Cambridge Studies in Biologic

M000080174

Mountain Gorillas

For the past three decades the mountain gorillas of Karisoke Research Center have been the subject of many studies focusing on their behavior and ecology. Long-term observations on known individuals, from birth to death, as well as data on social behavior within and between groups have led to an understanding of many aspects of the social system of gorillas. The findings have made significant contributions to models of comparative primate behavioral ecology. Mountain gorillas have also been the focus of intense conservation efforts, which have become a model for conservation programs elsewhere. While most of what we know about the genus *Gorilla* is based on mountain gorillas, data on the other two recognized subspecies have increased over the past 20 years. This book highlights and summarizes some of the behavioral, ecological, and conservation work on mountain gorillas, and makes comparisons with findings from other study sites. Aimed at graduate students and researchers in primatology and behavioral ecology, it will also appeal to all those interested in gorilla conservation, and represents the most up-to-date and diverse collection of information available on this endangered ape.

MARTHA M. ROBBINS is Research Associate at the Max Planck Institute for Evolutionary Anthropology, Leipzig. Her research currently focuses on the behavioral ecology and reproductive strategies of the gorillas of Bwindi Impenetrable National Park, Uganda.

PASCALE SICOTTE is Associate Professor of Primatology at the University of Calgary. Her research interests lie in male–male competition, male–female relationships, and female mate choice. She was Director of the Karisoke Research Center from 1993 to 1994.

KELLY J. STEWART is Research Associate at the University of California, Davis, where her research focuses on the socioecology and conservation of apes. She is editor of the annual *Gorilla Conservation News* and serves on the Scientific Advisory Council of the Dian Fossey Gorilla Fund International.

Cambridge Studies in Biological and Evolutionary Anthropology

Series Editors

HUMAN ECOLOGY
C. G. Nicholas Mascie-Taylor, University of Cambridge
Michael A. Little, State University of New York, Binghamton
GENETICS
Kenneth M. Weiss, Pennsylvania State University
HUMAN EVOLUTION
Robert A. Foley, University of Cambridge
Nina G. Jablonski, California Academy of Science
PRIMATOLOGY
Karen B. Strier, University of Wisconsin, Madison

Consulting Editors
Emeritus Professor Derek F. Roberts
Emeritus Professor Gabriel W. Lasker

Mountain Gorillas

Three Decades of Research at Karisoke

EDITED BY

MARTHA M. ROBBINS
Max Planck Institute for Evolutionary Anthropology, Leipzig

PASCALE SICOTTE
University of Calgary

KELLY J. STEWART
University of California, Davis

CAMBRIDGE UNIVERSITY PRESS
Cambridge, New York, Melbourne, Madrid, Cape Town, Singapore, São Paulo

Cambridge University Press
The Edinburgh Building, Cambridge CB2 2RU, UK

Published in the United States of America by Cambridge University Press, New York

www.cambridge.org
Information on this title: www.cambridge.org/9780521780049

© Cambridge University Press 2001

This publication is in copyright. Subject to statutory exception
and to the provisions of relevant collective licensing agreements,
no reproduction of any part may take place without
the written permission of Cambridge University Press.

First published 2001
This digitally printed first paperback version 2005

A catalogue record for this publication is available from the British Library

Library of Congress Cataloguing in Publication data
Mountain gorillas: three decades of research at Karisoke/edited by
Martha M. Robbins, Pascale Sicotte, Kelly J. Stewart.
  p. cm. – (Cambridge studies in biological and evolutionary
anthropology)
Includes bibliographical references (p.).
ISBN 0 521 78004 7
1. Gorilla – Rwanda – Congresses. 2. Karisoke Research Center –
Congresses. I. Robbins, Martha M., 1967– II. Sicotte, Pascale,
1961– III. Stewart, Kelly, J., 1951– IV. Series.

QL737.P96 M68 2001
599.884´0967571–dc21   00-066718

ISBN-13 978-0-521-78004-9 hardback
ISBN-10 0-521-78004-7 hardback

ISBN-13 978-0-521-01986-6 paperback
ISBN-10 0-521-01986-9 paperback

Contents

Contributors

Byrne, Richard W.   rwb@st-andrews.ac.uk
Scottish Primate Research Group, School of Psychology, University of
St. Andrews, St. Andrews, KY16 9JU, UK

Cranfield, Michael R.   mrcranfi@mail.bcpl.net
Medical Department, The Baltimore Zoo, Druid Hill Park, Baltimore, MD
21217, USA & Mountain Gorilla Veterinary Project, Druid Hill Park, Baltimore,
MD 21217, USA & Division of Comparative Medicine, Johns Hopkins
University, Baltimore, MD 21205, USA

Czekala, Nancy   czekala@sunstroke.sdsu.edu
Center of Reproduction of Endangered Species, San Diego Zoological Society,
PO Box 120551, San Diego, CA 92112, USA

Doran, Diane M.   ddoran@notes.cc.sunysb.edu
Department of Anthropology, State University of New York at Stony Brook,
Stony Brook, NY 11794–4364, USA

Eilenberger, Ute
Mountain Gorilla Veterinary Project, B.P. 1321, Kigali, Rwanda

Fletcher, Alison   a.fletcher@chester.ac.uk
Department of Biology, University College Chester, Parkgate Road, Chester,
CH1 4BJ, UK

Gerald-Steklis, Netzin   science@gorillafund.org
DFGF-International 800, Cherokee Ave, SE Atlanta, GA 30315-1440, USA

Harcourt, Alexander H.   ahharcourt@ucdavis.edu
Department of Anthropology, University of California, Davis, 330 Young Hall,
1 Sheilds Avenue, Davis, CA 95616, USA

Kahekwa, John
Parc National de Kahuzi-Biega, Institut Congolais pour Conservation de la
Nature, B.P. 895, Bukavu, République Démocratique du Congo

McNeilage, Alastair   mcneilage@aol.com
Institute of Tropical Forestry Conservation, PO Box 44, Kabale, Uganda

Mudakikwa, Antoine B.   mgvc@inbox.rw
Mountain Gorilla Veterinary Project, B.P. 1321, Kigali, Rwanda

Plumptre, Andrew J.   aplumptre@aol.com
Wildlife Conservation Society, 2300 Southern Boulevard, Bronx, NY 10460,
USA

Robbins, Martha M.   robbins@eva.mpg.de
Max Planck Institute for Evolutionary Anthropology, Inselstrasse 22, 04103
Leipzig, Germany

Sicotte, Pascale   sicotte@ucalgary.ca
Department of Anthropology, University of Calgary, 2500 University Drive NW,
Calgary, AB T2N IN4, Canada

Sleeman, Jonathan M.
Virginia–Maryland Regional College of Veterinary Medicine, University of
Maryland, 8075 Greenmead Drive, College Park, MD 20742, USA

Steklis, H. Dieter   mntgorilla@aol.com
Department of Anthropology, Rutgers – State University of New Jersey,
DFGF-International, Box 270, New Brunswick, NJ 08903–0270, USA

Stewart, Kelly J.   kjstewart@ucdavis.edu
Department of Anthropology, University of California, Davis, 330 Young Hall,
1 Sheilds Avenue, Davis, CA 95616, USA

Vedder, Amy   avedder@wcs.org
Wildlife Conservation Society, 2300 Southern Boulevard, Bronx, NY 10460,
USA

Watts, David P.   david.watts@yale.edu
Department of Anthropology, Yale University, PO Box 208277, New Haven, CT
06529–8277, USA

Weber, William A.   Wcsnap@aol.com
Wildlife Conservation Society, 2300 Southern Boulevard, Bronx, NY 10460,
USA

Williamson, Elizabeth A.   DFGFRWANDA@aol.com
DFGF-International, B.P. 1321, Kigali, Rwanda and Department of Psychology,
University of Stirling, UK

Yamagiwa, Juichi   yamagiwa@jinrui.zool.kyoto-u.ac.jp
Laboratory of Human Evolution Studies, Faculty of Science, Kyoto University,
Sakyo, Kyoto 606–8502 Japan

Acknowledgements

For initiating and facilitating research

There are a tremendous number of people that made research on mountain gorillas and their ecosystem possible. Without them, this book would not have been possible. We would like to thank them all. The danger is of course to forget some people, and it seems almost inevitable that we will.

First and foremost, the Office Rwandais du Tourisme et des Parcs Nationaux has allowed long-term research to be conducted on the gorilla research groups and generally in the Parc National des Volcans. We are grateful to them for their understanding of the need to collect long-term data on known individuals. The Institut Congolais pour la Conservation de la Nature has also allowed researchers based on the Rwandan side of the Virunga to cross freely the border to follow the gorillas. This has tremendously facilitated research in a situation that could potentially have been extremely difficult, since the forest encompasses the borders of three countries. We also thank the park staff at the Parc National des Volcans' headquarters in Kinigi and Ruhengeri for their cooperation and for the work that they have performed, sometimes during tumultuous times.

Dian Fossey paved the way for the long-term research that took place at Karisoke. She set up the Digit Fund to help maintain the activities of the research center, and its descendant, the Dian Fossey Gorilla Fund International, still continues to do so. The National Geographic Society funded Fossey's early work, and without this continued support, Karisoke probably would not have survived the few first crucial years. Several organizations have contributed for many years to gorilla conservation efforts in the Virungas. Some of them have helped fund regular censuses of the gorilla population in the Virungas, a vital exercise in order to evaluate whether conservation policies are working. Among them, the Wildlife Conservation Society, the World Wildlife Fund, the African Wildlife Foundation, the Fauna and Flora Preservation Society, the Morris Animal Foundation, and the Dian Fossey Gorilla Fund – UK.

The list of the people that have worked at Karisoke as census workers, research assistants, students, co-directors, and directors would take several

xii   *Acknowledgements*

pages and, again, inevitably, we would forget some people. They all faced different challenges depending on the times. In their own way, they all contributed to the efficient running of research at Karisoke. We are indebted to all of them.

Over the years, the Karisoke Research Center has employed several Rwandan trackers, research assistants, anti-poaching patrollers, and camp staff, without whom work would not have been possible. These men came mostly from the communities surrounding the Parc National des Volcans. Many have been not only co-workers, but have become friends. Over the years, many have suffered through difficult times: Dian Fossey's murder in the mid 1980s, and the insecurity that plagued the region since the early 1990s. Many have worked for the project for decades. We are grateful to them for their commitment to their work, for the fact that they became ambassadors for gorilla conservation in their community, and for their friendship: Balinda Michel, Banyangandora Antoine, Bapfakwita Vincent, Barabwiriza Faustin, Bigegra Gabriel, Bizumuremyi Jean Bosco, Harerimana Maurice, Hategekimana Jean Damascène, Hitayezu Emmanuel, Kamufozi Damascène, Kananira, Kanyorgano Leonidas, Kwiha Salatiel, Mbonigaba André, Mpiranya Mathias, Mugiraneza Fidèle, Munyejabo Enok, Munyambonera Bazira, Munyanganga Kana, Munyanshoza Leonard, Munyaziboneye Jean Damascène, Ndaruhebeye Jean Damascène, Ndibeshe Jean Pierre, Nemeye Alphonse, Ngaruyintwari François, Nkeramugaba Célestin, Nkunda Alphonse, Nsezeyintwali Gabriel, Nshogoza Fidèle, Ntahontuye Jean Damascène, Ntibiringirwa Félicien, Rukera Elias, Rwabukamba Ladislas, Rwampogazi, Rwelekana Emmanuel, Rwihandagaza Eliachem, Sebatware André, Sebazungu Leonard, Sekaryongo Jean, Uwimana Fidèle, Vatiri André.

For the book process
This idea for this book originated from the desire to get Karisoke researchers together to discuss the directions of future research on mountain gorillas. The very first such gathering took place in 1996, at the 16th Congress of the International Primatological Society in Madison, Wisconsin, where the Dian Fossey Gorilla Fund International organized a round table on this topic. Then, in early 1999, Christophe Boesch and the Max Planck Institute for Evolutionary Anthropology in Leipzig graciously supported "Mountain Gorillas of Karisoke: Research Synthesis and Potentials", the conference that was to become the stepping stone for this book. This book includes chapters by most participants of the Leipzig conference, but also includes chapters by Karisoke researchers who could not attend. We want to thank especially all those who contributed chapters

to this book. We are sincerely grateful to them, not only for their efforts and scholarship, but for their patience and responsiveness to our editing. This book is a result not only of the authors' research, but also of their dedication to Karisoke Research Center and the gorillas.

Several people reviewed chapters of this book. We would like to thank all of them for their enthusiastic and generous participation. In addition, most contributors to the book reviewed one or two chapters as well, for which we are grateful.

Martha Robbins thanks Christophe Boesch, Alastair McNeilage, Chuck Snowdon, and Karen Strier for advice and support during the editing of this book, and Leila Kunstmann and Serge Nsuli for secretarial and technical support. Pascale Sicotte thanks Tracy Wyman, Jill Ogle, and May Ives for technical support. The Kananaskis Field Station of the University of Calgary provided a lovely setting in the Canadian Rockies during the initial reviews of manuscripts. We thank Grace LeBel at the station for her help, and for making computers and other office equipment available. At Cambridge University Press, we thank Tracey Sanderson, Anna Hodson and Sarah Jeffery for their assistance.

To Dian Fossey

1 *Mountain gorillas of the Virungas: a short history*

KELLY J. STEWART, PASCALE SICOTTE
& MARTHA M. ROBBINS

Dian Fossey and Digit. (Photo by Kelly J. Stewart.)

Introduction

In January 1999, the Max Planck Institute for Evolutionary Anthropology hosted a conference in Leipzig, to celebrate more than three decades of research on wild mountain gorillas at the Karisoke Research Center. To be more specific, it was 32 years and 9 months since Dian Fossey had set up camp in the Rwandan sector of the Virunga Volcanoes (Figure 1.1). On September 24, 1967, using a marriage of Karisimbi and Visoke, the names of the two closest volcanoes, Fossey christened her site Karisoke. She could not have known that this would be her home for the rest of her life, or that the tent she pitched at 3000 m in that wet, montane forest would become one of the longest-running research sites in field primatology. Many of the direct descendants of the mountain gorillas she first contacted in 1967 are still being observed today.

The story of Karisoke is a chronicle of the development of behavioral and ecological research, intertwined with the growth of conservation efforts to save mountain gorillas. It has been played out against a back-drop of political instability and, over the past decade, devastating war. We present briefly this story below, to set the stage for the chapters that follow (for a more detailed description of the development and history of behavioral ecology, we recommend Strier, 1994 and Janson, 2000).

The intellectual setting

By 1967, primatology and anthropology were ripe for a long-term study of gorillas. It was four years after the publication of George Schaller's classic work, *The Mountain Gorilla*, a landmark study of remarkable detail that described the basics of the subspecies' social organization, life history, and ecology (Schaller, 1963). While western science had been aware of, and intrigued by, the gorilla since its discovery in Gabon in 1847, Schaller's work was based on more direct observations of wild gorillas and provided far more information than had ever been gathered before on any of the three gorilla subspecies (see Yerkes & Yerkes, 1929; Schaller, 1963, for a review of discovery and exploration).

Primatologists at the time were recognizing the value of long-term observations of known individuals, thanks to naturalistic studies of Japanese macaques and rhesus monkeys, initiated in the 1950s and early 1960s (e.g. Kawai, 1965; Koford, 1965). During the same period, anthropologists were beginning to use the behavior of living primates as a window into our hominid past. Baboons, by virtue of living in a habitat which was thought to be the one in which early hominids evolved, became a favorite species for modeling the behavior of our ancestors (Washburn & DeVore, 1961). The great apes, because of their close phylogenetic related-

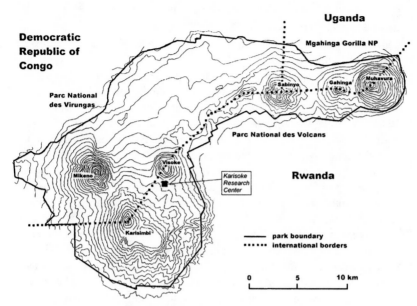

Figure 1.1. Map of the Virunga Volcano region.

ness to humans, also became models of human evolution (Rodman, 1994). Jane Goodall had been observing wild chimpanzees in Gombe Stream, Tanzania, since the early 1960s, and she was making discoveries that narrowed the traditional gaps between humans and apes (Goodall, 1968). It was Goodall's mentor, paleoanthropologist Louis Leakey, who, in his search for someone to conduct similar work on gorillas, helped launch Dian Fossey's project.

The ecological setting

Karisoke Research Center lies in the Rwandan portion of the Virunga Volcanoes (Figure 1.1), a chain of mountains stretching 77 km across the bottom of the Albertine Rift. The eight volcanic peaks (two still active) and their intervening saddle areas comprise one continuous ecosystem of largely moist montane forest (Schaller, 1963; Spinage, 1972). Today, the region encompasses the borders of three different countries: 211 km lie within the Parc National des Virungas, in the Democratic Republic of Congo (formerly known as Zaire); 125 km belong to Rwanda's Parc National des Volcans, and 44.5 km to Uganda's Mgahinga Gorilla Park (Figure 1.1). The Virungas are the "classic home" of mountain gorillas, *Gorilla gorilla beringei* (see Box 1.1), being the place where they were discovered by German officer, Oscar von Beringe, in 1902 (Schaller, 1963).

Box 1.1.   Gorilla taxonomy
Currently three subspecies of gorillas are recognized (Figure 1.3) (Groves, 1970). Western lowland gorillas (*Gorilla gorilla gorilla*) are the most widely distributed, with populations totaling approximately 110 000 gorillas occurring in Gabon, Democratic Republic of Congo (formerly Zaire), Congo, Cameroon, Equatorial Guinea, and Central African Republic (Harcourt, 1996). Eastern lowland gorillas (*G. g. graueri*) are found only in the Democratic Republic of Congo with approximately 17 000 gorillas being found in several isolated populations (Harcourt, 1996; Hall *et al.*, 1998). Mountain gorillas (*G. g. beringei*) number only approximately 600 individuals divided into two populations, the Virunga Volcanos of Rwanda, Uganda, and Democratic Republic of Congo, and Bwindi Impenetrable National Park, Uganda (Sholley, 1991; McNeilage *et al.*, 1998).

At a meeting held in February, 2000, members of the IUCN/SSC Primate Specialist Group proposed a revised consensus taxonomy for gorillas (J. Oates, personal communication). This proposed classification describes two species and four subspecies of gorillas. Western gorillas are divided into two subspecies, the western lowland gorilla (*G. g. gorilla*) and the Cross River gorilla (*G. g. diehli*). The Cross River gorilla, found in at least five small subpopulations in Nigeria, is critically endangered with approximately only 150–200 individuals remaining (Oates, 2000). The other species of gorilla, eastern gorilla, is divided into mountain gorillas (*G. beringei beringei*) and Grauer's gorilla (*G. b. graueri*). There is debate as to whether mountain gorillas should be further split into two subspecies by raising the Bwindi Impenetrable gorilla population to the level of subspecies. Limited morphological and ecological comparisons between the Virunga and Bwindi gorillas suggest that they should not be considered the same subspecies (Sarmiento *et al.*, 1996), but genetic evidence indicates that there is no difference between the two populations (Garner & Ryder, 1996; Jensen-Seaman, 2000). Further research on both populations, but particularly the Bwindi gorillas, is necessary to resolve this issue.

But why were mountain gorillas the choice for a long-term study and why have they dominated the gorilla scene ever since? They are hardly representative of the genus. *Gorilla gorilla beringei* is the rarest of the subspecies (Groves, 1970; Harcourt, 1996), and is widely thought to occur in only two populations of about 300 individuals each. The Virunga gorillas live at the ecological extreme of the gorilla's distribution (Schaller, 1963). Their habitat ranges up to 4507 m, far higher than any other gorilla habitat and its montane vegetation is notably lacking in fruit. Furthermore, it is the only gorilla site where gorillas do not overlap with their closest relatives, chimpanzees.

The nature of their habitat, however, is precisely what makes mountain gorillas more observable than other populations. This is as true today as it was back in 1967. Their highly folivorous, fruit-poor diet results in shorter daily ranges and presumably in smaller home ranges than those of gorillas in lowland tropical forest (Tutin, 1996; Remis, 1997; Doran & McNeilage, 1998). When the gorillas feed and travel in the dense undergrowth of montane forest, they leave behind a well-marked trail of trampled vegetation that enables humans to follow them relatively easily. In addition, the rugged terrain sometimes permits excellent visibility (for both humans and gorillas) while maintaining a fair distance with the observer that may help reduce the initial fear response of the gorillas (Fossey, 1983). Finally, the Virunga gorillas have not been traditionally hunted for food by humans, as have other gorilla populations (Schaller, 1963). These last two points have probably contributed to making the mountain gorillas more amenable to habituation than their lowland counterparts.

This combination of logistical factors no doubt contributed to the "gorilla-friendly" attitude of westerners towards the Virungas from the 1920s onwards, when expeditions began to focus on observing the animals rather than "collecting" them (Akeley, 1923). In fact, gorillas were a main reason for the creation in 1926 of Albert National Park, Africa's first national park. Thus, when Dian Fossey arrived in Rwanda in 1967, the gorillas were, at least on paper, legally protected, but known to be threatened; and history had shown that observing them was feasible.

The 1970s
Research

The 1970s opened the doors into the individual lives of gorillas. By 1972, Fossey and others at Karisoke had habituated three groups, enabling close-range observations of known animals. Using the systematic methods of data collection that had been developed in field primatology by this time (Altmann, 1974), Karisoke researchers documented and quantified the fundamentals of gorilla ecology, demography, and social organization.

A brief review of what was to become our "classic" understanding of gorilla socioecology follows. Most individuals dispersed from their natal group. Females always joined either a lone silverback or another breeding group, while males did not immigrate into breeding groups, but attracted females away from other silverbacks (Harcourt, 1978). The resulting social organization consisted of stable, cohesive groups held together by long-term bonds between adult males and females, while relationships among females were relatively weak (Harcourt, 1979a,b). Dominance relationships among adult females were generally unclear, and agonistic interactions relatively infrequent (Harcourt, 1979b). These findings could be tied to low levels of feeding competition, as well as to the dominant male's control of female–female aggression (Fossey & Harcourt, 1977; Harcourt, 1979a,b). In the groups that contained more than one adult male, rank differences between males were clear-cut and the dominant male appeared to do most of the mating (Harcourt, 1979c; Harcourt et al., 1980). The nature of courtship and mating supported the notion that male–male competition for females found its fullest expression in contests between, not within, groups (Harcourt et al., 1980). Indeed, relations between groups were not based on resource defense, since gorillas' home ranges overlapped extensively (Fossey & Harcourt, 1977). Rather, the nature of inter-group interactions was the result of intense mating competition between adult males.

As our understanding of gorilla socioecology increased, the growing number of studies on other primate species provided data for comparison. The "typical" primate, based on extensive studies on cercopithecines (particularly baboons and macaques), exhibited female philopatry, strong matrilineal kinship bonds, and highly structured dominance relationships (see Strier, 1994 for review). Mountain gorillas offered a sharp contrast to these findings.

Meanwhile, new field studies of orangutans in Borneo and Sumatra (Rodman, 1973; Rijksen, 1978; Galdikas, 1979), chimpanzees at other sites (Reynolds, 1965; Nishida, 1979), and bonobos (Kano, 1979) were revealing extraordinary diversity in ape social organizations. In most habitats, orangutans range in a solitary fashion. The home range of a male may overlap with the home ranges of several females. Chimpanzees and bonobos are found in fission–fusion social systems, where the composition of traveling parties within a community changes depending on the food resources available and on the reproductive state of females. Although the social system of mountain gorillas was seemingly quite different, these early studies nevertheless highlighted the fact that apes shared features, such as female dispersal and at least some degree of male philopatry, that distinguished them from many other primates (Harcourt, 1978).

Theoretical developments in the disciplines of animal behavior and ecology during the 1970s had a profound influence on the interpretation of gorilla behavior. While Hinde's conceptual framework described the relation between individuals' social interactions, their relationships, and the social system (Hinde, 1976), socioecological models from studies of birds and mammals related ecological variables to foraging strategies, range use, and social systems (Crook & Gartlan, 1966; Lack, 1968; Clutton-Brock & Harvey, 1977; Verencamp, 1979). Behavioral ecology blossomed in 1975 with the publication of *Sociobiology* (Wilson, 1975) providing the link between individuals' social behavior and evolutionary theory (Hamilton, 1964; Williams, 1966; Trivers, 1972, 1974). This approach held that individuals in a group could have divergent interests and therefore, that a behavior that was advantageous for an individual was not necessarily "good" for the group. Social systems resulted from the compromises among individuals in their strategies to gain resources and mates (Wrangham, 1979, 1980). The view of infanticide as an adaptive strategy is a good example of how an evolutionary approach influenced our interpretation of animal behavior. Working with langurs, Hrdy (1977, 1979) suggested that males may under certain circumstances benefit reproductively from killing unrelated infants. Females suffer reproductively from these killings, and she suggested that female counter-strategies to male infanticide should evolve. These have since been documented in a large number of species, including gorillas (Smuts & Smuts, 1993). Finally, the theory of sperm competition, a development of sexual selection theory stemming from male–male competition, put comparative data on the great apes' sexual morphology and mating behavior in an evolutionary perspective (Short, 1979; Harcourt *et al.*, 1981).

Conservation

Unfortunately, while research was thriving during the 1970s, the gorillas were not. The Virungas have a long history of human settlement on the edge of the gorillas' habitat, as well as incursions into the forest by people and cattle (Schaller, 1963; Curry-lindhal, 1969; Desforges, 1972; Spinage, 1972). For much of the 1970s, international conservation organizations were minimally involved in mountain gorillas, and Karisoke Research Center was the most constant and noticeable conservation presence in the forest. Its conservation activities, however, were focused on anti-poaching efforts in which patrols swept repeatedly through the forest to cut the snares that were set to catch antelopes and buffaloes (and to which gorillas are vulnerable), as well as to search for poachers' camps.

A Karisoke-based census of the entire Virunga ecosystem in the early 1970s showed that the gorilla population had declined since Schaller's estimate in the 1960s, from about 450 to about 275 animals (Figure 1.2) (Harcourt & Groom, 1972; Harcourt & Fossey, 1981; Weber & Vedder, 1983). One cause of this decline was habitat loss. Major loss of gorilla habitat occurred between 1958 and 1973, when more than 50% of the Parc National des Volcans (the Rwandan side of the Virungas) was cleared to

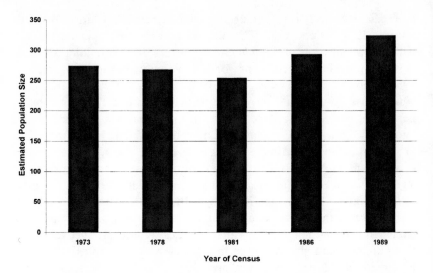

Figure 1.2. Changes in the population size of mountain gorillas in the Virunga
Volcanoes region as measured by periodic censuses.

allow human settlement and cultivation of cash crops (Weber, 1987, 1989).
Another cause of this decline in population was the incidental as well as
deliberate hunting of gorillas. In the mid 1970s, a gruesome trophy trade in
gorilla heads and skulls surfaced in Rwanda, with the main market being
foreign residents and visitors (Harcourt, 1986). This trade was behind the
poacher attacks in 1978 on Karisoke's longest-observed study group,
Group 4. Hunters killed two silverbacks, a female, and an infant, resulting
in the eventual disintegration of the breeding group (Fossey, 1983). The
massive publicity campaign in England and in the USA that followed these
killings resulted in the now famous Mountain Gorilla Project, a program
that became a model for gorilla conservation in other parts of Africa
(Harcourt, 1986; Weber, 1989).

The early 1980s
Research
While the demise of Group 4 and its aftermath were tragic, it taught us
much about the dynamics of group formation and dissolution, including
those of all-male groups (Fossey, 1983; Yamagiwa, 1987; Watts, 1991).
Perhaps most significantly, it underlined the role of adult males in protecting related infants from infanticide (Fossey, 1984; Watts, 1989). Soon after
the death of the leading silverback of Group 4, two infants were killed by
males who were not their fathers (Fossey, 1983). When females lose the

protection of the silverback of their group, their infants become vulnerable to infanticide (Watts, 1989).

The 1980s was also a time to start harvesting the fruits of long-term observations of known individuals. Gorillas who had first been seen as small infants in the study groups began reaching adulthood, and researchers could now examine variations in many aspects of behavior. Studies became more problem-oriented and considered the impact of factors such as group size, feeding competition, and kinship on social relationships and dispersal decisions (e.g. Watts, 1985, 1990, 1991, 1992; Stewart & Harcourt, 1987; Harcourt & Stewart, 1989). Our understanding of gorilla ecology increased significantly with detailed studies that measured food quality, distribution, and abundance, and related these to foraging patterns (Waterman *et al.*, 1983; Vedder, 1984; Watts, 1984). Finally, long-term ranging data enabled researchers to assess the relative impact of ecological and social factors on gorillas' habitat use (Yamagiwa, 1986; Watts, 1991, 1994*a*).

Within the discipline of behavioral ecology, models for the evolution of primate sociality provided a theoretical frame for the emerging pattern of gorilla behavior and ecology (Wrangham, 1979, 1980; van Schaik, 1989). But this picture of gorilla social evolution was still based on one subspecies. This situation began to change as observations on *G. g. graueri* in eastern Zaire, on groups habituated since the 1970s, became more consistent and systematic (Casimir, 1975; Yamagiwa, 1983). In addition, researchers were establishing long-term studies of the western lowland gorillas *G. g. gorilla*, first in Gabon in 1980 and then in Central African Republic and Congo (Tutin, 1996; Doran & McNeilage, 1998; see Figure 1.3 for location of field sites). The data emerging from this work confirmed hints from earlier studies (Sabater Pi, 1971; Calvert, 1985) as to major differences in diet and ranging behavior between lowland and mountain gorillas.

Conservation

The new conservation efforts initiated in the late 1970s in Rwanda appeared to work. Censuses during the 1980s showed a halt to the population decline after 1981, and a subsequent increase in numbers (Figure 1.2) (Harcourt, 1986; Vedder & Aveling 1987). A conservation program similar to the Mountain Gorilla Project was soon initiated on the Congo side of the Virungas (Aveling & Aveling, 1989), and efforts were mounted to improve the situation for gorillas in Uganda. In keeping with the expansion of the times, Karisoke extended its operations beyond the borders of the study site and the park, establishing cooperative links with

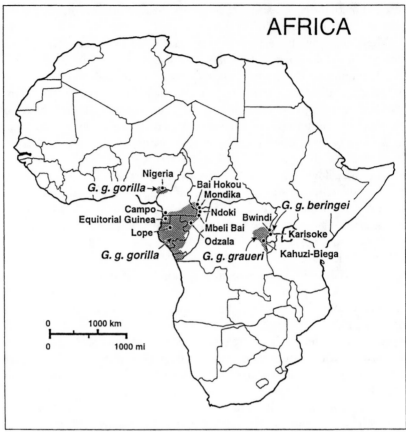

Figure 1.3. Map of Africa indicating the distribution of the three currently recognized subspecies of gorillas and the major research sites focusing on gorillas.

conservation projects in Rwanda, and later, Uganda and Congo, and with institutions such as l'Université Nationale du Rwanda. In addition, its conservation activities became closely integrated with those of the Rwandan park authorities, l'Office Rwandais du Tourisme et des Parcs Nationaux (ORTPN), specifically in well-organized joint patrols between the park guards and Karisoke field personnel. For the first time, sociological research was also directed towards the human population surrounding the park. Interviews of farmers revealed that the conservation education program put together by the Mountain Gorilla Project was having a positive impact on people's attitude towards conserving the park and its wildlife (Weber, 1987, 1989). Overall, there were grounds for cautious optimism.

Dian Fossey's murder

In December 1984, Dian Fossey was murdered in her cabin at Karisoke by unknown perpetrator/s. It is believed by many in the public domain that research at Karisoke ended with Fossey's death, but her legacy had far too much momentum to die with her. The Digit Fund, later to become the Dian Fossey Gorilla Fund International (DFGFI), was established to ensure that Karisoke Research Center would carry on its work. DFGFI continues to fund operation of the center today.

The late 1980s and early 1990s
Research

In the second half of the 1980s and the early 1990s, many research topics focused on intraspecific variation. Not only were behavioral ecologists as a whole recognizing the importance of intraspecific variation, but the Karisoke study groups were exhibiting variation in their composition. In general, the groups became larger, and included more related females and more adult males.

This variation in group composition challenged some of the long-held notions about gorilla social behavior. It became clear, for example, that under certain circumstances, nepotism among females could influence the nature of their social relationships (Harcourt & Stewart, 1989; Watts, 1991, 1994b, 1995). While these findings did not undermine the fundamental concept that male–female bonds are the bases of group cohesion, they have enabled us to draw a more complex picture of gorilla sociality. In addition, the long-term existence of groups with more than one silverback has led to a better understanding of within-group male–male competition (Watts, 1992; Sicotte, 1994; Robbins, 1995, 1996a) and brought the one-male mating system of gorillas into question (Robbins, 1999). These multimale groups have also allowed us to document behaviors rarely, or never, observed in one-male groups, namely herding of females during inter-group encounters (Sicotte, 1993) and interpositioning by females and infants to diffuse tension during male–male conflicts (Sicotte, 1995). These large groups finally meant that several females within a group could have offspring over a short time-span, which opened the way to more detailed investigation and comparison of immature development (Watts & Pusey, 1993; Fletcher, 1994; Doran, 1997).

Researchers also recognized that the current knowledge of gorilla feeding ecology was based on a small subset of their habitat. This led to a study encompassing a larger proportion of the Virungas (McNeilage, 1995). Studies on non-gorilla features of the ecosystem were also initiated, for example, the feeding ecology of large ungulates (bushbucks, buffaloes, elephants: Plumptre, 1993, 1995, 1996; Plumptre & Harris, 1995).

At the same period, new non-invasive technologies enabled researchers to examine underlying physiological mechanisms of behavior with hormone assays (Robbins & Czekala, 1997; Czekala & Sicotte, 2000). Genetic studies were also initiated, to address questions of taxonomy, genetic variability, and paternity (Garner & Ryder, 1996; Field *et al.*, 1998).

Conservation

It can be said that, at the end of the 1980s, conservation efforts had advanced to the stage where the future of mountain gorillas looked optimistic. The 1989 gorilla census in the Virungas showed a continued increase in population size and growth (Figure 1.2) (Sholley, 1991). Deliberate poaching had become extremely rare. Gorilla-based tourism was thriving in Rwanda and Democratic Republic of Congo (then Zaire) and was being developed in Uganda. Rwanda had a Veterinary Center devoted to monitoring the health of the gorilla population. Conservation education programs were in place in all three countries with mountain gorillas, and the two mountain gorilla sites in Uganda, Bwindi Impenetrable and Mgahinga forests, were on the brink of being made national parks. Unfortunately, this promising situation would not last long.

The early 1990s were a start to a tragic era for the people and the wildlife of the Virunga region. In October, 1990, war broke out in Rwanda and as a result, tourism, one of the mainstays to gorilla conservation and to the Rwandan economy, declined drastically. Researchers and conservation personnel coped with land mines, gunfire, and mortar shells near or in the park. To make matters worse, instability on the Congolese side of the Virungas, starting in 1991, was eroding the ability of the park system to manage its gorillas. Conservation organizations had to develop contingency plans for working in times of great uncertainty. The only good news was that in 1991, Uganda made both Bwindi Impenetrable Forest and the Mgahinga Game Reserve national parks, a boost to the fortunes of mountain gorillas. Despite the violence in Rwanda, research and conservation activities at Karisoke continued. In fact, while several students from the Université Nationale du Rwanda conducted short-term studies at Karisoke, data for four long-term research projects were collected (Fletcher, 1994; McNeilage, 1995; Doran, 1996, 1997; Robbins, 1996b).

Fighting in Rwanda escalated during 1993 and in February of that year, the Karisoke Research Center was looted and destroyed. Everyone was evacuated, but after a short period, the Rwandan field staff resumed their work, taking the long daily trek from their homes into the forest every day to monitor the study groups. Thanks largely to the dedication and skill of these men, there was barely a gap in the long-term monitoring of gorillas. Karisoke made a remarkable comeback, concurrent with a reduction of open hostilities in Rwanda. By August of 1993, DFGFI had completely rebuilt and modernized the research center. Research on gorillas continued, along with studies of other features of the ecosystem (e.g. tree hyrax: Milner & Harris, 1999*a,b*; sunbird–lobelia interactions: Burd, 1994, 1995). But the lull in the violence was misleading.

The late 1990s
Genocide
In April 1994, the Rwandan President Juvenal Habyarimana was killed in a plane crash. The hours that followed saw the start of the systematic extermination of hundreds of thousands of Tutsis and moderate Hutus by the militia associated with the extremist Hutus. Research activities were disrupted at Karisoke by the evacuation of all the expatriate staff, but the monitoring of the research groups by the field assistants and a contingent of students from the Université Nationale du Rwanda continued for a few months. In July of that year, in the aftermath of the successful takeover of the country by the Rwandan Patriotic Front, one million people, mainly civilians but also the soldiers and militiamen associated to the Habyarimana regime fled from Rwanda into neighboring Zaire (Prunier, 1995). The Karisoke staff and park personnel were among the refugees. Many of the people that fled to the North Kivu followed a path through the gorillas' habitat. The refugees were placed in three large camps, just north of Goma and only a few kilometers away from the boundary of the Parc National des Virungas. The director of the Center at the time, Pascale Sicotte, worked to organize a small camp in Zaire for the Karisoke staff and park and conservation personnel from Rwanda. This action saved many lives, as it enabled people to avoid the disease-ravaged camps of Goma. Nevertheless, since that time, research and conservation have lost far too many irreplaceable people.

Aftermath
The post-1994 Rwanda, with a new government installed, started the slow process of national reconciliation and began to rebuild itself. The Virunga region, however, with its three national borders, has suffered chronic destabilization since then. In 1996, the offensive against Zaire's Mobutu started in the Kivu region. The Parc National des Volcans and the neighboring Parc National des Virungas became a base for rebels, many of whom had belonged to the army of Rwanda's former Habyarimana

(A)

Figure 1.4. Titus represents one of many individuals whom researchers at Karisoke have studied throughout his life. (A) held by his mother, Flossie, on the second day of his life in 1974; (B) as a 6-year-old juvenile; and (C) as a dominant silverback in the 1990s and into the 2000s.

(B)

(C)

regime. A large number of civilians who had come back from Zaire in 1996 after the dismantlement of the camps were terrorized by these rebels, who saw them as collaborators with the new regime. Among them, several Karisoke workers were targeted and paid with their lives for the fact that they had resumed their work. In the forest, poaching activities were on the rise, as suggested by the fact that the price of bushmeat was considerably lower than before the war (Plumptre *et al.*, 1997). Many people regularly hid in the forest; some because they collaborated with the rebels, some because they were afraid of them. The Karisoke Research Center was, again, completely destroyed.

The present

If you go to the site of the research center today, there is little evidence that it was ever there. The lush forest has claimed back the spot, and it probably looks much the same as it did when Dian Fossey first set eyes on it in 1967.

But although Karisoke no longer physically exists, it has always been present in Rwanda. By February 1995, a temporary director, funded by DFGFI, was in place in Kigali. A year later, the long-term director, Liz Williamson, had arrived to take up her post, working closely with government, military, and park personnel, and with the Karisoke field team. Although this book contains data mostly collected before 1994, there has been only one period of 15 months during 1997–98 without regular monitoring of the Karisoke study groups. Amazingly, the gorillas have fared remarkably well, despite the violent incursions into their range. Total membership of the three study groups in 1995, for example, was 73. By the new century, it was 87. At the time this book went to press, monitoring programs in the Virungas determined that the gorilla population has reached a minimum of 360 individuals (J. Kalpers *et al.* personal communication). For this, and for the consistency of the long-term, demographic data, we owe an enormous debt to the Karisoke staff. In the face of unspeakable horrors, and the disintegration of their lives and those of their families, they have maintained their commitment to their work and to the mountain gorillas.

Focus of this book

This book is not meant to be a comprehensive account of all research that has ever been done at Karisoke. Rather, it aims to present an overall picture of our current knowledge about mountain gorillas. We take a comparative perspective, viewing the findings from long-term study of the Karisoke groups alongside data from other gorilla populations and other primate species.

Can we justify such a narrow approach in a book: research from just one field site, on just one subspecies, of just one great ape? Our limiting the focus of this book to the mountain gorilla subspecies arises partly from a fact of life: we simply do not have comparable detailed long-term data on other wild gorilla populations. This contrasts markedly with studies of chimpanzees on whom long-term observations have been carried out at several sites (Wrangham *et al.*, 1994; McGrew *et al.*, 1996; Boesch & Boesch-Achermann, 2000). Also, the great apes are so different from each other in so many ways, that they each deserve focused attention. Since the publication of Fossey's *Gorillas in the Mist* in 1983, our knowledge of the Virunga gorillas as well as other populations has increased significantly. Furthermore, their conservation status has changed markedly, as has the political climate in which conservation work must operate. It's time for an update.

The chapters
The first section of this book focuses on variation in the social system of gorillas. Robbins (chapter 2) and Sicotte (chapter 3) take different viewpoints to discuss how various aspects of sexual selection, namely male–male competition and female choice, influence the mountain gorilla social system. One aspect of the variability in the social structure of mountain gorillas is the number of adult males per group. Robbins argues that whether males stay or leave their natal group is the result of individual strategies, which are influenced by life history variation, social relationships, and mating strategies of both sexes. Male's role in defending offspring from infanticide by extra-group male is crucial in mountain gorillas, but as Sicotte points out, the basis on which females select a male in particular remains unclear. Her chapter describes the behaviors used by females to exercise mate choice, as well as male behavior that apparently aims at influencing mate choice. Several of these behaviors involve movements between groups or retention of individuals, and thus influence group composition.

The next two chapters present an overview of reproductive parameters and life history traits in eastern lowland gorillas (Yamagiwa, chapter 4), and an overview of our knowledge of the social organization of western lowland gorillas (Doran & McNeilage, chapter 5). Both chapters make comparisons with mountain gorillas. They also suggest that infanticide, which is a major influence on the social grouping patterns in mountain gorillas, may not play such an important role for the two others subspecies. For instance, Yamagiwa reports several cases where female eastern lowland gorillas have transferred to a new group with an infant, or where females have stayed together with other females and their offspring after

18   *Kelly J. Stewart, Pascale Sicotte & Martha M. Robbins*

the death of their silverback. These infants have not been the target of infanticide by extra-group males or by the males in their new group. Doran & McNeilage, summarizing data from several western lowland gorilla research sites, suggest that group spread is generally larger than in mountain gorillas, and that inter-group encounters may be less hostile. These elements, they argue, might decrease the likelihood of infanticide compared to its occurrence in mountain gorillas.

The next section presents in details various aspects of within-group social behavior. Fletcher (chapter 6) discusses the ontogeny of mother–infant relationships using the largest sample size to date. She considers variation between groups, as well as sex differences and maternal variables such as parity and reproductive state. As infants gradually attain independence from their mothers, they develop an attachment to adult males. Stewart (chapter 7) describes these relationships, considers the benefits that young animals gain from such association, and the possible influence of group composition. In many ways, the nature of immatures' relationships with mature males reflects those between adult females and silverbacks. Watts (chapter 8) draws a picture of the ecological and social factors that promote egalitarian female social relationships in mountain gorillas, but also shows that the quality of female–female relationships varies according to relatedness, residency, and that it may change over time, perhaps depending on group size. Finally, Harcourt & Stewart (chapter 9) take a functional approach to studying intra-group vocalizations by considering their social correlates and their possible meaning.

Part III includes papers that deal with feeding behavior. McNeilage (chapter 10) shows that there is more variation in the feeding ecology of mountain gorillas in the Virungas than previously thought. This could have important implications for management of the national parks, for instance in determining what is or is not "suitable" gorilla habitat. In contrast to this wide-angled view, Byrne (chapter 11) zooms in on the details of feeding behavior. Mountain gorilla's folivorous diet appears to present little mental challenge; it needs no tools, hunting skills, or phenological memory. Byrne, however, argues that the sequences of movements used by mountain gorillas in food processing is indicative of high cognitive abilities, because they are standardized, involve multiple stages, and are bimanually coordinated – a series of traits apparently only found in apes.

Part IV focuses on gorilla conservation and management. Chapter 12 (Czekala & Robbins) discusses how hormone assessment is a useful tool for monitoring reproductive function and stress levels in the wild and contributes to management of captive gorillas. The Virunga mountain gorillas have been fortunate enough to have a veterinary center focused just on their health; veterinary medicine as a conservation practice is the focus of chapter 13 (Mudakikwa *et al.*). Plumptre and Williamson (chapter 14) give an overview of the threats to mountain gorillas, and describe how research assists in conservation practices. Steklis and Gerald-Steklis (chapter 15) combine knowledge from population censuses, social dynamics, and known threats to gorillas to comment on the attempts that have been made at modeling the viability of the Virunga gorillas. Finally, Weber & Vedder (Afterword), who were largely responsible for the establishment of the Mountain Gorilla Project, take a look back at the research and conservation efforts on mountain gorillas to consider "lessons learned" and to offer suggestions for future integrated efforts to conserve and better understand mountain gorillas and their ecosystem.

The future

The first 30 years of research on the Karisoke mountain gorillas have revealed much about the complexity of their life trajectory and their social system. Most contributors highlight in their chapters the areas for future research that relate to their area of expertise. Among the most important is probably the continued investigation of individual variation. To take just a few examples, we do not know the effect, if any, of multiple transfers on female reproductive success, nor do we know how paternity is distributed among the males in multi-male groups; extensive paternity determination analysis is just now being conducted to address this question. The range of the variation in weaning age is not known either, nor are the factors that could be associated with this variation. Obviously, the analysis of the Karisoke long-term records will be essential to answer these questions, and these records must continue to be updated.

Another target of research should be the gorillas outside of the Karisoke study area. The Karisoke gorillas are found at one extreme of gorilla habitat, even within the Virungas, and we know little about the gorillas that live at the lower altitude areas of the ecosystem in the Democratic Republic of Congo. Despite Bwindi Impenetrable National Park being only 30 km away from the Virungas and having about one-third of its altitudinal range in common with the Virungas, the little-studied Bwindi gorillas have been the cause of taxonomic debate (Box 1.1) (Garner & Ryder, 1996; Sarmiento *et al.*, 1996). We need further genetic, ecological, and behavioral studies to resolve these issues. Preliminary data on the Bwindi gorillas suggests that unlike the Karisoke gorillas, fruit is an important component of their diet on a seasonal basis, and the extent to which this influences their sociality is currently being investigated (M.M. Robbins, unpublished data). More information on lowland gorillas is urgently needed to put mountain gorillas in proper perspective in comparative studies of the apes and primates. Finally, most non-gorilla features of the unique Virunga ecosystem have been understudied. We can only hope that improved security will allow the continuation of the research and conservation efforts in this region and in all gorilla habitats.

Acknowledgements
We thank Sandy Harcourt and Linda Fedigan for comments that helped improve an earlier draft of this manuscript. Alastair McNeilage provided Figure 1.1 and Diane Doran provided Figure 1.3.

References

Akeley, C E (1923) *In Brightest Africa*. Garden City: Garden City Publishers.
Altmann, J (1974) Observational study of behavior: sampling methods. *Behaviour*, **49**, 1–41.
Aveling, C & Aveling, R (1989) Gorilla conservation in Zaire. Oryx, **23**, 64–70.
Boesch, C & Boesch-Achermann, H (2000) *The Chimpanzees of the Tai Forest: Behavioural Ecology and Evolution*. Oxford: Oxford University Press.
Burd, M (1994) A probabilistic analysis of pollinator behavior and seed production in *Lobelia deckenii*. *Ecology*, **75**, 1635–46.
Burd, M (1995) Pollinator behavioral responses to reward size in *Lobelia deckenii*: no escape from pollen limitation of seed set. *Journal of Ecology*, **83**, 865–72.
Calvert, J J (1985) Food selection by western gorillas in relation to food chemistry. *Oecologia*, **65**, 236–46.
Casimir, M J (1975) Feeding ecology and nutrition of an eastern gorilla group in Mt. Kahuzi region (Republic of Zaire). *Folia Primatologica*, **24**, 81–136.
Cheney, D (1977) The acquisition of rank and the development of reciprocal alliances among free-ranging immature baboons. *Behavioral Ecology and Sociobiology*, **2**, 303–18.
Clutton-Brock, T H & Harvey, P H (1977) Primate ecology and social organization. *Journal of Zoology, London*, **183**, 1–39.
Crook, J H & Gartlan, J S (1966) Evolution of primate societies. *Nature*, **210**, 1200–3.
Curry-lindhal, K (1969) Disaster for gorilla. *Oryx*, **10**, 7.
Czekala, N M & Sicotte, P (2000) Reproductive monitoring of free-ranging female mountain gorillas by urinary hormone analysis. *American Journal of Primatology*, **51**, 209–15.
Desforges, A (1972) Defeat is the only bad news: Rwanda under Musiinga, 1896–1931. PhD thesis, Yale University.
Doran, D (1996) Comparative positional behavior of the African apes. In *Great Ape Societies*, ed. W C McGrew, L F Marchant & T Nishida, pp. 213–24. Cambridge: Cambridge University Press.
Doran, D (1997) Ontogeny of locomotion in mountain gorillas and chimpanzees.

Journal of Human Evolution, **32**, 323–44.

Doran, D M & McNeilage, A (1998) Gorilla ecology and behavior. *Evolutionary Anthropology*, **6**, 120–31.

Field, D, Chemnick, L, Robbins, M M, Garner, K & Ryder, O (1998) Paternity determination in captive lowland gorillas and orangutans and wild mountain gorillas by microsatellite analysis. *Primates*, **39**, 199–209.

Fletcher, A W (1994) Social development of immature mountain gorillas (*Gorilla gorilla beringei*). PhD thesis, University of Bristol.

Fossey, D (1983) *Gorillas in the Mist*. London: Hodder & Stoughton.

Fossey, D (1984) Infanticide in mountain gorillas (*Gorilla gorilla beringei*) with comparative notes on chimpanzees In *Infanticide: Comparative and Evolutionary Perspectives*, ed. G Hausfater & S Blaffer Hrdy, pp. 217–35. New York: Aldine Press.

Fossey, D & Harcourt, A H (1977) Feeding ecology of free ranging mountain gorilla. In *Primate Ecology*, ed. T H Clutton-Brock, pp. 415–47. London: Academic Press.

Galdikas, B M F (1979) Orangutan adaptations at Tanjung Puting Reserve: mating and ecology. In *The Great Apes*, ed. D A Hamburg & E R McCown, pp. 195–233. Menlo Park CA: Benjamin/Cummings.

Garner, K J & Ryder, O A (1992) Some applications of PCR to studies in wildlife genetics. *Symposia of the Zoological Society of London*, **64**, 167–81.

Garner, K J & Ryder, O A (1996) Mitochondrial DNA diversity in gorillas. *Molecular Phylogenetics and Evolution*, **6**, 39–48.

Goodall, J (1968) The behaviour of free-living chimpanzees in the Gombe Stream Reserve. *Animal Behaviour Monographs*, **1**, 161–311.

Groves, C P (1970) Population systematics of gorilla. *Journal of Zoology, London*, **161**, 287–300.

Hall, J S, White, L J T, Inogwabini, B I, Omari, I, Simons Morland, H, Williamson, E A, Saltonstall, K, Walsh, P, Sikubwabo, C, Bonny, D, Kiswele, K P, Vedder, A & Freeman, K (1998) Survey of Grauer's gorillas (*Gorilla gorilla graueri*) and eastern chimpanzees (*Pan troglodytes schweinfurthi*) in the Kahuzi-Biega National Park lowland sector and adjacent forest in eastern Democratic Republic of Congo. *International Journal of Primatology*, **19**, 207–35.

Hamilton, W D (1964) The genetical evolution of social behaviour. *Journal of Theoretical Biology*, **7**, 1–52.

Harcourt, A H (1978) Strategies of emigration and transfer by primates with particular reference to gorillas. *Zeitschrift für Tierpsychologie*, **48**, 401–20.

Harcourt, A H (1979a) Contrasts between male relationships in wild gorilla groups. *Behavioral Ecology and Sociobiology*, **5**, 39–49.

Harcourt, A H (1979b) Social relationships among adult female mountain gorillas. *Animal Behaviour*, **27**, 251–64.

Harcourt, A H (1979c) Social relationships between adult male and female mountain gorillas. *Animal Behaviour*, **27**, 325–42.

Harcourt, A H (1986) Gorilla conservation: anatomy of a campaign. In *Primates: The Road to Self-Sustaining Populations*, ed. K Benirschke, pp. 31–46. New

York: Springer-Verlag.

Harcourt, A H (1996) Is the gorilla a threatened species? How should we judge? *Biological Conservation*, **7**, 165–76.

Harcourt, A H & Fossey, D (1981) The Virunga gorillas: decline of an "island" population. *African Journal of Ecology*, **19**, 83–97.

Harcourt, A H & Groom, A F G (1972) Gorilla census. *Oryx*, **11**, 355–63.

Harcourt, A H & Stewart, K J (1989) Functions of alliances in contests within wild gorilla groups. *Behaviour*, **109**, 176–90.

Harcourt, A H, Stewart, K J, Fossey, D & Watts, D P (1980) Reproduction in wild gorillas and some comparisons with chimpanzees. *Journal of Reproduction and Fertility*, Supplement **28**, 59–70.

Harcourt, A H, Harvey, P H, Larson, S G & Short, R V (1981) Testis weight, body weight and breeding system in primates. *Nature*, **293**, 55–7.

Hinde, R A (1976) Interactions, relationships, and social structure. *Man*, **11**, 1–17.

Hinde, R A (1983) *Primate Social Relationships: An Integrated Approach*. Oxford: Blackwell Scientific Publications.

Hrdy, S B (1977) *The Langurs of Abu: Female and Male Strategies of Reproduction*. Cambridge MA: Harvard University Press.

Hrdy, S B (1979) Infanticide among animals: a review, classification, and examination of the implications for the reproductive strategies of females. *Ethology and Sociobiology*, **1**, 13–40.

Janson, C H (2000) Primate socio-ecology: the end of a golden age. *Evolutionary Anthropology*, **9**, 73–86.

Jensen-Seaman, M I (2000) Evolutionary genetics of gorillas. PhD thesis, Yale University.

Kano, T (1979) A pilot study on the ecology of pygmy chimpanzees, *Pan paniscus*. In *The Great Apes*, ed. D A Hamburg & E R McCown, pp. 123–35. Menlo Park CA: Benjamin/Cummings.

Kawai, M (1965) On the system of social ranks in a natural troop of Japanese monkeys. I. Basic rank and dependent rank. In *Japanese Monkeys*, ed. K Imanishi & S A Altmann, pp. 66–86. Chicago: Altmann.

Koford, C B (1965) Population dynamics of rhesus monkeys on Cayo Santiago. In *Primate Behavior*, ed. I DeVore, pp. 160–74. New York: Holt Rinehart & Winston.

Lack, D (1968) *Ecological Adaptations for Breeding in Birds*. London: Methuen.

Massey, A (1977) Agonistic aids and kinship in a group of pigtail macaques. *Ecology and Sociobiology*, **2**, 31–40.

McGrew, W C, Marchant, L F & Nishida T (1996) *Great Ape Societies*. Cambridge: Cambridge University Press.

McNeilage, A (1995) Mountain gorillas in the Virunga Volcanoes: ecology and carrying capacity. PhD thesis, University of Bristol.

McNeilage, A, Plumptre, A J, Brock-Doyle, A & Vedder, A (1998) Bwindi Impenetrable National Park, Uganda Gorilla and Large Mammal Census, 1997. Wildlife Conservation Society, Working Paper No. 14.

Milner, J M & Harris, S (1999a) Activity patterns and feeding behaviour of the tree hyrax, *Dendrohyrax arboreus*, in the Parc national des Volcans, Rwanda.

African Journal of Ecology, **37**, 267–80.

Milner, J M & Harris, S (1999*b*) Habitat use and ranging behaviour of tree hyrax, *Dendrohyrax arboreus*, in the Virunga Volcanoes, Rwanda. *African Journal of Ecology*, **37**, 281–94.

Missakian, E A (1972) Genealogical and cross-genealogical dominance relations kin in a group of free-ranging rhesus monkeys (*Macaca mulatta*) on Cayo Santiago. *Primates*, **13**, 169–80.

Nishida, T (1979) The social structure of chimpanzees of the Mahale Mountains. In *The Great Apes*, ed. D A Hamburg & E R McCown, pp. 73–121. Menlo Park CA: Benjamin/Cummings.

Oates, J F (2000) The Cross River gorilla: a neglected and critically endangered subspecies. Talk presented at *The Apes: Challenges for the 21st Century*, Brookfield Zoo, Chicago, May 10–13, 2000.

Plumptre, A J (1993) The effects of trampling damage by herbivores on the vegetation of the Parc National des Volcans, Rwanda. *African Journal of Ecology*, **32**, 115–29.

Plumptre, A J (1995) The chemical composition of montane plants and its influence on the diet of large mammalian herbivores in the Parc National des Volcans, Rwanda. *Journal of Zoology, London*, **235**, 323–37.

Plumptre, A J (1996) Modeling the impact of large herbivores on the food supply of mountain gorillas and implications for management. *Biological Conservation*, **75**, 147–55.

Plumptre, A J & Harris, S (1995) Estimating the biomass of large mammalian herbivores in a tropical montane forest: a method of faecal counting that avoids assuming a "steady state" system. *Journal of Applied Ecology*, **32**, 111–20.

Plumptre, A J, Bizumuremyi, J B, Uwimana, F & Ndaruhebeye, J D (1997) The effects of the Rwandan civil war on poaching of ungulates in the Parc National des Volcans. *Oryx*, **31**, 265–73.

Prunier, G (1995) *The Rwanda Crisis 1959–1994: History of a Genocide*. Kampala: Fountain.

Remis, M J (1997) Ranging and grouping patterns of a western lowland gorilla group at Bai Hokou, Central African Republic. *American Journal of Primatology*, **43**, 111–33.

Reynolds, V (1965) Some behavioral comparisons between the chimpanzee and the mountain gorilla in the wild. *American Anthropologist*, **67**, 691–706.

Rijksen, H D (1978) *A Field Study on Sumatran Orang Utans*. Wageningen: Veenman & Zonen.

Robbins, M M (1995) A demographic analysis of male life history and social structure of mountain gorillas. *Behaviour*, **132**, 21–47.

Robbins, M M (1996*a*) Male–male interactions in heterosexual and all-male wild mountain gorilla groups. *Ethology*, **102**, 942–65.

Robbins, M M (1996*b*) The social system of mountain gorillas: variation in male life history, social behavior, and steroid hormone profiles. PhD thesis, University of Wisconsin, Madison.

Robbins, M M (1999) Male mating patterns in wild multimale mountain gorilla groups. *Animal Behaviour*, **57**, 1013–20.

Robbins, M M & Czekala N M (1997) A preliminary investigation of urinary testosterone and cortisol levels in wild male mountain gorillas. *American Journal of Primatology*, **43**, 51–64.

Rodman, P S (1973) Population composition and adaptive organization among orang-utans of the Kutai Reserve. In *Ecology and Behaviour of Primates*, ed. R P Michael & J H Crook, pp. 171–209. London: Academic Press.

Rodman, P S (1994) The human origins program and evolutionary ecology in anthropology today. *Evolutionary Anthropology*, **2**, 215–24.

Sabater Pi, J (1971) Exploitation of gorillas *Gorilla gorilla gorilla* Savage & Wyman 1847 in Rio Muni, Republic of Equatorial Guinea, West Africa. *Biological Conservation*, **19**, 131–40.

Sarmiento, E E, Butynski, T M & Kalina, J (1996) Gorillas of Bwindi-Impenetrable Forest and the Virunga Volcanoes: taxonomic implications of morphological and ecological differences. *American Journal of Primatology*, **40**, 1–21.

Schaller, G B (1963) *The Mountain Gorilla: Ecology and Behavior*. Chicago: University of Chicago Press.

Seyfarth, R M (1977) A model of social grooming among adult female baboons. *Journal of Theoretical Biology*, **65**, 671–98.

Sholley, C R (1991) Conserving gorillas in the midst of guerillas. *American Association of Zoological Parks and Aquaria Annual Proceedings*, pp. 30–7.

Short, R V (1979) Sexual selection and its component parts, somatic and genital selection, as illustrated by man and the great apes. *Advances in the Study of Behavior*, **9**, 131–58.

Sicotte, P (1993) Inter-group encounters and female transfer in mountain gorillas: influence of group composition on male behavior. *American Journal of Primatology*, **30**, 21–36.

Sicotte, P (1994) Effect of male competition on male–female relationships in bimale groups of mountain gorillas. *Ethology*, **97**, 47–64.

Sicotte, P (1995) Interpositions in conflicts between males in bimale groups of mountain gorillas. *Folia Primatologica*, **65**, 14–24.

Smuts, B B & Smuts, R W (1993) Male aggression and sexual coercion of females in nonhuman primates and other mammals: evidence and theoretical implications. *Advances in the Study of Behavior*, **22**, 1–63.

Spinage, C A (1972) The ecology and problems of the Volcano National Park, Rwanda. *Biological Conservation*, **4**, 194–204.

Stewart, K J & Harcourt, A H (1987) Gorillas: variation in female relationships. In *Primate Societies*, ed. B B Smuts, D L Cheney, R M Seyfarth, R W Wrangham & T T Struhsaker, pp. 155–64. Chicago: University of Chicago Press.

Strier, K B (1994) Myth of the typical primate. *Yearbook of Physical Anthropology*, **37**, 233–71.

Trivers, R L (1972) Parental investment and sexual selection. In *Sexual Selection and the Descent of Man*, ed. B Campbell, pp. 136–79. London: Heinemann.

Trivers, R L (1974) Parent–offspring conflict. *American Zoologist*, **14**, 249–64.

Tutin, C E G (1996) Ranging and social structure of lowland gorillas in the Lopé

Reserve, Gabon. In *The Great Apes*, ed. W C McGrew, L F Marchant & T Nishida, pp. 58–70. Cambridge: Cambridge University Press.

van Schaik, C P (1983) Why are diurnal primates living in groups? *Behaviour*, **87**, 120–44.

Vedder, A & Aveling, C (1987) 1986 census of the Virunga population of *Gorilla gorilla beringei*. *Gorilla Conservation News*, **2**, 8–12.

Vedder, A L (1984) Movement patterns of a group of free-ranging mountain gorillas (*Gorilla gorilla beringei*) and their relation to food availability. *American Journal of Primatology*, **7**, 73–88.

Vehrencamp, S L (1979) The roles of individual, kin, and group selection in the evolution of sociality. In *Handbook of Behavioral Neurobiology*, vol. 3, ed. P Marler & J G Vandenbergh, pp. 351–79. New York: Plenum Press.

Washburn, S L & DeVore, I (1961) Social behavior of baboons and early man. In *The Social Life of Early Man*, vol. 31, ed. S L Washburn, pp. 91–105. New York: Viking Fund Publications in Anthropology.

Waterman, P G, Choo, G M, Vedder, A L & Watts, D P (1983) Digestibility, digestion inhibitors, and nutrients of herbaceous foliage from an African montane flora and comparison with other tropical flora. *Oecologia*, **60**, 244–9.

Watts, D P (1984) Composition and variability of mountain gorilla diets in the central Virungas. *American Journal of Primatology*, **7**, 325–56.

Watts, D P (1985) Relations between group size and composition and feeding competition in mountain gorilla groups. *Animal Behaviour*, **33**, 72–85.

Watts, D P (1989) Infanticide in mountain gorillas: new cases and a reconsideration of the evidence. *Ethology*, **81**, 1–18.

Watts, D P (1990) Ecology of gorillas and its relation to female transfer in mountain gorillas. *International Journal of Primatology*, **11**, 21–44.

Watts, D P (1991) Strategies of habitat use by mountain gorillas. *Folia Primatologica*, **56**, 1–16.

Watts, D P (1992) Social relationships of immigrant and resident female mountain gorillas. I. Male–female relationships. *American Journal of Primatology*, **28**, 159–81.

Watts, D P (1994a) The influence of male mating tactics on habitat use in mountain gorillas. *Primates*, **35**, 35–47.

Watts, D P (1994b) Social relationships of immigrant and resident female mountain gorillas. II. Relatedness, residence, and relationships between females. *American Journal of Primatology*, **32**, 13–30.

Watts, D P (1996) Comparative socio-ecology of gorillas. In *Great Ape Societies*, ed. W C McGrew, L F Marchant & T Nishida, pp. 16–28. Cambridge: Cambridge University Press.

Watts, D P (2000) Causes and consequences of variation in male mountain gorilla life histories and group membership. In *Primate Males*, ed. P Kappeler, pp. 169–79. Cambridge: Cambridge University Press.

Watts, D P & Pusey, A E (1993) Behavior of juvenile and adolescent great apes. In *Juvenile Primates*, ed. M E Pereira & L A Fairbanks, pp. 148–67. New York: Oxford University Press.

Weber, A W (1987) Socioecologic factors in the conservation of Afromontane

forest reserves. In *Primate Conservation in the Tropical Rain Forest*, ed. C W
Marsh & RA Mittermeier, pp. 205–29. New York: Alan R. Liss.

Weber, A W (1989) Conservation and development on the Zaire–Nile Divide: an
analysis of value conflicts and convergence in the management of afromontane forests in Rwanda. PhD thesis, University of Wisconsin, Madison.

Weber, A W & Vedder, A L (1983) Population dynamics of the Virunga gorillas:
1959–1978. *Biological Conservation*, **26**, 341–66.

Williams, G C (1966) *Adaptation and Natural Selection*. Princeton: Princeton
University Press.

Wilson, E O (1975) *Sociobiology*. Cambridge MA: Belknap Press of Harvard
University Press.

Wrangham, R W (1979) On the evolution of ape social systems. *Social Science
Information*, **18**, 334–68.

Wrangham, R W (1980) An ecological model of female-bonded primate groups.
Behaviour, **75**, 262–300.

Wrangham, R W, McGrew W C, de Waal, F B M & Heltne P G (1994) *Chimpanzee
Cultures*. Cambridge MA: Harvard University Press.

Yamagiwa, J (1983) Diachronic changes in two eastern lowland gorilla groups
(*Gorilla gorilla graueri*) in the Mt. Kahuzi Region, Zaire. *Primates*, **24**, 174–83.

Yamagiwa, J (1986) Activity rhythm and the ranging of a solitary male mountain
gorilla (*Gorilla gorilla beringei*). *Primates*, **27**, 273–82.

Yamagiwa, J (1987) Male life history and the social structure of wild mountain
gorillas (*Gorilla gorilla beringei*). In *Evolution and Coadaptation in Biotic
Communities*, ed. S Kawanao, J H Connell & T Hidaka, pp. 31–51. Tokyo:
University of Tokyo Press.

Yerkes, R M & Yerkes, A W (1929) *The Great Apes*. New Haven: Yale University
Press.

Part I
The social system of gorillas

2 *Variation in the social system of mountain gorillas: the male perspective*

MARTHA M. ROBBINS

Pablo and Ziz, second and first ranking males, respectively, of Group 5 from the mid 1980s until Ziz's death in 1993. (Photo by Martha M. Robbins.)

Introduction

A wide variety of social systems is observed across the animal kingdom. The degree of sociality and the type of social system exhibited by a species are influenced by many factors including phylogeny, physiology, ecology, life history, and behavior (Emlen & Oring, 1977; Wrangham & Rubenstein, 1986; van Hooff & van Schaik, 1992; Lee, 1994). The observed social system is the outcome of individual strategies to maximize fitness, and both inter- and intraspecific variation in social systems can be studied within this framework. Examining intraspecific social system variation improves our understanding of the flexibility in behavioral patterns of individuals, the proximate mechanisms that produce such variation, the adaptive significance of sociality, and the evolution of social structure (Lott, 1991; Lee, 1994). Currently, behavioral ecologists are no longer viewing species as having static patterns of behavior and social structure, but instead the variation in group size, age structure, sex ratio, degree of relatedness between individuals, etc. is being examined (Strier, 1994). In particular, looking at the factors that influence the number of adult males per social unit is interesting because it relates to reproductive strategies of individuals (Kappeler, 2000).

One of the greatest benefits of the long-term monitoring of several neighboring mountain gorilla groups for over 30 years is that it has enabled us to gain a good understanding of mountain gorilla socioecology and to observe the variation in their social system. Mountain gorillas feed on evenly distributed herbaceous vegetation and face low feeding competition (Watts, 1985, 1996). Due to the distribution of food resources, feeding competition is mainly of the scramble type and there is little benefit from forming alliances with kin to attain food resources (Wrangham, 1979; van Schaik, 1989; Watts, 1990*a*). Group formation in mountain gorillas seems to be in part due to females selecting males who provide protection against other potentially infanticidal males and a strategy of long-term mate guarding by males (Harcourt, 1981; Wrangham, 1986; Watts, 1989, 1996). Mountain gorillas exhibit female transfer and male dispersal, although individuals of both sexes may remain in their natal group (Harcourt *et al.*, 1976; Watts, 1996). Mountain gorillas have been classified as "non-female-bonded" (Wrangham, 1980; van Schaik, 1989) and "dispersal egalitarian" (Sterck *et al.*, 1997), and long-term male–female relationships are at the base of sociality for gorilla groups (Harcourt, 1979*a*; Stewart & Harcourt, 1987; Watts, 1996).

Despite that fact that multimale groups have been observed since mountain gorillas were first studied intensively in the late 1950s (Schaller, 1963), the classic description of a mountain gorilla group is that of one silverback,

30

Table 2.1. *Percent of gorilla populations consisting of multimale groups*

Gorilla population	Percent multimale groups
Mountain gorillas	
Virunga Volcanoes[a]	26%–40%
Bwindi Impenetrable Forest[b]	46%
Eastern lowland gorillas	
Kahuzi-Biega[c]	8%
Western lowland gorillas[d]	Observed in some but not all study areas; percentage unknown

[a] Harcourt & Groom, 1972; Weber & Vedder, 1983; Sholley, 1991.
[b] McNeilage et al., 1998.
[c] Yamagiwa et al., 1993.
[d] Remis, 1994; Goldsmith, 1996; Tutin, 1996; Doran & McNeilage, 1998; Magliocca et al., 1999.

several females, and their offspring. A comparison of various gorilla populations illustrates that multimale groups and variation in the social system are regular phenomena in gorillas (Table 2.1). However, a common interpretation of multimale groups assumes that because they are age-graded, they exist for only short periods of time, prior to the dispersal of maturing natal males (Eisenberg et al., 1972), when in fact males can coexist in a group for many years. We can therefore ask why do we see the observed proportions of one-male, multimale, and all-male groups in the Virunga population and what are the benefits of and constraints to the various group structures?

Many ecological, demographic, and behavioral factors act together to influence the variation in the social system of mountain gorillas. In this chapter, I first examine the demographic influences on male membership in groups. Then I examine several aspects of social relationships from the male's perspective, specifically male–male relationships, male–female relationships, and reproductive strategies. I also use the outcome of a group fission as a case study to further explore male–female relationships.

Demographic influences
Mountain gorillas have an age-graded social structure with the existence of and transition between solitary individuals and one-male, multimale, and all-male gorilla units (Figure 2.1) (Eisenberg et al., 1972; Yamagiwa, 1987a; Robbins, 1995). According to this model, a one-male group forms when a solitary silverback (adult male) acquires females from other groups. When male offspring reach maturity, which takes approximately

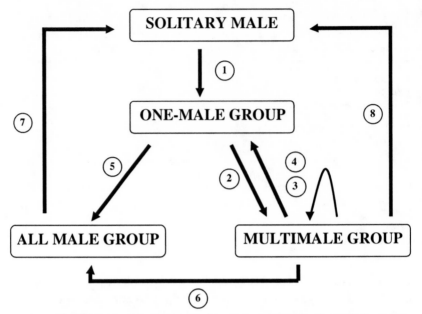

Figure 2.1. Transition between social units in mountain gorillas. (1) Solitary males acquire females to form one-male groups. (2) Maturing males remain in natal group and the group becomes multimale. (3) Death of a silverback in a multimale group may cause it to return to the one-male structure or retain its multimale structure. (4) A multimale group fission results in groups being either one-male or multimale. (5, 6) Males emigrate into an all-male group from either a one-male or multimale group. (7, 8) Males may emigrate to become solitary from either a heterosexual or all-male group. (Expanded from Yamagiwa, 1987a.)

13 years, the group can be considered a multimale group. If a silverback in a multimale group dies, depending on the number of silverbacks present, the group either returns to the one-male group structure or remains as a multimale group. If the remaining male(s) are incapable of maintaining the group, it may disband with females joining other groups. Multimale groups may also fission to form new groups with one or more males. Males from heterosexual groups may also form or join all-male groups and males may emigrate from heterosexual or all-male groups to become solitary.

Within the Karisoke study groups (Figure 2.2), we have observed three cases of one-male groups disintegrating upon the death of the silverback (Group 8, Nunkie's Group, Tiger's Group). Five groups have been observed to fluctuate between being one-male and multimale (Group 4, Group 5, Group 8, Beetsme's Group, Shinda's Group). A group fission was observed in 1993 when a multimale group (Group 5) split following the death of the dominant silverback, resulting in a one-male and a multimale group (Shinda's Group and Pablo's Group). An all-male group

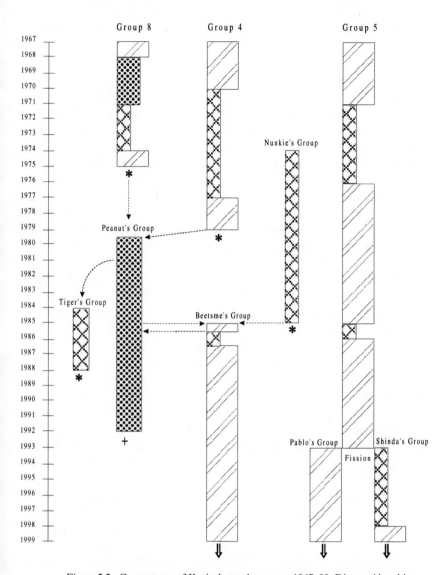

Figure 2.2. Group types of Karisoke study groups, 1967–99. Diagonal hatching denotes multimale group status, crosshatching denotes one-male group status, and dots indicate all-male group status. * denotes disintegration of the group due to death of leader silverback. + denotes disintegration of group due to all members emigrating. Dashed arrows indicate movement of monitored individuals following group disintegration (e.g. Peanut's Group acquired a solitary male who had been in Group 8 and three males from Group 4. Beetsme's Group formed through the merger of Peanut's Group and Nunkie's Group following the death of Nunkie. Three subadult males from Nunkie's Group later joined Peanut's Group.). Solid arrows indicate continuing status of groups in 1999. (Modified from Robbins, 1995.)

(Peanut's Group) gained its members from disintegrating groups (related males) and unrelated solitary males.

No cases of fully mature males immigrating into heterosexual groups have been observed and in only two cases have immature males been seen to successfully join heterosexual groups (Fossey, 1983; Sicotte, 2000). Therefore multimale groups usually consist of either father–son or brother/half-brother combinations, although sometimes groups may contain unrelated males (Robbins, 1995). The same mature males have coexisted in multimale groups for periods of over 10 years, indicating that multimale groups are not simply a temporary situation prior to male dispersal upon reaching maturity. Less than half ($n = 4$) of all males that reached maturity in heterosexual groups ($n = 11$) emigrated (Robbins, 1995; Watts, 2000). All males who were members of an all-male group (Peanut's Group) emigrated ($n = 12$) or died within the group ($n = 1$).

Given the variation in social units, does one particular group type (one-male or multimale) confer an advantage to male mountain gorillas at different stages of their life? In multimale groups there is a reduced risk of infanticide and immature males have an increased likelihood of remaining in a heterosexual group through maturity compared to males growing up in one-male groups (Robbins, 1995). If the silverback in a one-male group dies, the females move into other groups in which the new silverback is typically intolerant of offspring sired by another male, leading to infanticide of unweaned infants (Watts, 1989). Immature males in this situation face group eviction and typically end up in all-male groups. However, if a silverback dies in a multimale group, another silverback is often able to take over leadership and retain the integrity of the group, enabling offspring to remain in a stable group setting that can confer reproductive advantages later in life (Yamagiwa, 1987a; Robbins, 1995).

The benefits of multimale group membership continue into adulthood. Most importantly, males that adopt the strategy of "following" or staying, in which males remain in the group where they reach maturity, appear to have a higher success rate of obtaining reproductive opportunities compared to males who disperse and attempt to form new groups (Robbins, 1995; Watts, 2000), although more data on the fates of dispersing males would be helpful. This strategy is believed to be more advantageous in other species (waterbuck, *Kobus ellipsiprymnus*: Wirtz, 1982; howler monkey, *Alouatta seniculus*: Pope, 1990; gelada baboon, *Theropithecus gelada*: Dunbar, 1984; golden lion tamarin, *Leontopithecus rosalia*: Baker *et al.*, 1993). In addition, during inter-group encounters, males may form coalitions to defend the females in their group and prevent them from emigrating (Sicotte, 1993). Silverbacks may lose their dominance status as they age but they are not evicted from the group (Robbins, 1995). Presumably, it is beneficial for these males to stay in their groups because they can provide further protection and care to their offspring and they may have greater reproductive opportunities than if they were solitary males.

Paternity determination studies are necessary to study how differing male strategies affect lifetime reproductive success. A preliminary study of paternity determination in Group 5 has identified several useful genetic markers, but the results are limited due to the small number of individuals examined and the lack of samples from all possible sires (Field *et al.*, 1998). However, estimates of reproductive success from demographic data suggest two things (Robbins, 1995). First, male reproductive success appears to vary greatly; some males may never reproduce and others may sire as many as twelve offspring. Second, male reproductive success is positively correlated to the number of adult females in the group and the length of male reproductive tenure. This indicates that males in multimale groups may attract and/or retain more females, but genetically based paternity studies are needed to determine the exact reproductive costs and benefits to both dominant and subordinate males.

The many benefits to the multimale group structure leads to the question of why multimale groups are not more common. A study to examine how particular population level variables (e.g. habitat quality, sex ratio, age structure) can influence the occurrence of each group type would be useful. Currently we can identify both demographic and behavioral limitations to group structure. Perhaps most important, there have been no recorded cases of outsider silverbacks taking over or joining an established group (but see Yamagiwa, this volume). Therefore, the only route for a one-male group to become multimale is for the male offspring within the group to reach maturity, a process that takes approximately 13 years. From a behavioral perspective, male–male competition for reproductive opportunities is high (Harcourt, 1981; Watts, 1990*b*, 1996; Sicotte, 1994). As a result, male–male social relationships are usually weak compared to relationships between males and females. How is it that despite these limitations males can coexist in multimale groups?

Social relationships
Male–male relationships
In general, male primates form much weaker social relationships with one another than females do with each other or with males, and mountain gorillas are no exception (van Hooff & van Schaik, 1992, 1994; Kappeler, 2000). This is primarily because males compete for access to females, a resource that is not easily divided, compared to competition for access to

food resources which influences female–female social relationships. Male philopatry may encourage relationships among male relatives, but kinship is neither a prerequisite for nor a guarantee that strong male–male relationships will form (van Hooff & van Schaik, 1994; Kappeler, 2000).

Early studies of male–male relationships in mountain gorillas focused on relationships between silverbacks (adult males) and blackbacks (maturing males, age 8–12 years) because of the composition of the study groups (Harcourt, 1979*b*; Harcourt & Stewart, 1981). Silverbacks are always dominant over blackbacks. Blackbacks generally spend a great deal of time on the periphery of the group, perhaps to avoid conflict with the silverback(s) and adult females. These studies revealed that although silverback–blackback relationships are generally weak, there is variation in the strength of the relationship as measured by proximity patterns and levels of aggression and affiliation, which may be due to the relatedness or familiarity between the males. Harcourt & Stewart (1981) hypothesized that blackbacks who have strong relationships with the silverback were less likely to emigrate than males with weak social relationships, but this topic has not yet been examined in a systematic manner.

Other types of male social relationships, specifically those between silverbacks in heterosexual groups and those in all-male groups, have been examined as the composition of the study groups has permitted. In heterosexual groups, relationships between silverbacks are weak, with males spending little time in close proximity (< 5 m) and rarely affiliating with each other (Robbins, 1996). Dominance relationships are clearly defined and are not always positively correlated with age; younger males may ascend to the alpha position while the older male remains in the group. Despite obvious dominance relationships, aggression may be directed to the dominant silverback from the subordinate males, especially on days when females are in estrus (Sicotte, 1994). Rates of aggressive behavior between males will depend on a variety of factors, including the number and reproductive status of females, the age of males, and the stability of their relationships. Agonistic interventions are rarely observed among silverbacks (Robbins, 1996; Watts, 1997), partly because most multimale groups contain only two males. Reconciliation between adult males has not been observed (Watts, 1995*a*). Sicotte (1995) suggests that interpositions by females and infants between two competing males can be a proximate mechanism that facilitates male coexistence. Male behavior has not been studied shortly before emigration from heterosexual groups to determine if male dispersal is related to an increase in rates of aggression and intolerance on the part of the dominant male. Overall, it appears that in groups with stable dominance hierarchies, the males coexist through avoidance or tolerance of one another rather than by using frequent, active antagonism or by forming strong supportive relationships.

Relationships among males in all-male groups are quite different from those in heterosexual groups (Yamagiwa, 1987*b*; Robbins, 1996). Males are much more affiliative with each other based on the higher amount of time spent grooming, playing, and in close proximity. In two groups of five to six males each, the frequency of homosexual behavior was approximately one encounter per 10 hours of observation, and appeared to be partially based on the group composition, with the presence of subadults leading to more encounters. Dominance relationships are apparent between males of differing age classes, with silverbacks being dominant over all other individuals, but dominance relationships are less clear among individuals in the same age class, particularly among blackbacks. Although males in one all-male group exhibited higher rates of aggression than did males in two heterosexual groups, most of the aggression consisted of mild pig-grunting or moderate displays, rather aggression with physical contact. This is further shown by the higher incidence of wounds on the males in the heterosexual groups (Robbins, 1996). Interventions during agonistic encounters may play an important role in maintaining group stability in all-male groups, with younger males mediating in fights between older silverbacks (Yamagiwa, 1987*b*, 1992). The proximate factors leading to male dispersal from all-male groups are not well understood. In the case of one all-male group, the dominant male's unexpected emigration did not appear to be precipitated by any changes in aggressive or affiliative interactions with other group members (Robbins, 1996, unpublished data).

Although all-male groups provide no reproductive opportunities for males, they may furnish males with a better setting to develop adult social relationships than if they were solitary males because they can gain experience in aggressive and affiliative social interactions. Most males who reside in all-male groups eventually emigrate and some of them are successful at forming their own heterosexual groups. We do not yet have enough data to determine if males who have resided in all-male groups are more or less successful at group formation and reproduction than males who follow the solitary male strategy.

Overall, these studies show that there are weak male–male social relationships in mountain gorillas. As has been observed in all-male groups of other primate species (Pusey & Packer, 1987), males in all-male mountain gorilla groups who do not have any alternative social partners, nor reproductive opportunities over which to compete, have stronger relationships with each other than do males in heterosexual groups. The weak social relationships among males in heterosexual groups suggest that the competitive interactions over access to reproductive opportunities are a stronger influence on their social relationships than the benefits of cooperative defense of females from outsider males. Indeed, when an all-male group (Group PN) merged with the females and offspring of a heterosexual group following the death of the silverback (Group NK), the rate of aggression between the two unrelated silverbacks increased sharply until eventually one of the silverbacks was evicted (Watts, 1989).

Male–female relationships
Male–female relationships reflect the balance of costs and benefits to both sexes (van Hooff & van Schaik, 1992). In the case of mountain gorillas, it seems to be in a female's best interest to stay with a male that can provide protection against harassment and infanticide from extra-group males (Wrangham, 1986; Watts, 1989). Because of dispersal patterns, females often lack relatives in their group so males may be their most consistent long-term social partners. Males benefit from group formation by gaining long-term access to mates. Male behavior should be oriented towards acquiring new mates, preventing females from emigrating, and protecting offspring.

The most important early contribution to understanding male–female social relationships in mountain gorillas was a study of two one-male groups by Harcourt (1979a, c). Watts (1992) studied a multimale group and added much to what we know about variation in male–female relationships. Several factors, including female residence status, male age, and relatedness among individuals, influence male–female relationships. Despite the many studies of male–female relationships at Karisoke, we know little about the development of male–female relationships as individuals reach maturity, or when females immigrate into new social units (but see Watts, 1991b, 1992; Sicotte, 2000). Both of these topics are particularly important for understanding male dispersal patterns and female group choice.

Mountain gorillas spread out during feeding periods, presumably to reduce feeding competition, and then usually congregate near the silverback during rest sessions (Fossey & Harcourt, 1977; Harcourt, 1979a). Females can spend as much as 50% of their resting time and approximately 20% of their feeding time within 5 m of the silverback (Watts, 1992). Females appear to be more responsible than males for maintaining social proximity in one-male groups (Harcourt, 1979a). The reproductive status of females is an important factor in determining the amount of time females spend near silverbacks. Females with young infants spend significantly more time near silverbacks than do other females, presumably for protection and to familiarize the offspring with the silverback. There is an increase in the time spent in close proximity between males and females on days when females are in estrus (Harcourt, 1979a). In multimale groups, male–male competition influences the role males take in maintaining their relationships with females. Sicotte (1994) showed that in multimale groups, males were more responsible for maintaining proximity to females that were proceptive than to nonproceptive females and males followed proceptive females. Although there was no difference in the rate of "neigh-ing" (a vocalization emitted by males towards departing females) towards proceptive and nonproceptive females, perhaps because males already were maintaining close proximity to proceptive females, males were more likely to "neigh" to nonproceptive cycling females than to lactating fe-males (Sicotte, 1994).

Grooming, a typical measure of affiliation in primates, is not a common activity in mountain gorillas. Interestingly, there is no clear pattern of the directionality of grooming between the sexes. In some cases, particular silverbacks receive more grooming than they give and in other cases the opposite occurs (Harcourt, 1979a; Watts, 1992). Some males, especially subordinate silverbacks, may use grooming as an affiliative mating strat-egy (Watts, 1992).

Adult males are dominant over all females, but females are often dominant over blackbacks. It is unknown at what age this rise in domi-nance status occurs for males, but it probably correlates with the increase in size as males reach maturity. Silverbacks are more aggressive toward females than vice versa, with most aggression by females toward males consisting of pig-grunt vocalizations. Male aggression toward females is common, despite the lack of resource competition, but it consists mainly of displays rather than physically damaging attacks (Watts, 1992). This aggression could be a mating strategy or form of sexual coercion to prevent females from transferring to another social unit (Harcourt, 1981; Watts, 1992; Sicotte, unpublished data, this volume). During inter-group encoun-ters males may aggressively herd females away from the opposing group (Sicotte, 1993).

Harcourt (1981) suggested that within one-male groups the silverback does not need to increase the rate of aggression toward females when they are in estrus because the female has no choice in mates. However, in multimale groups where females do have a choice, if aggression is a mating tactic an increase in rate of aggression would be expected toward potential mates during the months when females are cycling and particularly on the days they are in estrus. To the contrary, Harcourt (1979a) observed lower rates of aggression directed toward females on the days they were in estrus compared to days when they were not. In multimale groups, Sicotte (unpublished data) found no difference in the rate of male displays directed at females on estrus versus nonestrus days. However, several studies have found that recent immigrant females (presumably cycling) received higher rates of displays and aggression than long-term resident females (Harcourt, 1979*a*; Watts, 1992; Sicotte, 2000). Comparing rates of aggression directed at subadult versus adult females would be useful for determining how male–female relationships develop as females reach maturity.

Females will reconcile with males after conflicts, further indicating that these relationships are important to the females (Watts, 1995*a*, *b*). Males intervene in many female–female conflicts but they rarely support either female (Harcourt & Stewart, 1987, 1989; Watts, 1997). These "control interventions" may prevent differences in competitive abilities from developing among the females, which could reduce the likelihood of female emigration caused by competition among females (Watts, 1997).

Case study: effect of male–female relationships on outcome of group fission

Given that male–female relationships appear to be more important than female–female relationships to the integrity of mountain gorilla groups, one would expect that following a group fission the composition of the resulting groups would not be a random outcome, but would be dependent on the relationship each male had with the females. One would expect the females to remain with the male with whom they had a stronger relationship. Specifically, the females should have higher levels of close spatial proximity and higher levels of affiliation with the male they remain than with the other male. Additionally, if male aggression toward females (sexual coercion) serves to influence female long-term mate choice, then one would also expect higher levels of aggression directed toward females that remain with each particular male. In 1993, following the death of the dominant silverback, Group 5 split into two groups. Unfortunately, no researchers were present at Karisoke for several months when the silverback died and the group split so male–female interactions shortly prior to and during the fission were not observed. My goal here is not to determine what caused the fission, but to see if variation in male–female relationships prior to the fission could be used as a determinant of which male each particular female joined.

At the time of the death of the dominant male (Zz), Group 5 contained three other silverbacks, ten adult females, and two subadult (sexually active) females. One of the resulting groups consisted of the second (Pb) and fourth (Ca) ranking males and six females (Group PB and hereafter

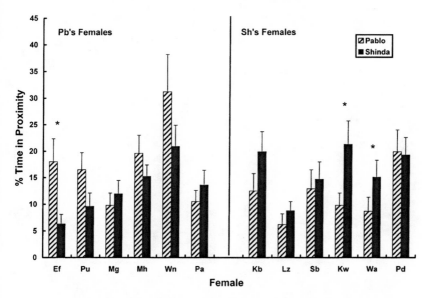

Figure 2.3. Comparison of time spent in proximity between the females and the two silverbacks in Group 5 prior to group fission. Bars represent means and standard errors. * $p < 0.05$.

called Pb's females). The other six females (referred to as Sh's females) remained with the previously third ranking male (Sh; Group SH). From 1990 to 1992, during over 80 hours of focal animal sampling on each adult male, all agonistic, affiliative (grooming and resting in physical contact), and sexual interactions between the males and females were noted. All aggressive interactions reported here were from the males directed at the females (females rarely aggress against adult males and it is usually mild pig-grunting), whereas affiliative interactions initiated by both males and females were combined for analysis. Hourly rates of aggression and affiliation per male–female dyad were calculated by dividing the number of interactions by the total amount of focal observation time. Wilcoxon Rank Sum tests were then used to compare hourly rates of aggression and affiliation that each subset of females had with each male. Also, every 15 minutes, scan samples were conducted to record all individuals within 5 m of the focal male. The mean time spent in proximity to each male was determined by calculating the percentage of time females were within 5 m of the male per focal period (number of occurrences in proximity divided by the total number of scans) and taking a mean of all the focal periods for each male. Mann–Whitney U tests were used to compare the mean time that each female spent in proximity to the two males.

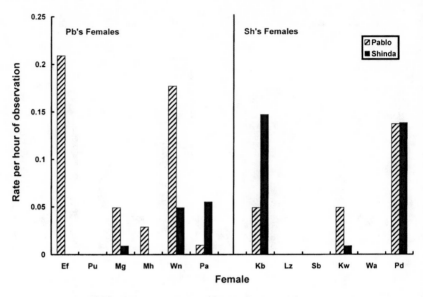

Figure 2.4. Comparison of rates of affiliation between females and the two silverbacks in Group 5 prior to group fission.

Four of the six females who remained with Pb were within 5 m proximity to him for a greater percentage of time than to Sh, but only one female spent significantly more time near Pb (Figure 2.3) (Ef: Mann–Whitney U Test, $n_1 = 29$, $n_2 = 28$, $U = 520.0$, $p = 0.031$). Five of the six Sh's females were in 5 m proximity to him more often than to Pb, but the values reached significance for only two of the females (Kw: $U = 268.5$, $p = 0.041$; Wa: $U = 277.0$, $p = 0.05$).

There were low rates of affiliation between both Pb and Sh and the females, but it does appear that there was a tendency for some females to join the male with whom they had more affiliative interactions (Figure 2.4). There was significantly more affiliation between Pb and his females than between Sh and Pb's females (Wilcoxon Rank Sum test, $n = 6$, $p = 0.014$), whereas there was no significant difference in the level of affiliation between Sh and his females than between Pb and Sh's females ($p = 0.593$). Sh did have a higher rate of affiliation than Pb with one of his females but overall, there were very few affiliative interactions involving Sh's females.

Both Pb and Sh were more aggressive toward four of six females that remained in his group than was the other male (Figure 2.5). However, neither silverback exhibited a significantly greater rate of aggression toward either the females who stayed with him or the females that joined the

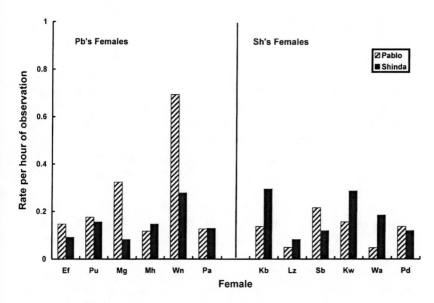

Figure 2.5. Comparison of rates of aggression between females and the two silverbacks in Group 5 prior to group fission.

other group (Wilcoxon Rank Sum test, Pb's females: $p = 0.173$; Sh's females: $p = 0.173$).

The number of matings observed was too low for statistical analysis. However, of the eight females that were observed mating, five of them mated with both Pb and Sh. One female, Pa, was observed to mate only with Pb and not with Sh (her maternal sibling), whereas other females, Pd and Sb, were seen to mate only with Sh. However, we cannot exclude the possibility that matings with these three females occurred when observers were not present.

Other variables, particularly the relatedness between females and males, probably also influenced the compositions of the two groups. Female Pa was a maternal sister to Sh, and despite her higher proximity value and rate of affiliation with Sh, she stayed in Pablo's Group. Four of the six females who went into Pablo's Group were related to each other and to the other silverback in the group, Ca (Ef was the mother of Pu, Mg, and Mh and the grandmother of Ca through Pu).

None of the variables examined (proximity, aggression, affiliation, sexual behavior, or relatedness) independently explain the outcome of the fission. However, by considering all of these variables together, it does appear that preexisting male–female relationships had an impact on the composition of the two groups following the fission.

Mating behavior

Mountain gorillas were originally considered to have a one-male mating system (Harcourt, 1981). This is not surprising given the groups studied in the 1970s were predominantly one-male groups. Initially little attention was paid to female choice or sexual coercion in mountain gorillas, probably because male–male competition was the focus of studies of sexual selection in animal behavior at the time. As the size of study groups increased, by both the number of silverbacks and females, the full complexity of reproductive strategies used by mountain gorillas began to emerge and enabled researchers to study various aspects of sexual selection within groups.

Male–male competition for access to mates is considered to be high both between and within groups (Harcourt, 1978, 1981; Watts, 1990*b*, 1996; Sicotte, 1993). One possible benefit of obtaining high dominance rank is to have higher reproductive success than that of subordinate individuals. Indeed, dominant male mountain gorillas have been observed to participate in the majority of matings in multimale groups, emphasizing a strong relationship between dominance rank and mating success, but subordinate males do mate, even at the likely time of conception (Watts, 1990*b*, 1991; Robbins, 1999). However, when using mating behavior as an indirect measure of reproductive success, it is important to consider the reproductive status of the females involved. In two heterosexual groups, dominant males were observed to mate more than subordinate males with cycling adult and pregnant females (Robbins, 1999). Subordinate males mate more with subadult females than do dominant males (Harcourt, 1981; Watts, 1990*b*, 1991; Robbins, 1999); such matings may pose little threat to the reproductive success of dominant males, which may explain the low levels of male–male aggression associated with certain matings.

Harassment was observed for 30% (35 of 115) and 22% (12 of 54) of matings in two groups and it usually consisted of only mild aggression (pig-grunt vocalizations or running at the mating pair). Both dominant and subordinate males harass each other while mating and their rank appears to influence their success in terminating copulations. For example, in Group 5 where dominant males harassed subordinate males more than vice versa, 63% of harassed copulations were stopped (20% of overall copulations), whereas in Beetsme's only 8% of harassed copulations were terminated. This was probably because most of the harassment was by the older deposed male and he was unsuccessful in stopping the younger dominant male (Robbins, 1999). Overall, the occurrence of harassment seems low given the relatively tight spatial group cohesion and high levels of male–male competition characteristic of mountain gorillas.

Interestingly, in two multimale groups studied the majority of females (11 out of 14) mated with more than one male, suggesting that female choice does play a role in the mating system of mountain gorillas (Robbins, 1999; Sicotte, this volume). Watts (1992) and Smuts & Smuts (1993) have suggested that two opposing but not mutually exclusive strategies, affiliation and coercion (aggression), are used by male mountain gorillas to influence female choice. However, we do not know if females exhibit partner preference, and if so, whether it is based on male dominance rank or the strength of social relationships with particular males (Watts, 1992), or if they simply choose to or are coerced to mate with all the males in the group. Females may also mate with more than one male in the group to increase the potential that these males will protect their infant against infanticidal attacks as suggested for Hanuman langurs (Borries *et al.*, 1999). Mate guarding could also be used to restrict female mate choice in multimale groups. According to early studies of one-male groups, male gorillas do not mate guard or form consort relationships with females (Harcourt *et al.*, 1980), but more recent observations of multimale groups suggest that mate guarding may occur (Sicotte, 1994; Watts, 1996).

These observations of mating behavior in the larger groups indicate that the mating system of mountain gorillas should perhaps be considered multimale (Robbins, 1999) or polygynandrous (Watts, 2000). Paternity determination studies are necessary to resolve the issue of reproductive costs and benefits of the multimale group structure to dominant and subordinate males. Further examination of male–female relationships in relation to within-group male–male competition, female reproductive status, and mating behavior in both one-male and multimale groups is necessary to understand strategies used by males and females to obtain mates.

It should also be noted that these recent findings on the variability of the mountain gorilla mating system do not fit well into the sperm competition hypothesis which correlates relative testes size with the mating systems of the great apes and primates in general (Short, 1979; Harcourt *et al.*, 1981, 1995). According to this hypothesis, intrasexual competition in mountain gorillas occurs mainly on the level of male–male competition that has resulted in extreme sexual dimorphism, the long-term monopolization of several females by one male, and therefore minimal opportunities for sperm competition, and no selection for increased testes size (Short, 1979; Harcourt *et al.*, 1981). Some possible explanations for the observed disparity, all of which would be difficult to empirically test, include: (1) the recent observations of a multimale mating system are actually maladaptive behavior, perhaps induced because of more males staying in their natal groups as a result of human pressures; (2) testes size in gorillas is a nonadaptive trait disassociated from other sexually selected traits including their extreme sexual dimorphism; and (3) sexually selected behavioral flexibility has not evolved in tandem with morphological traits. Other elements are important to consider, however, to understand the link between sexual dimorphism, testes size, and male mating behavior. For instance, female gorillas have short estrous periods (1–2 days on average) and relatively few females are in estrus simultaneously, which results in male gorillas mating relatively infrequently compared to other species such as chimpanzees. Interestingly, there are other species (e.g. lions, gelada baboons) that do not conform to the expected relationship of relatively larger testes in species exhibiting multimale mating systems (Dixson, 1998).

Discussion
Variation in the social system of mountain gorillas
The three decades of research at Karisoke have enabled researchers to witness much more variation in the social system of mountain gorillas than would be attainable through a short-term study. However, more long-term data are needed to fully address particular questions concerning life histories, reproductive strategies, and gorilla social dynamics. For example, we still do not know the average life span for gorillas, and particular phenomena (e.g. group formation and group fission) occur infrequently.

Based on our knowledge of ecological, demographic, and behavior patterns of the Karisoke mountain gorillas, I suggest a model to explain the social system of gorillas (Figure 2.6). Part A is the basic explanation for the observed group formation in mountain gorillas based on the general socioecological model for primates (Sterck *et al.*, 1997) and that specific to mountain gorillas (Watts, 1996). Part B investigates the factors that contribute to the variability observed in the social system, some of them having been discussed in this chapter. First, when considering emigration (for males) and transfer (for females) decisions, individuals need to assess the conditions of their current social unit against that of external conditions. Availability of food resources will need to be sufficient for survival and allow for a tolerable level of feeding competition. This will vary depending on group size as well as the density of gorillas in an area of particular habitat quality. Next, individuals will need to consider the demographic conditions, specifically the number, age, and relatedness of potential mates and same-sex group members. However, an assessment of actual demographic conditions may be tempered by particular morphological and behavioral aspects of sexual selection (e.g. variability of body size, male fighting ability, female choice, etc.). A temporal element

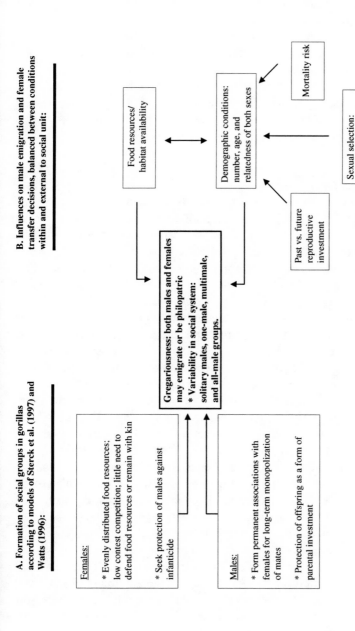

A. Formation of social groups in gorillas according to models of Sterck et al. (1997) and Watts (1996):

B. Influences on male emigration and female transfer decisions, balanced between conditions within and external to social unit:

Females:

* Evenly distributed food resources; low contest competition; little need to defend food resources or remain with kin

* Seek protection of males against infanticide

Males:

* Form permanent associations with females for long-term monopolization of mates

* Protection of offspring as a form of parental investment

Gregariousness: both males and females may emigrate or be philopatric
* Variability in social system: solitary males, one-male, multimale, and all-male groups.

Food resources/ habitat availability

Demographic conditions: number, age, and relatedness of both sexes

Mortality risk

Sexual selection: competition, choice, and coercion

Past vs. future reproductive investment

Figure 2.6. Schematic model of factors influencing the variable social system in mountain gorillas. (A) Basic explanation for group formation based on the general socioecological model for primates (Sterck *et al.*, 1997) and that specific to mountain gorillas (Watts, 1996); (B) factors that contribute to the variability in the social system.

also contributes to individual strategies, as individuals need to weigh their past versus future parental/reproductive investment as well as lifetime variation in mortality risk (e.g. age). Stochastic factors (e.g. a skewed sex ratio at birth or differential mortality caused by disease or poaching) may also cause a sudden shift in individual strategies that, in turn, may impact the overall distribution of group types across the population. In addition, one must consider whether individuals are acting "optimally" or they are "making the best of a bad job"; such a compromise may be due to conflicting strategies within and between the sexes.

Areas for future research

To date, the socioecological research conducted at Karisoke has focused primarily on the within-group aspects of food availability and feeding competition, demographic conditions, and behavioral patterns, with little emphasis placed on these variables from the perspective of between-group interactions (but see Sicotte, this volume) or from the level of the entire Virunga gorilla population and many questions still remain unanswered. The observations of the study groups in recent years lead to the question of whether multimale groups perpetuate multimale groups, as their presence may make it particularly difficult for a male to succeed in forming a new one-male group. Also, little is known about the relationship between group type and female reproductive success. For example, Gerald (1995) found that females in multimale groups begin to reproduce significantly earlier than do females in one-male groups, but we do not know how group type affects female lifetime reproductive success. It appears that females are more likely to join multimale groups than is expected by chance (Watts, 2000), but many factors influence female transfer decisions and group choice (Sicotte, this volume). Females may choose between smaller groups with less competition among females or larger, multimale groups that offer greater within-group mate choice and increased protection against harassment from outsider males (Watts, 1990b).

Future work on conditions external to particular study groups would be beneficial. In particular, it will be interesting to see how an increase in population size, which will lead to an increased density of gorillas and increased competition for habitat, will affect the variability of the social system. Yamagiwa (1999) suggests that a change in population density of gorillas may influence the degree of between-group male–male competition and opportunities for female transfer. In baboons, the cost of dispersal for males, which impacts male fitness, is strongly influenced by population density and predation risk (Alberts & Altmann, 1995).

Comparisons with other gorillas and apes

This model (Figure 2.6) can be extended to other populations and sub-species of gorillas. Currently other researchers are considering how food distribution and abundance, population density, and the risk of infanticide influence the social system in other gorilla subspecies (Yamagiwa, 1996, 1999, this volume; Doran & McNeilage, 1998, this volume). Research on the behavioral ecology and reproductive strategies of the only other population of mountain gorillas, in Bwindi Impenetrable National Park, Uganda, will also provide a test of this model (Robbins, unpublished data). Increased habituation and long-term monitoring of groups from these areas will shed light on the relative importance of ecological, demographic, and behavioral influences on social variability in gorillas.

Mountain gorillas present an interesting case when comparatively studying primate socioecology because both sexes disperse and they exhibit such variation in their social system. A comparison of the four species of great apes reveals a wide variety of male behaviors associated with the variation in social systems (Table 2.2). On the one extreme, orangutans generally associate very little with each other due to constraints imposed on them by their nutritional needs (van Schaik & van Hooff, 1996). At the other extreme of male social bonding, we find multimale–multifemale communities of chimpanzees with philopatric males who develop strong social ties. Although males compete for access to mates, they have elaborate systems for alliance formation and for reconciling after conflicts (Goodall, 1986; Furuichi & Ihobe, 1994; Boesch, 1996). Bonobos are more similar to chimpanzees than to either orangutans or mountain gorillas, but it appears that male–female relationships may play an important role in the maintenance of male dominance rank (Furuichi & Ihobe, 1994; White, 1996; Hohmann et al., 1999). Mountain gorillas fit somewhere in the middle of this continuum with the possibility of coexistence in groups, but weak male–male social relationships. One obvious omission to this discussion is the social system of western lowland gorillas because little is known about their behavior patterns. The distribution of fruit and terrestrial herbaceous vegetation and the consumption of these foods by lowland gorillas, relative to that of mountain gorillas, may lead to differences in their socioecology. For example, lowland gorillas may experience more within-group competition for food resources, a greater likelihood to subgroup, and possibly lower limits on group size, although they likely will still be influenced by the threat of infanticide (Remis, 1994; Goldsmith, 1996; Tutin, 1996; Doran & McNeilage, 1998, this volume).

Table 2.2. *Comparison of variables related to the influence of males on the social systems of the great apes*

	Orangutan[a]	Mountain gorilla[b]	Chimpanzee[c,d,e,f]	Bonobo[e,g,h]
Grouping pattern	Solitary; very loose communities	Solitary males; one-male, multimale, all-male groups	Multimale, multifemale fission–fusion communities	Multimale, multifemale fission–fusion communities
Dispersal patterns	Mainly female philopatry	Both sexes exhibit philopatry and dispersal	Male philopatry; occasional female philopatry	Male philopatry
Mating system	One-male, multimale	One-male, multimale	Multimale; extra-goup reproduction observed	Multimale
Infanticide as sexual selection	Not observed	Yes	Yes	Not observed
Male affiliation	Almost none	Low; more common in all-male groups	Common	Less common than in chimpanzees
Male dominance relationships	Decided, but not always transitive	Decided; less apparent in all-male groups	Decided among high and low ranking males; less apparent in mid ranking males	Decided among high ranking males; less apparent in low ranking males
Male–male alliances during within-group interactions	None	None	Yes	Occasional
Male–male alliances during between-group encounters	None	Some, in multimale groups	Yes	None
Male–male conflict resolution	None	None	Very common; formalized behavior pattern	Less common than in chimpanzees

[a] van Schaik & van Hooff, 1996.
[b] This chapter and references therein.
[c] Goodall, 1986.
[d] Boesch, 1996.
[e] Furuichi & Ihobe, 1994.
[f] Gagneux et al., 1997.
[g] White, 1996.
[h] Hohmann et al., 1999.

Comparisons with other primates and mammals

Looking to other primate species, the social systems of langurs offer an interesting comparison with variability in female transfer patterns, female dominance relationships, number of males in groups, the occurrence of all-male groups, and the incidence of infanticide (Newton & Dunbar, 1994; Sterck, 1997, 1999; Sterck & van Hooff, 2000). For example, Hanuman langurs (*Presbytis entellus*) are found in one-male, multimale, and all-male groups and females seek protection against infanticide from males (Newton, 1988), however males disperse and females form philopatric matrilines. In addition, all-male groups may take over heterosexual groups by evicting leader males (Rajpurohit *et al.*, 1995). In comparison, Thomas langurs (*Presbytis thomasi*) exhibit low levels of feeding competition and females disperse (Sterck, 1997; Sterck & Steenbeek, 1997). Males protect females against infanticide, group takeovers do occur, and the age-graded structure appears to benefit dominant males (Sterck, 1997; Steenbeek *et al.*, 2000).

Some parallels of the variation between one-male, multimale, and all-male groups observed in mountain gorillas can also be made with non-primate species (e.g. horses, *Equus caballus*: Berger, 1986; Stevens, 1990; Feh, 1999; oribi, *Ourebia ourebi*: Arcese, 1999). In multimale groups of equids and oribi, the males are often related, but not always. Dominant males are more likely to retain both their territories and females with the assistance of subordinate males than if alone. The reproductive costs of having auxiliary males in these species has not been empirically studied, but there is mating by subordinate males as well as mate guarding by the dominant males. Variation in the mating strategies and social system of many ungulate species is usually related to habitat quality, population density, population age structure, and female distribution (Thirgood *et al.*, 1998).

Management and conservation

Understanding the flexibility in the social system of mountain gorillas is useful for the long-term management and conservation of the small populations of this endangered species. In particular, the effective population size and amount of gene flow within the population may be greater than initially thought if more than one male is siring offspring per reproductive unit (Durant & Mace, 1994). However, we cannot ignore the possibility that human pressure has had some impact on the mountain gorilla social system as has been postulated for other species with a variable social system (langurs: Sterck, 1998, 1999). Differential poaching for silverbacks would have an obvious impact on the number of males in social units and

52   *Martha M. Robbins*

the rate of infanticide as has been shown in other species (Greene *et al.*, 1998). Additionally, due to the sharp division between protected forest and farmland, individuals from groups along the park border may have reduced dispersal options (fewer neighboring groups) than those groups in the interior of the protected area. Future conservation efforts should not overlook the interplay between natural and human influences on the variability of genetics, life history patterns, behavior, and ecology of gorillas.

Summary

The long-term monitoring of several well-habituated mountain gorilla groups at Karisoke has enabled researchers to observe extensive variability in their social system. Groups fluctuate between being one-male, multi-male, or all-male in structure. Male–male social relationships in heterosexual groups can be best described as competitive and weak, whereas males in all-male groups have more affiliative relationships. Male–female relationships are beneficial to both sexes with both males and females investing more in maintaining relationships with each other than with same-sex individuals. This is exemplified by considering how male–female relationships influenced the outcome of a group fission. Observations of mating behavior in multimale groups has revealed that mating strategies are more complex than initially thought, but the reproductive costs and benefits to males of varying social ranks awaits paternity determination studies. Overall, the variability in the social system of mountain gorillas appears to be due to complex interactions between ecology, demographic and life history patterns, social relationships, and mating strategies.

Acknowledgements

I thank l'Office Rwandais du Tourisme et des Parcs Nationaux and l'Institut Zairois pour la Conservation de la Nature for permission for all Karisoke researchers to work in the Parc des Volcans. I thank the many researchers who contributed to the long-term records of the Karisoke Research Center over the past three decades and the Dian Fossey Gorilla Fund International for permission to use these records. This work would also have not been possible without the dedicated efforts of the Karisoke field assistants. This manuscript benefited greatly from the comments of C. Boesch, B. Bradley, R. Dunbar, P. Sicotte, and K. Stewart.

The mountain gorilla social system: the male perspective   53

References

Alberts, S C & Altmann, J (1995) Balancing costs and opportunities: dispersal in male baboons. *American Naturalist* **145**, 279–306.

Arcese, P (1999) Effect of auxiliary males on territory ownership in the oribi and the attributes of multimale groups. *Animal Behaviour*, **57**, 61–71.

Baker, A M, Dietz, J M & Kleiman, D G (1993) Behavioural evidence for monopolization of paternity in multimale groups of golden lion tamarins. *Animal Behaviour*, **46**, 1091–1103.

Berger, J (1986) *Wild Horses of the Great Basin*. Chicago: University of Chicago Press.

Boesch, C (1996) Social grouping in Tai chimpanzees. In *Great Ape Societies*, ed. W C McGrew, L F Marchant & T Nishida, pp. 101–13. Cambridge: Cambridge University Press.

Borries, C, Launhardt, K, Epplen, C, Epplen, J T & Winkler, P (1999) Males as infant protectors in Hanuman langurs (*Presbytis entellus*) living in multimale groups – defense pattern, paternity and sexual behaviour. *Behavioral Ecology and Sociobiology*, **45**, 350–6.

Dixson, A F (1998) *Primate Sexuality*. Oxford: Oxford University Press.

Doran, D M & McNeilage, A (1998) Gorilla ecology and behavior. *Evolutionary Anthropology*, **11**, 120–31.

Dunbar, R I M (1984) *Reproductive Decisions: An Economic Analysis of Gelada Baboon Reproductive Strategies*. Princeton: Princeton University Press.

Durant, S M & Mace, G M (1994) Species differences and population structures in population viability analysis. In *Creative Conservation: Interactive Management of Wild and Captive Animals*, ed. P J S Olney, G M Mace & A T C Feistner, pp. 67–91. London: Chapman & Hall.

Eisenberg, J F, Muckenhirn, N A & Rudran, R (1972) The relation between ecology and social structure in primates. *Science*, **176**, 863–74.

Emlen, S T & Oring, L W (1977) Ecology, sexual selection, and the evolution of mating systems. *Science*, **197**, 215–23.

Feh, C (1999) Alliances and reproductive success in Camargue stallions. *Animal Behaviour*, **57**, 705–13.

Field, D, Chemnick, L, Robbins, M, Garner, K & Ryder, O (1998) Paternity determination in captive lowland gorillas and orangutans and wild mountain gorillas by microsatellite analysis. *Primates*, **39**, 199–209.

Fossey, D (1983) *Gorillas in the Mist*. Boston MA: Houghton Mifflin.

Fossey, D & Harcourt, A H (1977) Feeding ecology of free-ranging mountain gorillas. In *Primate Ecology*, ed. T H Clutton-Brock, pp. 415–49. New York: Academic Press.

Furuichi, T & Ihobe, H (1994) Variation in male relationships in bonobos and chimpanzees. *Behaviour*, **130**, 211–28.

Gagneux, P, Woodruff, D S & Boesch, C (1997) Furtive mating in female chimpanzees. *Nature*, **387**, 358–9.

Gerald, C N (1995) Demography of the Virunga mountain gorilla (*Gorilla gorilla beringei*). MSc thesis, Princeton University.

Goldsmith, M L (1996) Ecological influences on the ranging and grouping behavior of western lowland gorillas at Bai Hokou in the Central African Republic. PhD thesis, State University of New York, Stony Brook.

Goldsmith, M L (1999) Ecological constraints on the foraging effort of Western Gorillas (*Gorilla gorilla gorilla*) at Bai Hokou, Central African Republic. *International Journal of Primatology*, **20**, 1–23.

Goodall, J (1986) *The Chimpanzees of Gombe: Patterns of Behavior.* Cambridge MA: Harvard University Press.

Greene, C, Umbanhower, J, Mangel, M & Caro, T (1998) Animal breeding systems, hunter selectivity, and consumptive use in wildlife conservation. In *Behavioral Ecology and Conservation Ecology*, ed. T Caro, pp. 271–305. Oxford: Oxford University Press.

Harcourt, A H (1979a) Social relationships between adult male and female mountain gorillas in the wild. *Animal Behaviour*, **27**, 325–42.

Harcourt, A H (1979b) Contrasts between male relationships in wild gorilla groups. *Behavioral Ecology and Sociobiology*, **5**, 39–49.

Harcourt, A H (1979c) Social relationships among adult female mountain gorillas. *Animal Behaviour*, **27**, 251–64.

Harcourt, A H (1981) Intermale competition and the reproductive behavior of the great apes. In *Reproductive Biology of the Great Apes*, ed. C E Graham, pp. 301–18. New York: Academic Press.

Harcourt, A H & Groom, A F G (1972) Gorilla Census. *Oryx*, **10**, 355–63.

Harcourt, A H & Stewart, K J (1981) Gorilla male relationships: can differences during immaturity lead to contrasting reproductive tactics in adulthood? *Animal Behaviour*, **29**, 206–10.

Harcourt, A H & Stewart, K J (1987) The influence of help in contests on dominance rank in primates: hints from gorillas. *Animal Behaviour*, **35**, 182–90.

Harcourt, A H & Stewart, K J (1989) Functions of alliances in contests within wild gorilla groups. *Behaviour*, **109**, 176–90.

Harcourt, A H, Stewart, K J & Fossey, D (1976) Male emigration and female transfer in wild mountain gorilla. *Nature*, **263**, 226–7.

Harcourt, A H, Fossey, D, Stewart, K J & Watts, D P (1980) Reproduction in wild gorillas and some comparisons with chimpanzees. *Journal of Reproduction and Fertility*, Supplement **28**, 59–70.

Harcourt, A H, Harvey, P H, Larson, S G & Short, R V (1981) Testis weight, body weight and breeding system in primates. *Nature*, **293**, 55–7.

Harcourt, A H, Purvis, A & Liles, L (1995) Sperm competition: mating system, not breeding season, affects testes size of primates. *Functional Ecology*, **9**, 468–76.

Hohmann, G, Gerloff, U, Tautz, D & Fruth, B (1999) Social bonds and genetic ties: kinship, association and affiliation in a community of bonobos (*Pan paniscus*). *Behaviour*, **136**, 1219–35.

Kappeler, P M (2000) *Primate Males.* Cambridge: Cambridge University Press.

Lee, P C (1994) Social structure and evolution. In *Behavior and Evolution*, ed. P J B Slater & T R Halliday, pp. 266–303. Cambridge: Cambridge University Press.

Lott, D F (1991) *Intraspecific Variation in the Social Systems of Wild Vertebrates.*

Cambridge: Cambridge University Press.

Magliocca, F, Querouil, S & Gautier-Hion, A (1999) Population structure and group composition of western lowland gorillas in north-western Republic of Congo. *American Journal of Primatology*, **48**, 1–14.

McNeilage, A, Plumptre, A J, Vedder, A & Brock-Doyle, A (1998) *Bwindi Impenetrable National Park, Uganda Gorilla and Large Mammal Census, 1997*. Wildlife Conservation Society, Working Paper No. 14.

Newton, P N (1988) The variable social organization of hanuman langurs (*Presbytis entellus*), infanticide, and the monopolization of females. *International Journal of Primatology*, **9**, 59–77.

Newton, P N & Dunbar, R I M (1994) Colobine monkey society. In *Colobine Monkeys: Their Ecology, Behaviour and Evolution*, ed. A G Davies & J F Oates, pp. 311–46. Cambridge: Cambridge University Press.

Pope, T R (1990) The reproductive consequences of male cooperation in the red howler monkey: paternity exclusion in multi-male and single-male troops using genetic markers. *Behavioral Ecology and Sociobiology*, **27**, 439–46.

Pusey, A E & Packer, C (1987) Dispersal and philopatry. In *Primate Societies*, ed. B B Smuts, D L Cheney, R M Seyfarth, R M Wrangham & T T Struhsaker, pp. 250–66. Chicago: University of Chicago Press.

Rajpurohit, L S, Sommer, V & Mohnot, S M (1995) Wanderers between harems and bachelor bands: male Hanuman langurs (*Presbytis entellus*) at Jodhpur in Rajasthan. *Behaviour*, **132**, 255–99.

Remis, M J (1994) Feeding ecology and positional behavior of Western Lowland Gorillas (*Gorilla gorilla gorilla*) in the Central African Republic. PhD thesis, Yale University.

Robbins, M M (1995) A demographic analysis of male life history and social structure of mountain gorillas. *Behaviour*, **132**, 21–47.

Robbins, M M (1996) Male–male interactions in heterosexual and all-male wild mountain gorilla groups. *Ethology*, **102**, 942–65.

Robbins, M M (1999) Male mating patterns in wild multimale mountain gorilla groups. *Animal Behaviour*, **57**, 1013–20.

Rubenstein, D I (1986) Ecology and sociality in horses and zebras. In *Ecological Aspects of Social Evolution*, ed. D I Rubenstein & R W Wrangham, pp. 282–302. Princeton: Princeton University Press.

Schaller, G B (1963) *The Mountain Gorilla: Ecology and Behavior*. Chicago: University of Chicago Press.

Sholley, C R (1991) Conserving gorillas in the midst of guerrillas. *American Association of Zoological Parks and Aquaria Annual Proceedings*, pp. 30–7.

Short, R V (1979) Sexual selection and its component parts, somatic and genital selection, as illustrated by man and the great apes. *Advances in the Study of Behavior*, **9**, 131–58.

Sicotte, P (1993) Inter-group encounters and female transfer in mountain gorillas: influence of group composition on male behavior. *American Journal of Primatology*, **30**, 21–36.

Sicotte, P (1994) Effect of male competition on male–female relationships in bi-male groups of mountain gorillas. *Ethology*, **97**, 47–64.

Sicotte, P (1995) Interpositions in conflicts between males in bimale groups of mountain gorillas. *Folia Primatologica*, **65**, 14–24.

Sicotte, P (2000) A case study of mother–son transfer in mountain gorillas. *Primates*, **41**, 93–101.

Smuts, B B & Smuts, R W (1993) Male aggression and sexual coercion of females in nonhuman primates and other mammals: evidence and theoretical implications. *Advances in the Study of Behavior*, **22**, 1–63.

Steenbeek, R, Sterck, E H M, de Vries, H & van Hooff, J A R A M (2000) Costs and benefits of the one-male, age-graded and all-male phase in wild Thomas's langur groups. In *Primate Males*, ed. P M Kappeler, pp. 130–45. Cambridge: Cambridge University Press.

Sterck, E H M (1997) Derminants of female dispersal in Thomas langurs. *American Journal of Primatology*, **42**, 179–98.

Sterck, E H M (1998) Female dispersal, social organization, and infanticide in langurs: are they linked to human disturbance? *American Journal of Primatology*, **44**, 235–54.

Sterck, E H M (1999) Variation in langur social organization in relation to the socioecological model, human habitat alternation, and phylogenetic constraints. *Primates*, **40**, 199–213.

Sterck, E H M & Steenbeek, R (1997) Female dominance relationships and food competition in the sympatric Thomas langur and long-tailed macaque. *Behaviour*, **134**, 749–74.

Sterck, E H M & van Hooff, J A R A M (2000) The number of males in langur groups: monopolizability of females or demographic processes? In *Primate Males*, ed. P M Kappeler, pp. 120–9. Cambridge: Cambridge University Press.

Sterck, E H M, Watts, D P & van Schaik, C P (1997) The evolution of female social relationships in nonhuman primates. *Behavioral Ecology and Sociobiology*, **41**, 291–309.

Stevens, E F (1990) Instability of harems of feral horses in relation to season and presence of subordinate stallions. *Behaviour*, **112**, 149–61.

Stewart, K J & Harcourt, A H (1987) Gorillas: variation in female relationships. In *Primate Societies*, ed. B B Smuts, D L Cheney, R M Seyfarth, R M Wrangham & T T Struhsaker, pp. 155–64. Chicago: University of Chicago Press.

Strier, K B (1994) Myth of the typical primate. *Yearbook of Physical Anthropology*, **37**, 233–71.

Thirgood, S, Langbien, J & Putman, R J (1998) Intraspecific variation in ungulate mating strategies: the case of the flexible fallow deer. *Advances in the Study of Behavior*, **28**, 333–61.

Tutin, C E G (1996) Ranging and social structure of lowland gorillas in the Lope Reserve, Gabon. In *Great Ape Societies*, ed. W C McGrew, L F Marchant & T Nishida, pp. 58–70. Cambridge: Cambridge University Press.

van Hooff, J A R A M & van Schaik, C P (1992) Cooperation in competition: the ecology of primate bonds. In *Coalitions and Alliances in Humans and Other Animals*, ed. A H Harcourt & F B M de Waal, pp. 357–89. Oxford: Blackwell Scientific Publications.

van Hooff, J A R A M & van Schaik, C P (1994) Male bonds: affiliative relationships among nonhuman primate males. *Behaviour*, **130**, 309–37.

van Schaik, C P (1989) The ecology of social relationships amongst female primates. In *Comparative Socioecology: The Behavioural Ecology of Humans and Other Mammals*, ed. V Standen & R A Foley, pp. 195–218. Oxford: Blackwell Scientific Publications.

van Schaik, C P & van Hooff, J A R A M (1996) Toward an understanding of the orangutan's social system. In *Great Ape Societies*, ed. W C McGrew, L F Marchant & T Nishida, pp. 3–15. Cambridge: Cambridge University Press.

Watts, D P (1985) Relations between group size and composition and feeding competition in mountain gorilla groups. *Animal Behaviour*, **33**, 72–85.

Watts, D P (1989) Infanticide in mountain gorillas: new cases and a reconsideration of the evidence. *Ethology*, **81**, 1–18.

Watts, D P (1990*a*) Ecology of gorillas and its relation to female transfer in mountain gorillas. *International Journal of Primatology*, **11**, 21–45.

Watts, D P (1990*b*) Mountain gorilla life histories, reproductive competition, and sociosexual behavior and some implications for captive husbandry. *Zoo Biology*, **9**, 185–200.

Watts, D P (1991*a*) Mountain gorilla reproduction and sexual behavior. *American Journal of Primatology*, **24**, 211–25.

Watts, D P (1991*b*) Harassment of immigrant female mountain gorillas by resident females. *Ethology*, **89**, 135–53.

Watts, D P (1992) Social relationships of immigrant and resident female mountain gorillas. I. Male–female relationships. *American Journal of Primatology*, **28**, 159–81.

Watts, D P (1995*a*) Post-conflict social events in wild mountain gorillas (Mammalia, Hominoidea). I. Social interactions between opponents. *Ethology*, **100**, 139–57.

Watts, D P (1995*b*) Post-conflict social events in wild mountain gorillas. II. Redirection, side direction, and consolation. *Ethology*, **100**, 158–74.

Watts, D P (1996) Comparative socioecology of gorillas. In *Great Ape Societies*, ed. W C McGrew, L F Marchant & T Nishida, pp. 16–28. Cambridge: Cambridge University Press.

Watts, D P (1997) Agonistic interventions in wild mountain gorilla groups. *Behaviour*, **134**, 23–57.

Watts, D P (2000) Causes and consequences of variation in male mountain gorilla life histories and group membership. In *Primate Males*, ed. P M Kappeler, pp. 169–79. Cambridge: Cambridge University Press.

Weber, A W & Vedder, A (1983) Population dynamics of the Virunga gorillas: 1959–1978. *Biological Conservation* **46**, 341–66.

White, F J (1996) Comparative socio-ecology of *Pan paniscus*. In *Great Ape Societies*, ed. W C McGrew, L F Marchant & T Nishida, pp. 29–41. Cambridge: Cambridge University Press.

Wirtz, P (1982) Territory holders, satellite males and bachelor males in a high density population of waterbuck (*Kobus ellipsiprymnus*) and their associations with conspecifics. *Zeitschrift für Tierpsychologie*, **58**, 277–300.

Wrangham, R W (1979) On the evolution of ape social systems. *Social Science*

58   *Martha M. Robbins*

Information, **18**, 335–68.

Wrangham, R W (1980) An ecological model of female-bonded primate groups. *Behaviour*, **75**, 262–300.

Wrangham, R W (1986) Ecology and social relationships in two species of chimpanzee. In *Ecological Aspects of Social Evolution: Birds and Mammals*, ed. D I Rubenstein & R W Wrangham, pp. 352–78. Princeton: Princeton University Press.

Wrangham, R W & Rubenstein, D I (1986) Social evolution of birds and mammals: ecological aspects of social evolution. In *Ecological Aspects of Social Evolution: Birds and Mammals*, ed. D I Rubenstein & R W Wrangham, pp. 452–70. Princeton: Princeton University Press.

Yamagiwa, J (1987a) Male life history and the social structure of wild mountain gorillas (*Gorilla gorilla beringei*). In *Evolution and Coadaptation in Biotic Communities*, ed. S Kawano, J H Connell & T Hidaka, pp. 31–51. Tokyo: University of Tokyo Press.

Yamagiwa, J (1987b) Intra- and inter-group interactions of an all-male group of Virunga mountain gorillas (*Gorilla gorilla berengei*). *Primates*, **28**, 1–30.

Yamagiwa, J (1992) Functional analysis of social staring behavior in an all-male group of mountain gorillas. *Primates*, **33**, 523–44.

Yamagiwa, J (1999) Socioecological factors influencing population structure of gorillas and chimpanzees. *Primates*, **40**, 87–104.

Yamagiwa, J, Mwanza, N, Spangenberg, A, Maruhashi, T, Yumoto, T, Fischer, A & Steinhauer, B B (1993) A census of the eastern lowland gorillas *Gorilla gorilla graueri* in Kahuzi-Biega National Park with reference to the mountain gorillas *G. g. beringei* in the Virunga Region, Zaire. *Biological Conservation*, **64**, 83–9.

3 *Female mate choice in mountain gorillas*

PASCALE SICOTTE

Male displays towards females can occur during inter-group encounters, where
they can be part of the sequence of behaviors that males use to herd females
away from the other group or male, or they can occur in within group
interactions. (Photo by Pascale Sicotte.)

Introduction

Mountain gorillas exhibit pronounced sexual dimorphism. The males are twice as large as the females and possess several traits, such as long canines, associated with fighting ability (Harvey *et al.*, 1978; Stewart & Harcourt, 1987; Plavcan & van Schaik, 1992). Across species, sexual dimorphism is associated with intense male competition for access to females and polygynous mating (Rodman & Mitani, 1987), although the relationship is not straightforward across taxa (van Hooff & van Schaik, 1994).

In mountain gorillas, the most conspicuous male competition takes place during inter-group encounters (Harcourt, 1981), and males seek these encounters because they are the occasion to attract females (Sicotte, 1993; Watts, 1994*a*). Indeed, females transfer between groups in gorillas, and only do so during inter-group encounters (Harcourt, 1978). During these encounters, males display at each other by beating their chest and sometimes fight with their opponents (Harcourt, 1978; Sicotte, 1993). The aggression displayed by the males is more intense when the number of females that can transfer between the two units is high (Sicotte, 1993). Males that succeed in forming and maintaining a group can have a long tenure and will have mating access to several females over a long period. Others are not so successful in attracting and retaining females (Stewart & Harcourt, 1987; Robbins, 1995; Watts, 2000).

This intense male–male competition to attract and retain females, as well as males' differential success in this competition, does not remove the possibility for active female mate choice (Smuts, 1987). Indeed, researchers studying the highly dimorphic baboons have shown early on that evidence for female mate choice exists in these species; females actively sought proximity and mated preferentially with specific males (hamadryas: Bachmann & Kummer, 1980; geladas: Dunbar, 1984; savanna baboons: Smuts, 1985). Gorillas are often not the first species that come to mind when one thinks of examples of female mate choice. However, the early work on mountain gorillas did suggest that females exercised mate choice. For instance, the first published observations of transfer between groups established that females transferred by their own volition as opposed to being "kidnapped", or "stolen" by males (Harcourt *et al.*, 1976; Harcourt, 1978).

This chapter will review female behaviors that suggest female mate choice in gorillas, and male behaviors that apparently influence this choice. My aim is to demonstrate that female gorillas do exercise choice in selecting their mate and their group of residence, despite the many elements (such as the marked sexual dimorphism, lack of female–female coalitions, and attempts by males to influence female choice, sometimes aggressively) that may suggest the contrary.

60

Overview of the chapter and definitions

In the past decade, female mate choice has become a growing focus for primate research (Huffman, 1987; Small, 1989, 1993; Kuester & Paul, 1992; Manson, 1992). Data now indicate that it can profoundly affect reproductive outcomes for males (Gagneux *et al.,* 1997). In this chapter, the term "female mate choice" refers to behaviors that indicate intra-group mating preference by females in multimale groups and behaviors that relate to residence decisions and influence the long-term association between a male and a female. I am thus dealing with both short-term and long-term mate choice. Whether these behaviors influenced paternity and subsequent survival of the offspring could not be assessed. More longitudinal data and paternity tests using DNA techniques are needed to answer these questions.

In the first section of this chapter, I review the behaviors that female gorillas utilize to exercise mate choice (Figure 3.1). The first and most obvious is that female gorillas can leave their group to find a mate in a new group. Following Pusey & Packer (1987), natal dispersal refers to dispersal from the natal group, and secondary dispersal refers to subsequent move-ment between groups. A transfer is a dispersal directly into another group. Natal individuals are individuals residing in their natal group as adults. In gorillas, the timing of female transfer is constrained by two factors; the age of a female's latest offspring and the occurrence of inter-group encounters. Females generally transfer only after their offspring is weaned, and only when another group is in the vicinity (Harcourt, 1978).

The possibility of exercising mate choice does not necessarily disappear when females stay in their group. Females can either solicit or resist mating with specific males. In theory, females could also stay in their unit, while trying to copulate with extra-group males when possible. Extra-group copulations are rare in mountain gorillas compared to other species (Goodall, 1986; Smuts, 1987). I report two instances where it apparently occurred and further describe the data that would be necessary to evaluate the potential for extra-group mating in gorillas.

In the second section of this chapter, I present aggressive and non-aggressive behaviors that could be evidence of males' attempts to influence female mate choice (Figure 3.1). Behaviors taking place during inter-group encounters, such as male display and male herding, are summarized. I also discuss the importance of male infanticide in influencing female transfer decision. These behaviors all relate to situations of potential transfer and possibly influence the choice of a long-term mate. I then discuss some within-group behaviors that can be interpreted as male attempts at in-fluencing female choice (male displays and male attempts to maintain

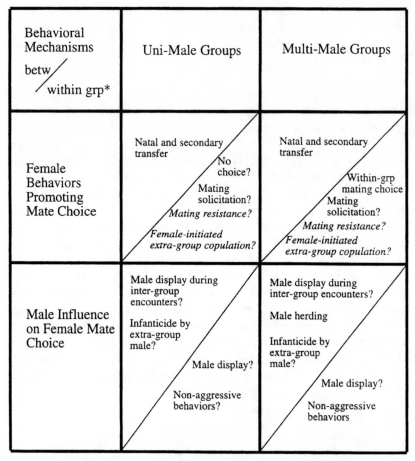

Figure 3.1. Summary of the behaviors promoting/influencing female mate choice in mountain gorillas. * Between groups: i.e. promoting long-term mate choice; within group; i.e. promoting short-term mate choice. Terms in roman characters refer to likely means by which female mate choice is exercised or influenced; terms in italics refer to means that are unlikely in mountain gorillas. ? indicates that data available remain insufficient to draw firm conclusions.

proximity with females). These behaviors contribute to the establishment and the maintenance of relationships between males and females, and may influence short-term mating decisions.

There are two points that I would like the reader to keep in mind throughout this chapter. Firstly, the discussion of behaviors promoting or influencing female choice in gorillas takes place in a context where the basis for the choice itself is not well understood. Male gorillas do not defend the ecological resources in an area (Watts, 1990*b*), and do not actively care for infants (Stewart, this volume); these factors apparently do not influence female mate choice. On the other hand, males defend females from other males during inter-group encounters (Harcourt, 1978). This observation led to the suggestion that the need to gain protection against extra-group males may have been a key element influencing female sociality in apes, because extra-group males can be potentially infanticidal (Wrangham, 1979). Further observations in that population confirmed that infanticide by extra-group males was indeed a threat (Fossey, 1984; Watts, 1989). But it remains unclear to this day what qualities make a male a good protector, which means that it is difficult to assess whether males vary in their capacity to protect females from infanticidal males. In fact, the presence of an adult male in the group seems sufficient in most cases, because infanticide often follows the death of a group leader (Watts, 1989). Beyond his mere presence in the group, it is unknown to what extent the individual qualities of a male (such as his size, strength, or intensity of displays) are important in influencing female choice, as opposed to qualities that can be described as external or contextual (such as his age, or the number of males in his group).

Secondly, our knowledge on the decision-making process leading to female transfer is limited, because most analyses of female transfer took into account only cases of actual transfer (Harcourt *et al.*, 1976; Harcourt, 1978; Watts, 1990*a*, 1991*a*), rather than considering all cases in which a female could potentially have transferred. Whether females transfer to a new group on the first occasion that presents itself, or whether they usually go through a few encounters with various groups before leaving their unit is unknown. If the latter tended to happen, females could be choosing between units of destination as opposed to simply making a choice between the unit of residence and the unit of destination.

Behaviors females use to exercise mate choice in gorillas
Female dispersal
The most conspicuous manifestation of female mate choice in gorillas is the fact that they can transfer between groups. Mountain gorillas are folivorous, and the wide and relatively even distribution of their low-quality food resources promotes low site fidelity (Watts, 1990*b*; Sterck *et al.*, 1997). Female transfer in gorillas is voluntary (Harcourt, 1978), and does not result from abduction by extra-group males (such as in hamadryas baboons: Abbeglen, 1984) or from eviction from the group by same-sex individuals (such as in howlers: Pusey & Packer, 1987; Glander, 1992; gibbons: Leighton, 1987; or ringtailed lemurs: Pereira & Kappeler, 1997).

In order to transfer, a female simply leaves her unit during an inter-group encounter, and enters the other unit. Transfers happen when another group is in the vicinity, and as a result, females do not travel on their own for very long; it is a matter of minutes before a transferring female reaches the other group. The thickness of the vegetation makes it easy for the female to avoid the males, who are busy displaying at each other (Harcourt, 1978; Sicotte, 1993). Other than the absence of a dependent offspring, it is difficult to point to another defining characteristic of trans-ferring females. As a result, a transfer can be a "slightly surprising" event, as females seem to "make sudden decisions to leave, and why they did not do so during a previous inter-unit encounter, or wait until another one is not known" (Harcourt, 1978:408).

When the silverback of a group dies, the situation is different because females travel on their own or with other females until they meet a group or a lone silverback (Harcourt, 1978; Watts, 1989). In rare cases, females have left their group and traveled together despite the fact that the silver-back of their group was still alive. For instance, Pp and Pe traveled together for nearly a week, spending time successively with a lone silver-back and with another group (K.J. Stewart, personal communication; Karisoke long-term records). This suggests some potential for flexibility in female behavior, perhaps especially if the possibility of traveling with another female is present (Yamagiwa, this volume).

Despite some similarities, particularly the fact that transfer is voluntary, the process of transfer in gorillas and in chimpanzees is quite different. In chimpanzees, females visit neighboring communities for a few days or a few weeks, usually when they are in estrus, and then return to their group (Goodall, 1986; Boesch & Boesch, 2000). Immigrant chimpanzee females can receive aggression from resident females, and to a lesser extent from males as well (Pusey, 1980; Nishida & Hiraiwa-Hasegawa, 1985; Nishida *et al.*, 1985; Takahata & Takahata, 1989), which may explain that they sometimes return in their group of origin. In mountain gorillas, these visits to other groups do not usually take place. Immigrant gorilla females sometimes receive aggression from other females in their new group (Watts, 1994*b*) or aggressive display from the males (Sicotte, 2000). These immigrant females are not typically in a position to go back to their group of origin even if they wanted to, however, because the groups move apart after the inter-group encounter. Some females do go back to their group of origin in a later encounter with that group, however (Harcourt, 1978; Karisoke long-term records).

Inter-group encounters and possibilities of transfer

The number of neighboring units probably influences the likelihood of female transfer. A large number of neighboring units not only provides a wide range of choice for the females in terms of group composition, but it also leads to a higher frequency of inter-group encounters, which create occasions for female transfer. Karisoke researchers consider that groups interact when they are within 500 m of each other because at this distance, it is likely that the groups hear or see each other. The males may choose to approach the other unit, or they may choose to simply display at a distance, or to avoid the other group (Sicotte, 1993). The main Karisoke research groups in the 1980s (Groups 5, NK, and BM) were involved in over 145 inter-group encounters in 1981–89. Groups NK and BM were in existence for only about half of this 8-year period, and Group 5 was in existence throughout. There are thus approximately 16 years of data collection on inter-group encounters, which translates into less than one encounter a month.

The distribution of these encounters is unlikely to be random for several reasons, so this figure of "less than one encounter per month" should not be used to actually evaluate transfer possibilities. Firstly, lone silverbacks and newly established groups are more motivated than well-established groups to enter inter-group interactions (Watts, 1991*b*, 1994*a*; Sicotte, 1993), so their number in an area will influence the rate and the duration of encounters. Lone males consistently show larger home range size than what would be expected on the basis of their nutritional requirements, strongly suggesting that they travel over a large area in order to interact with bisexual units (Watts, 1994*a*, 1998). Group size and home range size are also likely to influence the rate of contacts between bisexual groups (Yamagiwa, 1999), although not necessarily in a linear fashion. In fact, Watts (1998) showed that following a wave of transfer that considerably increased its size, Group 5 became less involved in inter-group encounters than before the transfers. This suggests that the young silverback leading the group during that period could have been seeking other units in order to provoke female transfers towards his group, but that he stopped doing so once several females joined him.

Ecological factors may also have an influence on the rate of inter-group encounters. For instance, if several groups share a bamboo zone in their home range, it could increase the frequency of inter-group encounter during the bamboo season, as it is the only seasonal food for mountain gorillas and is heavily used when available (Watts, 1984). It seems that ecological conditions may influence the occurrence of inter-group encounters more heavily in western lowland gorillas than in mountain gorillas, however, as groups are reported to interact at fruiting trees (Tutin, 1996), and in large swamps where gorillas use aquatic herbaceous vegetation (Olejniczak, 1994; Doran & McNeilage, this volume).

Why do females tranfer?

What motivates female transfer in gorillas? The length of male tenure is a factor associated with female dispersal in a cross-specific comparison (Clutton-Brock, 1989). Male gorillas can have a tenure of well over 10 years (Stewart & Harcourt, 1987; Robbins, 1995, this volume; Watts, 2000). This is longer than the median age at sexual maturity for females (6.3 years), and longer that the median age at first parturition (10 years) (Harcourt *et al.*, 1980; Watts, 1991*a*). Female natal dispersal is thus probably motivated by the lack of novel mates and by inbreeding avoidance. In fact, female natal dispersal in gorillas seems to be the rule when there are no males in the group other than the female's putative father. Several natal females have reproduced, but in all cases, the females had access to and mated with a male other than their putative father. Nevertheless, the fact that secondary transfer takes place in gorillas clearly shows that inbreeding avoidance is not the only reason underlying female dispersal in this species (Stewart & Harcourt, 1987; Watts, 1990*a*).

Another factor that has been suggested in transfer decision in gorillas is the size of the group. Earlier reports found that females transfer towards newly-formed units, which are smaller than established groups (Harcourt, 1978). This is consistent with the notion that the cost of scramble competition increases with group size (Watts, 1988; van Schaik, 1989). However, this tendency was not found in a larger sample, where a very large group unexpectedly attracted a number of females (Watts, 1990*b*). It appears that the size of the group is not the only important factor to consider, because females also transfer preferentially towards groups that include more than one adult male (Watts, 2000). This suggests that, from a female's point of view, the ideal group includes more than one male, and has a small number of adult females.

This preference for multimale groups might be associated with a higher reproductive success for females. Females in multimale groups may have a higher reproductive output than females in one-male groups because they are less likely to suffer from infanticide (Watts, 2000), but also possibly because females in multimale groups tend to reproduce at an earlier age (Steklis & Steklis-Gerald, this volume). It is also possible that they tend to have a shorter time to conception once they have resumed cycling because they have access to a higher supply of sperm, which may facilitate conception (Small, 1988). With endocrinological techniques allowing noninvasive testing of pregnancy in the wild (Czekala & Sicotte, 2000; Czekala & Robbins, this volume), it will become easier to distinguish females that have long inter-birth interval because they do not conceive from those that conceive but whose pregnancy is interrupted. These data should allow to test whether a high supply of sperm (i.e. a high number of copulations per cycle) decreases the time to conception. The preference that females show for multimale groups does have another consequence in that it insures possible mate choice within the group, a topic to which I will return in a later section.

Secondary dispersal

Females of all ages have transferred more than once between gorilla groups (Harcourt, 1978; Stewart & Harcourt, 1987; Watts, 1990a, 1991a). Secondary dispersal is unusual in mammals, and it is in fact rarer in female-dispersal species than in species showing predominantly male dispersal (Greenwood, 1980). It is also unusual in primates (Moore, 1984). The few primates where female secondary dispersal has been reported are red colobus (Marsh, 1979; Starin, 1981), hamadryas (Moore, 1984), and some langurs (Sterck, 1998; Steenbeek, 1999; Steenbeek et al., 1999).

The fact that female mountain gorillas can transfer more than once indicates that inbreeding avoidance is not the only factor motivating female transfer. Two of the most commonly cited reasons to explain female transfer in gorillas (reduction of feeding competition and improved protection against infanticide) can obviously also explain female secondary transfer. Females could also presumably use secondary transfer to increase the genetic variability of their offspring, or they could pursue a "bet hedging" strategy where they produce offspring in different groups that may vary in terms of mortality rates (Moore, 1993; see also Watts, 1990b). None of these ideas has been tested so far, partly because the analyses of female transfer have combined natal and secondary dispersals.

Despite the observation that secondary transfer is a regular occurrence in mountain gorillas, there is little information on the actual frequency of secondary female dispersal in that population. Harcourt (1978) showed that out of 11 females that transferred, seven dispersed more than once, and this is excluding returns to the natal group that sometimes happen. Watts (1991a) presented a larger, pooled, sample and reported that about 30% of transfers are likely to be secondary transfers. His data suggest that at least three females in the study population have transferred between groups three times, and at least another one has transferred five times.

I extracted information on female secondary dispersal from the Karisoke long-term records, to establish the exact occurrence of secondary transfers in this population. The females that I use were followed for at least 10 years during their reproductive life (n = 24). This corresponds roughly to one-third of a female's reproductive life. The number of transfers are the minimum number of transfers from one unit to another, and does not include group changes following the death of a group's silverback, or following a group fission. This is a conservative approach, where I record a secondary transfer only in those cases where a female actively leaves a cohesive unit to join another one. I did include returns to the natal group as instance of secondary dispersal because they are strong evidence of female choice, either against their new group of residence, or for their natal group.

The median number of secondary transfer for these females is 0.83 (Table 3.1). One female (Mo) was involved in at least three secondary transfers, while a few others were observed to participate in a secondary transfer twice (Lz, Fl, Pc, and Sb). A few more were involved in one secondary transfer. In total, nine females out of these 24 have transferred at least once more after their initial natal transfer (38%).

Females do not simply transfer more often as a result of being observed for a longer period of time. There is no correlation between the number of years a female was observed and her number of secondary transfers (Spearman rank correlation, n = 24, z corrected for ties = 0.538, n.s.). This could suggest that females are responding to local conditions in making their decision whether to transfer again or not. These local conditions could relate to the composition of their group, and possibly to the quality of their relationship with their silverback.

However, a crucial test of this idea relates to whether or not the females that did not show secondary dispersal had the possibility to transfer in the first place. The only time a female can transfer is when she is cycling (pregnant females are not known to transfer; Watts, 1990a). Female gorillas usually take only three or four cycles to conceive (Harcourt *et al.*, 1980; Watts, 1991a). This means that a female can be in a situation where she can transfer only during 3–4 months, once every 4 years. As shown earlier, inter-group encounters occur perhaps every month or every other month in any given unit. Therefore, only a few inter-group encounters might take place during the time a female is cycling. Some of these inter-group encounters are likely to be with the same group. If the opposing unit, for some reason, is not appropriate, or if the female is prevented from transferring by the male in her group, she may not be in position to transfer again for another 4 years. These limitations on choice could contribute to explain, for instance, the unusual situation where a female put her infant at risk for infanticide by transferring before he was weaned (Sicotte, 2000).

Table 3.1. *Natal and secondary transfer in female mountain gorillas*

Female	Natal transfer[a]	Secondary transfer	Number of years followed[b]	Date first seen and age class[c]	Death or disappearance[d]
Ef	?	None known	27	1967: Ad P	Died 1994
Fl	?	2	22	1967: Ad P	Died 1989
Lz	?	2	26	1967: Ad P	Died 1993
Ma	?	None known	13	1967: Ad P	Died 1980
Cl	Y	1	22	b. 1971	Now in Susa
Fu	Y	0	23	1976: Ad N	
Gi	N	0	13	b. 1980	
Jn	N	0	12	b. 1981	
Kw	Y	0	15	1984: Ad N	
Mo	Y	3	11	1967: Subad N	Died 1978
Mg	N	0	13	b. 1980	
Mw	N	0	11	b. 1982	
Mu	Y	1	16	b. 1977	Now in Susa
Pd	Y	0	23	1976: Ad	
Pa	N	0	29	1967: Juv	Died 1999
Pp	Y	1	29	1967: Juv	
Pe	Y	1	17	1967: Subad N	1984
Pc	?	2	13	1984: Ad P?	Now in Susa
Po	Y	0	17	b. 1976	Now in Susa
Pu	N	0	25	b. 1968	
Sg	N	0	13	b. 1980	
Sb	Y	2	25	b. 1968	Died 1999
Tu	Y	0	21	b. 1972	
Wa	Y	0	14	1985: Ad	

[a] When a female first appeared in a research group as nulliparous following a transfer, I take the conservative approach to consider this a natal transfer.
[b] Number of years followed after sexual maturity or after first sighting as adult either until death, disappearance, transfer to non-research group, or present. All the females have been followed at least 10 yrs in reproductive life. Sexual maturity is at 6 years old.
[c] N = Nulliparous, P = Parous
[d] Unless stated otherwise, female still in one of the research groups as of 1999.

Why female secondary transfer in some species and not in others?

In theory, females should show secondary dispersal less frequently in species in which dispersal involves the establishment of a new home range, since it probably involves costs related to ecological knowledge, to predation pressure, or to competition from female conspecifics. In species where females do not establish an individual home range, secondary transfers should be more likely. A comparison between gorillas and chimpanzees is useful to consider further this idea. In chimpanzees, several studies from different populations have reported that females use an area preferentially within the range of the community (Gombe: Wrangham & Smuts, 1980; Goodall, 1986; Mahale: Hasegawa, 1990; Kibale: Chapman & Wrangham, 1993). Therefore, when females transfer to a new community, they have to establish a home range within the range of the community. In mountain gorillas, females do not have to establish their own home range, as they adopt the home range of the group itself.

Accordingly, the rate of secondary transfers should be lower in chimpanzees than in mountain gorillas. I showed that secondary dispersal in mountain gorillas is a regular occurrence, despite the fact that opportunities are relatively few. In a review of female dispersal patterns in several chimpanzee populations, Christophe and Hedwige Boesch (2000) showed that few chimpanzee females exhibit secondary dispersal. There are three populations for which longitudinal information on female transfer is available; Mahale, Gombe, and Tai. Between these three sites, more hours of observations have probably been logged on female chimpanzees than on the mountain gorillas of Karisoke, which suggests that the lack of female chimpanzee secondary dispersal is a real phenomenon as opposed to reflecting a lack of data.

Mahale is the only population in which some females transferred more than once, and this only happened after males in a community began to disappear (Nishida *et al.*, 1990). When I compiled secondary dispersal data in the Karisoke population, I did not record as secondary dispersals the movements between groups caused by group disintegration. Therefore, for the sake of this comparison, these movements between groups in Mahale could be argued not to constitute secondary dispersal in a strict sense. Gombe is a small population with only a few communities, so the absence of female secondary dispersal could reflect a lack of choice in the communities of destination. In the Tai forest, the availability of neighboring communities is not a limiting factor, and yet, females do not show secondary dispersal. As more data accumulate from sites such as Kibale and Budongo, where several communities coexist in large forests, researchers will be able to confirm whether this finding from Tai can be generalized to the whole species.

At this stage, it does seem that there is a real difference between chimpanzees and mountain gorillas in the rate of female secondary dispersal. Is it possible to make sense of this difference? Later in this chapter, I will suggest that whether or not secondary dispersal is present in these species is perhaps better understood when it is juxtaposed with the presence or absence of extra-group copulations.

Evidence for within-group choice: mating solicitation and resistance to mating

There is a large degree of plasticity in mountain gorilla group composition, and a large proportion of groups includes more than one adult male (Weber & Vedder, 1983; Yamagiwa 1987; Robbins, 1995, this volume). In practice, this means that many females have mating access to more than one male during the same estrus period (Watts, 1990*a*, 1991*a*; Robbins, 1999). Females apparently show a preference for groups with more than one male, but do females show mating preference when they have a choice in mating partners?

The solicitation of copulations by females is usually taken as indicative of female mating preference (Smuts, 1987; Small, 1990). In gorillas, females solicit copulations by approaching the male, sometimes with pursed lips, establishing eye contact with him, and attracting his attention by touching him or by hitting the ground in front of him. Females can also sometimes give a vocalization called the train grunt (Fossey, 1972), which is usually given by males in the context of mating solicitation. Females initiated the majority of the observed copulations in data sets including both one-male and multimale groups (Fossey, 1982; Watts, 1991*a*). However, data collected only on a one-male group showed that the male initiated the majority of the copulations (Nadler, 1989).

These different trends may be due to the difference in the group composition. When a female is in a one-male group, she is in a situation where she has no mating choice in the short term. The situation is different if she belongs to a multimale group. Most females in multimale groups mate with more than one male during a given estrus period (Watts, 1991*a*; Robbins, 1999; personal observation). But do females solicit copulations from more than one male in these groups? If they do, it would suggest that females in multimale group are to a certain degree promiscuous, which seems hard to reconcile with the idea that females are selective (Small, 1989, 1990). In fact, females in multimale groups often do not solicit more than one male. Data from Group 5 and Group BM in 1988–89 show that some females in these groups solicited only one male (5 out of 9 in Group 5 and 3 out of 7 in Group BM; Table 3.2). This was the case even if the majority of the females mated with more than one male in their group. The male that received the majority of female solicitations was sometimes the dominant male, but not in all cases. This trend needs to be confirmed with a larger sample of groups, but it does suggest that females living in multimale groups may have preferential mating partners.

In theory, females could also resist mating with specific males as a way to exercise mate choice. Female resistance to mating, or refusal to mate, can

Table 3.2. *Solicitation of copulation by females*

Female	Parity	Natal	Male A	Male B
Group 5				
Ef	P		6/12	1/2
Fl	P		1/3	0
Kw	P		1/2	2/4
Mg	N	Y	0	1/4
Pd	P		1/5	1/2
Pa	P	Y	0/3	0
Pc	P		0/1	1/2
Pu	P	Y	1/1	0
Sb	P		2/2	0
Group BM				
Fu	P		0/1	0
Gi	N		2/5	2/5
Jn	N		2/4	2/5
Mw	N		1/2	0
Pp	P		0/1	3/4
Sh	N		1/1	0
Tu	P		5/12	2/13

The proportions are calculated using only copulations for which the solicitation was known. Male A is the dominant male in the group, Male B is one of the followers. In Grp 5, Male B can be the young SB or a blackback. In Grp BM, Male B used to be the leader of the group, but he is (in 1989) in the process of being outranked.

be successful in many species (Smuts, 1987), but it is likely that the degree of sexual dimorphism exhibited by a species, and the intensity of female coalitions, are two factors that influence whether resistance to mating is successful or not. For instance, in orangutans, which have marked sexual dimorphism and lack female coalitions, females are rarely successful when they resist mating (Rodman & Mitani, 1987). During 1988–89, I recorded two instances where female gorillas showed active resistance to mating, and when it happened, the resistance was not successful. Case 1: The male Bm approached the female Tu with pursed lips. Tu avoided Bm, but Bm grabbed Tu by the thighs and mated with her. In this case, the female Tu was a new immigrant in Group BM. Case 2: The male Zz grabbed the female Pa and pulled her towards him. Pa tried to get away, but Zz mounted her and successfully copulated. In this case, Pa was a natal female in Group 5. She and Zz were maternal siblings. Watts (1990a) also reports two instances of forced copulations in this population.

Despite the fact that the Karisoke groups have been followed for

30 years, there is an interesting lack of data about whether or not natal females in one-male groups mate before emigrating. Watts (1990*a*) reported that transfer by nulliparous females is associated with the absence of a mating partner other than the putative father in the natal group. However, whether these females mated with their putative father before their transfer is not known. In theory, this would be the setting under which female resistance, or at least avoidance, to mating is the most likely to be observed. It is worth pointing out, however, that only two groups in the study population can be described as having been strictly one-male (in all other groups, there was at least a blackback male), and both lasted for a relatively short period. Group TG lasted less than 3 years, and Group NK was in existence for 13 years (1972–85). Given the short life span of Group TG, it is only in Group NK that there was a possibility for a natal female to mate with her father. Of the 16 infants born in that group, none was sexually mature when the silverback died; the four oldest females were juveniles. Even if the silverback had lived a few years longer, these natal females would not have been in a situation where he was their only possible mate, because their half-brothers were also growing up, and would have been likely mating partners (Watts, 1990*a*).

Therefore, even if the gorilla social system tends to be one-male/multi-female, it is highly unlikely that a natal female will have only one possible mating partner. The first 10–12 years of the life of a single-male unit is the only period when immigrant females have access to only one possible mate. It is noteworthy that this phase, which was once thought to be at the basis of the formation of all gorilla groups, is now seen as only one of the processes by which gorilla groups can form (Yamagiwa, 1987; Robbins, 1995, this volume). It was also suggested that males who leave their group to form a new unit are in fact less successful reproductively than those who remain in their natal group and eventually inherit the females of that group (Robbins, 1995; Watts, 2000).

Extra-group mating

Extra-group copulations have been reported in several primate species (chimpanzees: Goodall, 1986; Wrangham, 1997; gibbons: Palombit, 1994; Reichard, 1995; Reichard & Sommer, 1997; guenons: Cords, 1987; multi-male multifemale cercopithecines: Smuts, 1987). Recent data collected on chimpanzees suggest that extra-group mating may have a profound influence on paternity in a community (Gagneux *et al.*, 1997), and further support the notion that social unit and reproductive unit are not fully overlapping entities in many species (Cords, 1987). The potential for extra-group copulations in a given species probably depends on the fluidity of its social structure. If individuals have the option of traveling on their own for some time, as is the case for chimpanzees with their fission–fusion social system, it is conceivable that females can engage in extra-group mating without the males of their group being aware of it (Goodall, 1986; Wrangham, 1997; Gagneux *et al.*, 1999). Alternatively, if group spread is important, females at the edge of the group may mate with extra-group males without attracting the attention of the resident males (such as in multimale multifemale groups of macaques and baboons: Smuts, 1987). Mountain gorillas have a cohesive social structure and a relatively small group spread. These two factors may not create many opportunities for extra-group mating. Also, when a lone male approaches a group, he is usually rapidly detected by the silverback of the group, and the two males engage in chest-beating display (personal observation).

The only likely instance where extra-group copulations could take place in mountain gorillas would be during mingling episodes in inter-group encounters. It is during inter-group encounters that extra-pair copulations are reported in gibbons (Reichard & Sommer, 1997). In gorillas, individuals from the two different units sometimes mingle, and juveniles from the two groups play together. So far, there have been at least two reports of a male mounting a young adult female during a mingling episode (Czekala & Sicotte, 2000; Y. Warren, personal communication). In the two cases, a young male (a blackback or a young silverback) mounted a nulliparous female from the opposing group. The case that I observed took place in July 1990. A young silverback belonging to an all-male group mounted a nulliparous female from Group BM. The two individuals were born in Group NK, but had been separated for 5 years, after Nk died and his group dispersed. The young female exhibited a sexual swelling at the time (Czekala & Sicotte, 2000). The mount lasted less than one minute, so it seems unlikely that it led to ejaculation. During the same inter-group encounter, a silverback from Group BM herded the young female away from the other group (Sicotte, 1993). Also, in the few days following the inter-group encounter, this female mated with one of the silverbacks in her group of residence on at least two occasions. The other male interfered in one of these copulations. These elements (the herding, the mating in the group of residence, and the male interference in the copulation) suggest that the males in Group BM did show sexual interest in this female, despite her young age. The fact that the female was involved in a case of attempted extra-group mating thus cannot be simply explained by a lack of interest from the males in her group.

In order to evaluate the potential for extra-group copulations in mountain gorillas, the episodes of mingling during inter-group encounters need to be quantified. Encounters where only mingling takes place, without any aggressive behavior, are a rare occurrence (7% of 58 inter-group encounters: Sicotte, 1993). However, bouts of mingling may take place during a generally aggressive inter-group interaction, so the occasions for extra-group mating may be more frequent than the frequency of inter-group interactions with mingling only would lead one to believe. The two examples described above also suggest that focusing observations on the young females and blackbacks during inter-group encounters may reveal more cases of extra-group mating in mountain gorillas. It is generally difficult to record all the social interactions taking place during inter-group encounters, because of the large number of individuals involved, their excitability, the dense vegetation, and the short duration of many of these interactions. A good illustration of this situation is that it is only recently that males were observed herding females to prevent them from transferring, despite years of observation of inter-group encounters (Sicotte, 1993).

At this point, however, it seems that extra-group mating in mountain gorilla is rare. As well, there are no reports of adult individuals (males or females) engaging in extra-group mating. Female chimpanzees, on the other hand, have the option of mating outside their group of residence. They also only rarely exhibit secondary dispersal. Therefore, I would argue that extra-group mating is a functional equivalent to secondary dispersal by providing avenues to exercise mate choice outside of the current group of residence. Extra-group mating could be characteristic of species in which females maintain an individual home range, and where, as a result, the ecological and possibly the social costs of transfer are high (species such as mantled howlers, multimale multifemale cercopithecines, gibbons, chimpanzees), while secondary dispersal would be found in species in which females do not typically face much social or ecological costs in transferring [species such as mountain gorillas, langurs that show "female split–merger" social system (Sterck, 1997), red colobus].

Before concluding this section, it should be noted that the potential for extra-group mating may be higher in western lowland gorillas than in the mountain gorillas. Indeed, western lowland gorillas seem to travel either in subgroups (Remis, 1994) or to have a larger group spread than mountain gorillas (Tutin, 1996; Doran & McNeilage, this volume). Furthermore, there are reports of group minglings in a large swamp where gorillas feed on aquatic vegetation (Olejniczak, 1994).

Behaviors used by males to influence female mate choice
Male displays and aggression during inter-group encounters

Male displays, which involve strutting and chest-beating (Schaller, 1963), are a behavior typical of males during inter-group encounters (although it also occurs in within-group interactions). When males display during inter-group encounters, it can be a means by which they exhibit their fighting ability to attract females, but they may also try to prevent extra-group males from approaching their unit. Male display may appear ritualized (Schaller, 1963), but it can nevertheless lead to contact aggression (Harcourt, 1978; Sicotte, 1993). The intensity of male aggression during inter-group encounters does vary with the number of potential migrants in the two units involved (Sicotte, 1993), which suggests that male aggression during inter-group encounters is related to the defense and acquisition of females rather than to the defense of food resources or of a territory.

If males vary in their capacity to defend females, females could assess male fighting abilities by observing their displays during inter-group encounters. It follows that females should transfer more towards males that display with the highest intensity, the longest duration, or towards the male that effectively displaces the other unit. These predictions remain untested, mainly because the number of inter-group encounters during which a transfer was observed is small.

Female herding during inter-group encounters

Males can directly influence whether females can transfer or not by herding them during inter-group encounters (Sicotte, 1993). Herding involves displaying towards a female, charging, and sometimes biting her when she moves in the direction of the other group (Figure 3.2). Male herding was documented in groups that include more than one male, presumably because males primarily display toward the other group during encounters; only when a second male is present can attention be devoted to the females. Herding appears to be effective in influencing female choice because in only one case did the female manage to transfer despite being herded. In fact, this case was the only one where herding was documented in a one-male group (1/13). All cases of herding in groups including more than one male were successful in preventing female transfer (Sicotte, 1993). If females transfer out of multimale groups less often than out of one-male groups (Watts, 2000), it may be partly because female herding occurs in the former.

Figure 3.2.(*opposite*) Herding during an inter-group encounter involving Group BM and PN. Male Bm targeted the female in low-center of the photograph and maneuvered her back into his group.

Infanticide by males

Males in several primate species sometimes kill unrelated infants. As a result, females tend to resume estrus sooner than if their infant had survived, and often mate with the infanticidal male, especially in species where infanticide follows male takeover (Leland & Struhsaker, 1987). Male takeover does not happen in mountain gorillas, but the threat of infanticide has been suggested to be an important evolutionary pressure forcing female gorillas into a permanent association with a male in order to gain his protection against other males (Wrangham, 1979; Watts, 1989; Smuts & Smuts, 1993). This implies that the presence of a male is crucial in the protection against infanticide, and this is confirmed by the fact that most cases of infanticide occur following the death of a group leader (Watts, 1989). In a few cases, infanticide was committed during an inter-group encounter, and these cases have led to the suggestion that male gorillas kill infants in order to induce females to transfer to them (Smuts, 1995; Wrangham & Peterson, 1996). The logic of this argument is not entirely clear, however, as it is not because a male can kill an infant during an inter-group encounter that he will be able to protect an infant from another male, as the first step involves mainly being aggressive towards a female and her infant, and the second step involves being aggressive towards an adult male. Nevertheless, investigating the timing of infanticide in relation to female transfer is crucial to shed light on these questions. It is also important to distinguish whether these cases of infanticide and female transfer took place after the death of the leader of the female's group or if the male was still alive.

There are ten cases of infanticide in mountain gorillas for which there is information on the timing of the infanticide in relation to the female transfer (Watts, 1989), and one case of attempted infanticide (Sicotte, 2000). In five cases out of these 11, the transfer and the infanticide happened on the same day, but in none of the cases were observers present to determine the actual sequence of events. I thus consider these cases incomplete.

From the six remaining cases, three have the transfer precede the infanticide. In these cases, the infants (all males who were close to weaning) followed their mother into her new group. In two cases, the leader of the group in which the infant was born had died and the infant became vulnerable to infanticide as a result. In the third case, which is the case of attempted infanticide (Sicotte, 2000), the infant became vulnerable to infanticide by following its mother in her new group. In this case, the mother's choice of a new male was presumably based on factors other than protection against infanticide, because this female was able to raise her infant to weaning age, which should indicate that the males in her former group of residence were good protectors.

In the last three cases, the infanticide occurred before the transfer. These cases thus correspond to the situation that would support the suggestion that males kill infants in order to provoke female transfer. In two of these cases, the transfer did not occur immediately after the infanticide, but occurred 5 and 2 months after the infanticide, and, in at least one of these cases, and possibly two, the female did not transfer to the infanticidal male's group.

There is thus a single case out of six complete documented cases, that fit the "males kill infants to induce females to transfer to them" scenario, in which infanticidal males are being chosen by females (cf. Smuts, 1995; Wrangham & Peterson, 1996). Two cases support the notion that males that are unsuccessful at preventing infanticide are deserted by females. There is thus little support for the idea that females select infanticidal males, although infanticidal males may gain reproductively by killing the offspring of their competitors.

Male attempts to influence female choice in within-group interactions

In this section, I discuss aggressive and non-aggressive behaviors that can be interpreted as male attempts to influence female mate choice. These behaviors have yet to be studied in one-male groups. I look at whether or not male displays at females in within-group interactions influence female short-term mating, and I review attempts by males to maintain proximity with females in light of the influence it may have on mating access.

Aggression towards females

Male gorillas regularly direct aggression to females in the form of aggressive displays, and these displays do not appear to be motivated by feeding competition or by intolerance of proximity (Harcourt, 1979; Watts, 1992). Male displays are suggested to serve as courtship aggression, and thus to promote short-term mating (Watts, 1992). However, estrous females do not receive a higher proportion of male display than anestrous females. In fact, most copulations take place without male display. In the few cases where the display was temporally associated with copulation, it occurred after the copulation. These results suggest that male display does not aim at promoting mating in the short term (P. Sicotte, unpublished data).

On the other hand, following a display, females often appease the male by performing a sequence of behavior that involves approaching, establishing contact with the male, and grumbling vocalizations (Harcourt *et al.*, 1993; Watts, 1994*b*). This reaction from females occurs significantly more often after a display than after any other type of male interaction, whether it is an approach or another type of aggression (P. Sicotte, unpublished data). This reaction seems exaggerated in light of the fact that females are rarely wounded as a result of male displays. The immediate motivation for females to appease the males may be to stop the display aggression, prevent its escalation, or reduce the likelihood of its reoccurrence. However, these possibilities do not explain the tendency for female gorillas to show little or no appeasement to males following other types of aggression. They also do not explain the tendency for males to display towards females in the first place. Watts (1995) suggested that the female appeasement reaction could function as a reaffirmation of allegiance towards the male. If this suggestion is valid, it implies that by displaying at a female, a male is testing her association with him and actively creating an opportunity for the female to express her choice.

Data from Group 5 indicated that the females receiving the least proportion of displays were the relatively recent immigrants that had a dependent infant (P. Sicotte, unpublished data). In other words, the females with no dependent offspring and the long-term residents received a higher proportion of male display. The females with no dependent offspring are the females that can potentially emigrate in the event of an inter-group encounter. By displaying at a potential migrant, the male may be evaluating the quality and the intensity of her appeasement reaction, and hence, the possibility that she will leave during an inter-group encounter.

Clearly, more work is needed to understand the role of within-group male aggression towards females, and its relation to short-term and long-term mate choice. It does seem that male display towards females is not simply acting as courtship aggression. The extent to which it influences residence decisions is unclear, as no data exist so far on the link between the frequency of male displays and the intensity and the frequency of female appeasement reaction on the one hand, and on the link between male behavior towards females during inter-group encounters and female transfer decisions on the other hand.

Non-aggressive behaviors towards females
Males use non-aggressive means to maintain proximity with females in multimale groups, and this may be an attempt to influence female short-term mating decisions (Sicotte, 1994). Males have a higher index of responsibility than females in the maintenance of proximity with proceptive females (excluding interactions immediately leading to copulations). They

also use a vocalization, the neigh, to negotiate proximity with the females. Males emit this vocalization most often when a female is leaving their proximity. It leads to the maintenance of the contact between male and female, because the female either stops her departing movement, the male follows her or the male follows her with his gaze, as if to know her location. The neigh is directed more often towards cycling females than towards pregnant or lactating females. Whether this maintenance of contact is associated with an increased mating access when the female is in estrus remains to be tested.

If the function of the neighing vocalization is to negotiate proximity with females, it should be heard less often in one-male groups, where females have only one possible male partner. I compared the hourly rate of emission of the neighing vocalization by the dominant male in Group 5 (11 females, of which four were cycling) and Group SH (five females, of which two were cycling). Many of the females in the two groups were the same, as Group SH was formed after Group 5 fissioned. Group 5 included two silverbacks and two blackbacks, whereas Group SH was one-male. I considered all the neighing vocalization emitted by the males, whether they were clearly directed at a female or not. Zz, in the multimale group, neighed 149 times in 108.5 hours of focal observation in 1989 (hourly rate of 1.37). Sh, in the one-male group, neighed 10 times in 30.2 hours of focal observation in 1996 (hourly rate of 0.3). A rate adjusted for the number of females in each group gives 0.12 for Zz and 0.06 for Sh, suggesting that there is indeed a difference in the rate at which dominant males neigh depending on the presence or absence of other males in the group. Needless to say, this remains to be confirmed with a large sample of males and of groups.

Summary

I reviewed the behaviors used by females that suggest female mate choice in gorillas, and male behaviors that seem to influence this choice. Females choose the male(s) with whom they associate by emigrating from their natal group, and by sometimes making secondary transfers. When females are in a multimale group, it appears that they may not solicit all males in the group, despite the fact that they often mate with all of them. This suggests some preferential mating relationships.

Males attempt to influence female short-term mating decisions and long-term residence decisions. They display during inter-group encounters, when female transfer can take place, perhaps to exhibit their fighting ability. Males sometimes actively prevent females from transferring by herding them away from the opposing group/male. Extra-group males are potentially infanticidal, and female grouping pattern is influenced by this threat. It is yet unclear what makes a male a good protector against infanticide, however. In within-group interactions, male displays should not be interpreted as courtship aggression, as displays do not seem to influence short-term mating access. Males often try to maintain proximity with proceptive females. It remains to be seen if these attempts translate into greater mating access, or if they are instances of mate guarding where one male tries to prevent the other male in the group from mating with the female. For future studies, it will be crucial to establish whether the behaviors that were identified as promoting or influencing female mate choice in gorillas do translate in mating access to specific males. In particular, data are lacking for behaviors that may influence short-term mating decisions. Also, whether mating access translates into paternity and in the long-term survival of the offspring will be important to determine. Finally, more comparisons between one-male and multimale groups are needed.

Acknowledgements

I thank D. Doran, M. Robbins, K. Stewart, and T. Wyman for critical reading of earlier versions of this chapter. Discussions with B. Chapais sharpened my thinking on the notion of the timing of female transfer in relation to infanticide. My work with gorillas was supported by the Natural Sciences and Engineering Research Council of Canada, the L.S.B. Leakey Foundation, DFGFI, and the University of Calgary. I thank DFGFI for permission to use the Karisoke long-term records.

References

Abbeglen, J J (1984) *On Socialization in Hamadryas Baboons.* Cranbury NJ: Associate University Press.
Bachmann, C & Kummer, H (1980) Male assessment of female choice in hamadryas baboons. *Behavioural Ecology and Sociobiology,* 6, 315–21.
Boesch, C & Boesch, H (2000) *The Chimpanzees of the Tai Forest: Behavioural Ecology and Evolution.* Oxford: Oxford University Press.
Chapman, C A & Wrangham, R W (1993) Range use of the forest chimpanzees of Kibale: implications for the understanding of chimpanzee social organization. *American Journal of Primatology,* 31, 263–73.
Clutton-Brock, T H (1989) Female transfer and inbreeding avoidance in social animals. *Nature,* 337, 70–2.
Cords, M (1987) Forest guenons and patas monkeys: male–male competition in one-male groups. In *Primate Societies,* ed. B B Smuts, D L Cheney, R M Seyfarth, R W Wrangham & T T Struhsaker, pp. 98–111. Chicago: University

of Chicago Press.

Czekala, N & Sicotte, P (2000) Reproductive monitoring of free-ranging female mountain gorillas by urinary hormone analysis. *American Journal of Primatology*, **51**, 209–15.

Dunbar, R I M (1984) *Reproductive Decisions.* Princeton: Princeton University Press.

Fossey, D (1972) Vocalizations of the mountain gorilla. *Animal Behaviour*, **20**, 36–53.

Fossey, D (1982) Reproduction among free-living mountain gorillas. *American Journal of Primatology*, Supplement, 97–104.

Fossey, D (1983) *Gorillas in the Mist.* Boston MA: Houghton Mifflin.

Fossey, D (1984) Infanticide in mountain gorillas (*Gorilla gorilla beringei*). In *Infanticide: Comparative and Evolutionary Perspectives*, ed. G Hausfater & S B Hrdy, pp. 217–35. New York: Aldine Press.

Gagneux, P, Boesch, C & Woodruff, D S (1997) Furtive mating by female chimpanzees. *Nature*, **387**, 327–8.

Gagneux, P, Boesch, C & Woodruff, D S (1999) Female reproductive strategies, paternity, and community structure in wild West African chimpanzees. *Animal Behaviour*, **57**, 19–32.

Glander, K E (1992) Dispersal patterns in Costa Rican mantled howling monkeys. *International Journal of Primatology*, **13**, 415–36.

Goodall, J (1986) *The Chimpanzees of Gombe: Patterns of Behaviour.* Cambridge: Cambridge University Press.

Greenwood, P J (1980) Mating systems, philopatry, and dispersal in birds and mammals. *Animal Behaviour*, **28**, 1140–62.

Harcourt, A H (1978) Strategies of emigrations and transfer by primates, with particular reference to gorilla. *Zeitschrift für Tierspsychologie*, **48**, 401–20.

Harcourt, A H (1979) Social relationships among adult female mountain gorillas. *Animal Behaviour*, **27**, 251–64.

Harcourt, A H (1981) Intermale competition and the reproductive behaviour of the great apes. In *Reproductive Biology of the Great Apes: Comparative and Biomedical Perspectives*, ed. C Graham, pp. 301–17. New York: Academic Press.

Harcourt, A H, Stewart, K J & Fossey, D (1976) Male emigration and female transfer in wild mountain gorilla. *Nature*, **263**, 226–7.

Harcourt, A H, Stewart, K J, Fossey, D & Watts, D P (1980) Reproduction in wild gorillas and some comparisons with chimpanzees. *Journal of Reproduction and Fertility*, **28**, 59–70.

Harcourt, A H, Stewart, K J & Hauser, M (1993) Functions of wild gorilla "close" calls. I. Repertoire, context, and interspecific comparison. *Behaviour*, **124**, 89–122.

Harvey, P H, Kavanagh, M & Clutton-Brock, T H (1978) Canine tooth size in female primates. *Nature*, **276**, 817–18.

Hasegawa, T (1990) Sex differences in ranging patterns. In *The Chimpanzees of the Mahale Mountains: Sexual and Life History Strategy*, ed. T Nishida, pp. 99–114. Tokyo: University of Tokyo Press.

Huffman, M (1987) Consort intrusion and female mate choice in Japanese macaques (*Macaca fuscata*). *Ethology*, **75**, 221–34.

Kuester, J & Paul, A (1992) Influence of male competition and female mate choice on male mating success in Barbary macaques (*Macaca sylvanus*). *Behaviour*, **120**, 192–217.

Leighton, D R (1987) Gibbons: territoriality and monogamy. In *Primate Societies*, ed. B B Smuts, D L Cheney, R M Seyfarth, R W Wrangham & T T Struhsaker, pp. 135–45. Chicago: University of Chicago Press.

Leland, L & Struhsaker, T T (1987) Colobines: infanticide by males. In *Primate Societies*, ed. B B Smuts, D L Cheney, R M Seyfarth, R W Wrangham & T T Struhsaker, pp. 83–97. Chicago: University of Chicago Press.

Manson, J (1992) Measuring female mate choice in Cayo Santiago Rhesus macaques. *Animal Behaviour*, **44**, 405–16.

Marsh, C W (1979) Female transference and mate choice among Tana River red colobus. *Nature*, **281**, 568–86.

Moore, J (1984) Female transfer in primates. *International Journal of Primatology*, **5**, 537–89.

Moore, J (1993) Inbreeding and outbreeding in primates: what's wrong with "the dispersing sex"? In *The Natural History of Inbreeding and Outbreeding*, ed. N W Thornhill, pp. 392–426. Chicago: University of Chicago Press.

Nadler, R D (1989) Sexual initiation in wild mountain gorillas. *International Journal of Primatology*, **10**, 91–2.

Nishida, T & Hiraiwa-Hasegawa, M (1985) Responses to a stranger mother–son pair in the wild chimpanzee: a case report. *Primates*, **26**, 1–13.

Nishida, T, Hiraiwa-Hasegawa, M, Hasegawa, T & Takahata, Y (1985) Group extinction and female transfer in wild chimpanzees in the Mahale National Park, Tanzania. *Zeitschrift für Tierspsychologie*, **67**, 284–301.

Nishida, T, Takasaki, H & Takahata, Y (1990) Demography and reproductive profiles. In *The Chimpanzees of the Mahale Mountains: Sexual and Life History Strategy*, ed. T Nishida, pp. 63–97. Tokyo: University of Tokyo Press.

Olejniczak, C (1994) Report on pilot study of western lowland gorillas at Mbeli Bai, Nouabale-Ndoki Reserve, northern Congo. *Gorilla Conservation News*, **8**, 9–11.

Palombit, R A (1994) Extra-pair copulations in a monogamous ape. *Animal Behaviour*, **47**, 721–3.

Pereira, M E & Kappeler, P M (1997) Divergent systems of agonistic behaviour in lemurid primates. *Behaviour*, **134**, 225–74.

Plavcan, J M & van Schaik, C P (1992) Intrasexual competition and canine dimorphism in anthropoid primates. *American Journal of Primatology*, **27**, 461–77.

Pusey, A E (1980) Inbreeding avoidance in chimpanzees. *Animal Behaviour*, **28**, 543–52.

Pusey, A E & Packer, C (1987) Dispersal and philopatry. In *Primate Societies*, ed. B B Smuts, D L Cheney, R M Seyfarth, R W Wrangham & T T Struhsaker, pp. 250–66. Chicago: University of Chicago Press.

Reichard, U (1995) Extra-pair copulations in a monogamous gibbon (*Hylobates lar*). *Ethology*, **100**, 99–112.

Reichard, U & Sommer, V (1997) Group encounters in wild gibbons (*Hylobates lar*): Agonism, affiliation, and the concept of infanticide. *Behaviour*, **134**, 1135–74.

Robbins, M M (1995) A demographic analysis of male life history and social structure of mountain gorillas. *Behaviour*, **132**, 21–47.

Robbins, M M (1996) Male–male interactions in heterosexual and all-male wild mountain gorilla groups. *Ethology*, **120**, 942–65.

Robbins, M M (1999) Male mating patterns in wild multimale mountain gorilla groups. *Animal Behaviour*, **57**, 1013–20.

Rodman, P S & Mitani, J C (1987) Orangutans: sexual dimorphism in a solitary species. In *Primate Societies,* ed. B B Smuts, D L Cheney, R M Seyfarth, R W Wrangham & T T Struhsaker, pp. 146–54. Chicago: University of Chicago Press.

Schaller, G (1963) *The Mountain Gorilla: Ecology and Behavior.* Chicago: University of Chicago Press.

Sicotte, P (1993) Intergroup interactions and female transfer in mountain gorillas: influence of group composition on male behaviour. *American Journal of Primatology*, **30**, 21–36.

Sicotte, P (1994) Effect of male competition on male–female relationships in bi-male groups of mountain gorillas. *Ethology*, **97**, 47–64.

Sicotte, P (2000) A case study of mother–son transfer in mountain gorillas. *Primates*, **41**, 95–103.

Small, M F (1988) Female primate sexual behaviour and conception. *Current Anthropology*, **29**, 81–100.

Small, M F (1989) Female choice in nonhuman primates. *Yearbook of Physical Anthropology*, **32**, 103–27.

Small, M F (1990) Promiscuity in barbary macaques (*Macaca sylvanus*). *American Journal of Primatology*, **20**, 267–82.

Small, M F (1993) *Female Choices: Sexual Behaviour of Female Primates.* Ithaca NY: Cornell University Press.

Smuts, B B (1985) *Sex and Friendship in Baboons.* Hawthorne NY: Aldine Press.

Smuts, B B (1987) Sexual competition and mate choice. In *Primate Societies,* ed. B B Smuts, D L Cheney, R M Seyfarth, R W Wrangham & T T Struhsaker, pp. 385–99. Chicago: University of Chicago Press.

Smuts, B B (1995) The evolutionary origins of patriarchy. *Human Nature*, **6**, 1–32.

Smuts, B B & Smuts, R W (1993) Male aggression and sexual coercion of females in nonhuman primates and other mammals: evidence and theoretical implications. *Advances in the Study of Behaviour*, **22**, 1–63.

Starin, E D (1981) Monkey moves. *Natural History*, **90**, 36–43.

Steenbeek, R (1999) Tenure related changes in wild Thomas's langurs. I. Between-group interactions. *Behaviour*, **136**, 595–625.

Steenbeek, R, Assink, P & Wich, S A (1999) Tenure related changes in wild Thomas's langurs. II. Loud calls. *Behaviour*, **136**, 627–50.

Sterck, E H M (1997) Determinants of female dispersal in Thomas langurs. *American Journal of Primatology*, **42**, 179–98.

Sterck, E H M (1998) Female dispersal, social organization, and infanticide in langurs: are they linked to human disturbance? *American Journal of Primatology*, **44**, 235–54.

Sterck, E H M, Watts, D P & van Schaik, C P (1997) The evolution of female relationships in nonhuman primates. *Behavioural Ecology and Sociobiology*, **41**, 297–309.

Stewart, K J & Harcourt, A H (1987) Gorillas: variation in female relationships. In *Primate Societies*, ed. B B Smuts, D L Cheney, R M Seyfarth, R W Wrangham & T T Struhsaker, pp. 155–64. Chicago: University of Chicago Press.

Takahata, H & Takahata, Y (1989) Inter-unit group transfer of an immature male of the common chimpanzee and his social interactions in the non-natal group. *African Studies Monographs*, **9**, 209–20.

Tutin, C (1996) Ranging and social structure of lowland gorillas in the Lope Reserve, Gabon. In *Great Ape Societies*, ed. W C McGrew, L F Marchant & T Nishida, pp. 58–70. Cambridge: Cambridge University Press.

van Hooff, J A R A M & van Schaik, C P (1994) Male bonds: affiliative relationships among nonhuman primate males. *Behaviour*, **130**, 307–37.

van Schaik, C P (1989) The ecology of social relationships amongst female primates. In *Comparative Socioecology*, ed. C Standen & R Foley, pp. 195–218. Oxford: Blackwell Scientific Publications.

Watts, D P (1984) Composition and variability of mountain gorilla diets in the Central Virungas. *American Journal of Primatology*, **7**, 323–56.

Watts, D P (1988) Environmental influences on mountain gorilla time budgets. *American Journal of Primatology*, **15**, 1–17.

Watts, D P (1989) Infanticide in mountain gorilla: new cases and a reconsideration of the evidence. *Ethology*, **81**, 1–18.

Watts, D P (1990a) Mountain gorilla life histories, reproductive competition, and sociosexual behaviour and some implications for captive husbandry. *Zoo Biology*, **9**, 185–200.

Watts, D P (1990b) Ecology of gorillas and its relation to female transfer in mountain gorillas. *International Journal of Primatology*, **11**, 21–45.

Watts, D P (1991a) Mountain gorilla reproduction and sexual behaviour. *American Journal of Primatology*, **24**, 211–25.

Watts, D P (1991b) Strategies of habitat use by mountain gorillas. *Folia Primatologica*, **56**, 1–16.

Watts, D P (1992) Social relationships of immigrant and resident female mountain gorillas. I. Male–female relationships. *American Journal of Primatology*, **28**, 159–81.

Watts, D P (1994a) The influence of male mating tactics on habitat use in mountain gorillas (*Gorilla gorilla beringei*). *Primates*, **35**, 35–47.

Watts, D P (1994b) Agonistic relationships between female mountain gorillas (*Gorilla gorilla beringei*). *Behavioural Ecology and Sociobiology*, **34**, 347–58.

Watts, D P (1995) Post-conflict social events in wild mountain gorillas (Mammalia, Hominoidea). I. Social interactions between opponents. *Ethology*, **100**, 139–57.

Watts, D P (1998) Long-term habitat use by mountain gorillas (*Gorilla gorilla beringei*). I. Consistency, variation, and home range size and stability. *International Journal of Primatology,* **19**, 651–80.

Watts, D P (2000) Causes and consequences of variation in male mountain gorilla life histories and group membership. In *Primate Males,* ed. P M Kappeler, pp. 169–79. Cambridge: Cambridge University Press.

Weber, A W & Vedder, A (1983) Population dynamics of the Virunga gorillas: 1959–1978. *Biological Conservation,* **26**, 341–66.

Wrangham, R W (1979) On the evolution of ape social systems. *Social Sciences Information,* **18**, 334–68.

Wrangham, R W (1997) Behaviour: subtle, secret female chimpanzees. *Science,* **277**, 774–5.

Wrangham, R W & Peterson, D (1996) *Demonic Males: Apes and the Origins of Human Violence.* Boston MA: Houghton Mifflin.

Wrangham, R W & Smuts, B B (1980) Sex difference in the behavioural ecology of chimpanzees in the Gombe National Park, Tanzania. *Journal of Reproduction and Fertility,* **28**, 13–31.

Yamagiwa, J (1987) Intra- and inter-group interactions of an all-male group of Virunga mountain gorillas (*Gorilla gorilla beringei*). *Primates,* **28**, 1–10.

Yamagiwa, J (1999) Socioecological factors influencing population structure of gorillas and chimpanzees. *Primates,* **40**, 87–104.

4 Dispersal patterns, group structure, and reproductive parameters of eastern lowland gorillas at Kahuzi in the absence of infanticide

JUICHI YAMAGIWA & JOHN KAHEKWA

Eastern lowland gorillas: silverback Nindja and infants. Chai, 3-year-old infant, put a twig between his lips to stare at us. He transferred to the Nindja Group with his mother and survived. (Photo by Juichi Yamagiwa.)

Introduction

Gorillas are classified into three subspecies (*Gorilla gorilla gorilla, G. g. graueri, G. g. beringei*) and are distributed in two widely separated forest habitats. Mitochondrial DNA sequence analyses have found a clear distinction between the western subspecies (*g. gorilla*) and the two eastern subspecies (*g. graueri and g. beringei*) that is equivalent to a species-level distinction (Ruvolo *et al.*, 1994; Garner & Ryder, 1996).

Recent studies have shown marked differences in ecological features among these three subspecies, such as diet and ranging. Mountain gorillas are terrestrial folivores, while western lowland gorillas regularly feed on fruits and insects, and eastern lowland gorillas seasonally show frugivorous characteristics even in montane forests (Watts, 1984; Tutin & Fernandez, 1992, 1993; Yamagiwa *et al.*, 1994, 1996; Kuroda *et al.*, 1996; Doran & McNeilage, this volume; McNeilage, this volume). More temporary and permanent shifts in ranging occur in western and eastern lowland gorillas and lead to their larger annual home range than that of mountain gorillas (Casimir & Butenandt, 1973; Remis, 1994; Tutin, 1996; Watts, 1998). This is presumably caused by the abundance or availability of high-quality foods (e.g. fruits) and may possibly influence the social organization of gorillas (Doran & McNeilage, 1998, this volume). Indeed, smaller-sized groups are expected under stronger within-group competition for foods (Wrangham, 1979). The fluid social units showing frequent subgrouping reported for western lowland gorillas at Ndoki (Mitani, 1992) and at Bai Hokou (Remis, 1994; Goldsmith, 1999) may be caused by the sparse distribution of fruits. The size of home range and length of day journey are expected to influence the frequency of inter-unit encounters, which provide female gorillas the opportunity to transfer into other units (Yamagiwa, 1999; Doran & McNeilage, this volume; Sicotte, this volume). Therefore, these ecological variations are regarded as important factors influencing group size and composition through female dispersal.

Most of our knowledge on the social organization of gorillas comes from the long-term data on mountain gorillas (*g. beringei*) gathered by the Karisoke Research Center in the Virunga region. Mountain gorillas usually form cohesive groups (Schaller, 1963; Fossey, 1983). About 40% of the groups contain more than one adult male, these being a father and his sons or brothers in most cases (Harcourt, 1978; Robbins, 1995). Females tend to leave their natal groups before maturity and to join other groups or solitary males (Harcourt *et al.*, 1976; Harcourt, 1978; Watts, 1991). The low ecological cost of female transfer is explained by low variation in habitat quality, which leads to extensive overlap of home range between units and low site fidelity of females (Watts, 1996). However, ecological factors are not the only determinant of female dispersal. Although group size relating to within-group feeding competition may influence female reproduction (birth rate and infant mortality), immigration and emigration rates are not dependent only on group size (Watts, 1990).

Social factors, such as avoidance of inbreeding or infanticide, may constitute the main reasons underlying female transfer. When females reach sexual maturity, they tend to avoid mating with their putative fathers and are probably motivated to seek novel mates outside their natal groups (Stewart & Harcourt, 1987; Watts, 1990; Sicotte, this volume). However, inbreeding avoidance does not explain their secondary transfer. Infanticide by males is another important factor influencing female transfer decisions. In the period of 1968–87, when an average of three groups was regularly monitored by the Karisoke researchers at any given time, infanticide occurred at a rate of 1 per 22.7 group–years or 1 per 17 group–years when a suspected case was included (Watts, 1989). It is regarded as a reproductive tactic adopted by extra-group males to resume the cyclicity of suckling females (Fossey, 1984; Watts, 1989). Although the threat of infanticide may limit female transfer decisions when they have infants, females usually seek a good protector male and occasionally transfer into large groups with two or more males in which the risk of infanticide is low (Watts, 1989, 1996; Robbins, 1995).

Thus, female reproduction is strongly influenced by inbreeding avoidance and infanticide. Females tend to have their first birth in the group into which they transfer, although some remain in their natal groups and reproduce with their putative brothers (Stewart & Harcourt, 1987; Watts, 1990, 1991). Inbreeding avoidance (with putative fathers) possibly affects these decisions by females to transfer from their natal groups, and transferring delays their first parturition (Watts, 1991). They usually transfer alone at inter-unit encounters in order to avoid infanticide, but sometimes transfer with other females and immatures after the death of their leading silverback (Fossey, 1983; Sicotte, 1993, this volume). But most of these infants that accompanied their mothers at transfer were killed by extra-group males (Watts, 1989) or received male aggression in their new groups (Sicotte, 2000). This may explain the tendency of female transfer without dependent infants when a leading male is present.

Infanticide causes 37% of infant mortality in the Virunga mountain gorilla population (Watts, 1991) and can be a major cause of infant mortality in other primate species (Hrdy *et al.*, 1995). Infanticide reduces female reproductive success, and females should have evolved strategies that reduce infanticide risk (Hrdy, 1979; Treves & Chapman, 1996; van Schaik, 1996; Sterck *et al.*, 1997). Observations on mountain gorillas provide good evidence to support the "male-protection model" that explains female transfer and prolonged association between a particular male and female in polygynous and monogamous primate species (van Schaik & Dunbar, 1990; Watts, 1996; Palombit, 1999).

However, the validity of this model has not been tested for eastern and western lowland gorillas, who differ considerably at least in diet from the mountain gorillas. Group sizes of western and eastern lowland gorillas seem to be similar to that of mountain gorillas, and most groups are polygynous (Harcourt *et al.*, 1981*b*; Yamagiwa *et al.*, 1993; Tutin, 1996). As observed for mountain gorillas, home range overlap is extensive among neighboring groups, and inter-group encounters occasionally involve agonistic interactions between silverbacks of both groups (Remis, 1994; Tutin, 1996; Yamagiwa *et al.*, 1996). Observations in captivity show similar behavioral and physiological features in reproduction of western lowland gorillas to those of mountain gorillas (Nadler, 1980; Sievert *et al.*, 1991).

Among eastern and western lowland gorillas, at present the eastern lowland gorilla population (*G. g. graueri*) inhabiting the Kahuzi-Biega National Park is the only population that provides some demographic and behavioral data for comparison with mountain gorillas. Like mountain gorillas, female eastern lowland gorillas commonly transfer between social units before maturity and have their first birth after transfer. However, they show marked differences in patterns of transfer and association. No infanticide has been reported at Kahuzi, and at least three infants transferred with their mothers between groups. Although such demographic data are fragmented, it is worth considering individual dispersal and reproduction through comparison between the Kahuzi and Virunga populations in order to understand intraspecific variation in the social organization of gorillas in relation to infanticide and various ecological constraints.

Kahuzi-Biega National Park and eastern lowland gorillas
Eastern lowland gorillas are unevenly distributed in eastern Democratic Republic of Congo. They appear in isolated habitats with high-concentration aggregations (Emlen & Schaller, 1960). The Kahuzi-Biega National Park (6000 km^2), established in 1970 (lowland sector in 1980), constitutes the largest continuous habitat which supports the largest population of eastern lowland gorillas. It is divided into the highland sector covered with montane forests (Kahuzi region) and the lowland sector covered with lowland tropical forests. A recent survey by Hall *et al.* (1998) estimates a total of 7670 (4180–10 830) weaned gorillas surviving in the lowland sector.

The Kahuzi region is located on the west side of Kivu Lake and covers

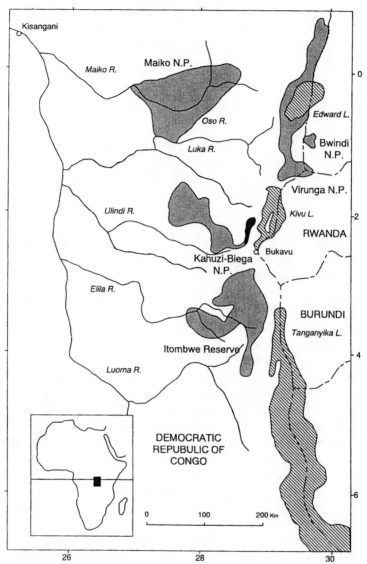

Figure 4.1. Map showing the location of national parks and reserved in the distribution of eastern lowland gorillas (Maiko, Kahuzi-Biega, and Itombwe) and mountain gorillas (Virunga and Bwindi). The black area shows the highland sector of Kahuzi-Biega National Park where our study has been conducted.

Table 4.1. *Size and composition of gorilla populations from three censuses taken in the Kahuzi-Biega National Park*

	1978[a]	1990[b]	1996[c]
Number of groups	14	25	25
Number of solitary males	5	9	2
Total number of gorillas	223	258	247
Mean group size	15.6	10.8	9.8
Percentage of infants	17.0%	8.4%	12.7%

[a] From Murnyak (1981).
[b] From Yamagiwa *et al.* (1993).
[c] From the population census conducted in 1996; unpublished data.

an area of 600 km² at an altitude of 1800 to 3308 m in the Kahuzi-Biega National Park (Figure 4.1). It is made up of bamboo forest, primary montane forests, secondary montane forests, *Cyperus* swamp, and sub-alpine vegetation (Goodall, 1977; Murnyak, 1981). The eastern lowland gorillas are widely distributed within the park and are separated from the mountain gorilla population in the Virunga region by 200 km of active volcanoes, lava forests, intensive agriculture, and human settlements. The vegetation of Kahuzi is similar to that of the Virungas, except that the herbaceous zone is absent in Kahuzi. The herbaceous zone is distributed just above the bamboo forest and provides important foods for gorillas in the Virunga region (Fossey & Harcourt, 1977), while the bamboo forest forms the upper limit of the Kahuzi gorillas' habitat.

In the highland sector (10% of the whole area) of Kahuzi-Biega, population censuses conducted in 1978 and 1990 found about the same population size (223–258) as that of the Virunga mountain gorillas surviving in an area of 600 km² of montane forest (Table 4.1). Two groups (Maeshe Group and Mushamuka Group) were habituated to accept human visitors in the early 1970s, and demographic changes in these groups have been recorded since the late 1970s (Yamagiwa, 1983; Mankoto *et al.*, 1994). A third group (Mubalala Group) was habituated in 1987 (Mankoto, 1988). A fourth group (Nindja Group) was formed by a young silverback, who was born in the Mushamuka Group, with several females and immatures from both the Mushamuka and the Maeshe Groups. They started to travel independently in 1989 and began accepting human visitors in 1990. This process of new group formation is regarded as a group fission.

One of the authors (J. Yamagiwa) made an 8-month survey of the habituated groups in 1978–79. He observed and identified most members of both the Maeshe and Mushamuka Groups during this survey. The other author (J. Kahekwa) started to work as a guide for gorilla tourism in 1983 and has had nearly daily contact with these gorillas. He named all individuals of the habituated groups and recorded the demographic changes in the four groups. Although a team consisting of a guide and several trackers has visited each group every day since the early 1970s, the recording of demographic changes in each group only started in 1983. Since 1984, Yamagiwa has visited the park every year and confirmed these changes with Kahekwa.

We reconstructed the demographic history of these four habituated groups from fragmented records on each individual. Although the exact dates of birth, death, immigration, and emigration of some individuals were recorded, most of the data on their demography were only collected weekly or monthly. Therefore, we calculated ages and other events by intervals of a month. Statistical analyses are two-tailed.

Major changes in group composition
Table 4.2 shows yearly changes in size and composition of each habituated group until 1998. Both Maeshe and Mushamuka Groups consisted of approximately 20 gorillas in 1972 (Casimir & Butenandt, 1973; Goodall, 1977). Marked social changes occurred in both groups in 1975 (Figure 4.2). The leader male named Casimir indulged in fierce displays and fights with a solitary silverback male, fell ill from an infected wound, and died. Maeshe, a young silverback, took over the leadership of the group. Although old trackers insist that Maeshe was Casimir's son, MacKinnon (1978:60), who visited the park for a short period in 1975, reported that the solitary male (named Dieter at that time) became the new leader. This is the only case that could suggest a group takeover by an extra-group male in Kahuzi. However, relationships between Maeshe and other group members remain anecdotal.

This social change resulted in transfer of several females from the Maeshe Group into the Mushamuka Group. The size of the Mushamuka Group reached 42 in 1978, while that of the Maeshe Group decreased to nine.

In the 1980s, the Maeshe Group frequently acquired females from neighboring groups and gradually increased in size. On the other hand, as Mushamuka, the leading silverback of the Mushamuka Group, grew older, females emigrated from the group one by one. Maturing males also emigrated one after another from the Mushamuka Group.

In 1986, Mubalala, a 13-year-old maturing silverback, left the Mushamuka Group and then took away three females and three juveniles

Table 4.2. *Yearly changes in age and sex composition of habituated groups*

	Silverback male >13yrs	Blackback male 10–12yrs	Adult female >10yrs	Subadult 7–9yrs	Juvenile 4–6yrs	Infant 0–3yrs	?	Total
Mushamuka Group								
1972	2	4	4		2	5	3	20
1976	1		12				18	31
1978	1	4	17		9	11		42
1983	1	4	6	2	2	6		21
1986	1	1	6	2	5	6		21
1989	1	1	6	3	3	5		19
1991	1	2	6	3	5	2		19
1993	1	2	3	3	2	2		13
1995	2	1	2	2	1	1		9
1996	1	2	2	2	1	1		9
1997	0	2	5	1	2	2		12
1998	0	1	5	1	2	2		11
Maeshe Group								
1972	1		2		4	2	9	18
1976	1	1	3	⌐——7——⌐				12
1978	1	2	3	0	3	0		9
1983	1	1	4	1	5	2		14
1986	1	0	8	4	1	11		25
1989	1	0	9	2	4	9		25
1991	1	0	5	3	6	2		17
1993	0	0	4	8	3	0		15
1995	0	1	4	5	2	0		12
1996	1	1	6	3	0	0		11
1997	1	1	9	2	1	2		16
1998	1	0	10	1	1	2		15
Mubalala Group								
1986	1	0	2	1	3	0		7
1989	1	0	6	3	0	2		12
1991	1	1	9	2	0	3		16
1993	1	0	10	0	2	2		15
1995	1	0	12	1	1	4		19
1996	1	0	10	2	2	4		19
1997	1	0	13	2	2	5		23
1998	1	1	10	1	3	4		20
Nindja Group								
1989	1	0	4	2	3	2		10
1991	1	0	8	5	0	3		17
1993	1	0	12	1	3	6	1	24
1995	1	0	13	0	3	7		24
1996	1	0	13	3	4	5		26
1997	1	0	11	3	8	2		25
1998	0	0	11	2	8	2		23

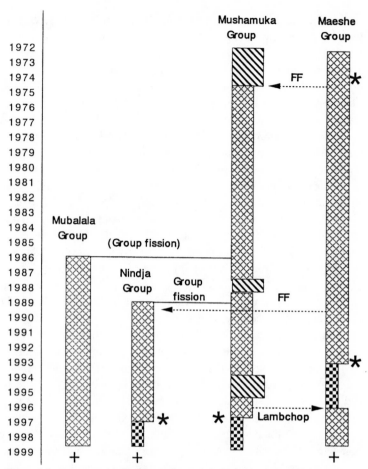

Figure 4.2. Group types of habituated groups in Kahuzi, 1972–99. Diagonal hatching denotes multimale group status, checkers hatching denotes one-male group status, and dots indicate a group consisting of females and immatures without any mature male. * denotes death of the leading male. + denotes disintegration of group due to slaughter. Dashed arrows indicate transfer of more than two females or a mature male between habituated groups. Group fission in parentheses is a suspected case based on trackers' observations.

to form his own group. In 1987, Nindja, a 12-year-old maturing silverback, started separate travels from the bulk of the group for hours with several females and immatures. In 1989, he frequently made prolonged subgrouping for days and for months and finally lured away five gorillas (including an infant) from the Mushamuka Group to establish his own group in July. These cases are regarded as group fission. Group fission was also reported in the Virungas (Robbins, this volume).

After the fission, the Nindja Group frequently encountered the Maeshe Group. After several agonistic interactions during encounters, Maeshe was seriously wounded and remained in a small area for two weeks in December, when Nindja acquired three females and three immatures who emigrated together from the Maeshe Group. In 1991 and 1992, the Nindja Group acquired other females from the unhabituated groups and rapidly increased in size.

On November 3, 1993, Maeshe disappeared and was later known to have been killed by poachers. Females and immatures moved together without any mature male for 29 months. During these periods, several (at least five) solitary silverbacks or their night nests were sometimes seen in the vicinity (within 200 m) of these groups. Maturing silverbacks (Lamb-chop and Mintsauce) of the Mushamuka Group were also seen to follow the Maeshe Group after Maeshe died. Sounds of chest-beating displays made by these males were occasionally heard near the groups in such cases. No physical contact or social interaction between group members and solitary males was observed, but the proximity of these extra-group males suggests that the groups were aware of their presence and it could have led to inter-group encounters. All three groups were involved in possible encounters with extra-group males once to several times per month. How-ever, we did not follow these extra-group males and it was difficult to detect their stealthy moving, so the frequency of encounters is unknown. Despite these possible encounters with extra-group silverbacks, the Maeshe Group did not disintegrate. Finally, Lambchop joined the group as the new leading silverback in April 1996. Mushamuka and Nindja were killed by soldiers and poachers during the civil war in April and October 1997, respectively. As with the Maeshe Group, females and immatures of both groups moved by themselves without any silverback until August 1998, when the civil war escalated in this area and none of the habituated groups could continue to be monitored.

Female dispersal in the Kahuzi population
A movement of a known individual between habituated groups with continuous stay (for more than 1 month) was defined as transfer, an appearance of new individual in the habituated group with continuous stay was defined as immigration, and disappearance of a known individual from a habituated group without confirmation of death was defined as emigration. A simultaneous dispersal of more than one individual was scored as one case. A total of 22 cases of female dispersal (4 transfers, 11 immigrations, and 7 emigrations) were confirmed in the four habituated groups from 1989 to 1998. A total of 38 adult and subadult females, 4

weaned juveniles, and 9 suckling infants were involved in these female dispersals. Other cases of female dispersal occurring before 1989 were excluded from the analysis because the time and accompanying individuals of each dispersal were unclear.

Two or more females emigrated from or immigrated into habituated groups in nine cases (Table 4.3). A total of 24 adult and subadult females were involved in these cases, in which two females dispersed twice. The mean number of females who dispersed together was 2.9 ($n = 9$, range: 2–4). Among these cases, 4 juveniles and 8 infants were involved in five cases.

The first case of multifemale dispersal occurred at a group fission in 1989 (Table 4.3, Case 1). Nindja, a son of Mushamuka, grew into a silverback and started to form another subgroup with several females in 1987. He began sleeping at separate nest sites more than 50 m away from the Mushamuka Group in 1989. He finally left Mushamuka in July 1989 and formed his own group with an adult female with a 2-month-old female infant, two subadult females, and one juvenile female.

From July 1989 to January 1990, the Nindja Group occupied the northern part of Mushamuka's home range and frequently encountered the Maeshe Group. Maeshe was wounded on his face and chest during severe fights with Nindja and could not move quickly until January 1990. Three adult females, two juvenile females, and a 7-month-old male infant transferred from the Maeshe Group into the Nindja Group in December 1989 (Case 2). The simultaneous transfer of these six individuals into the Nindja group was observed by John Kahekwa and trackers.

In other cases, females emigrated from or immigrated into the habituated groups together. These dispersals were recognized by the daily nest count firstly and were confirmed by direct observations (of their presence or absence) on the same or the next day. Three adult females and a subadult female (daughter of one of these females) emigrated from the Mushamuka Group in 1992 (Case 4). Their presence has not been confirmed in any group since then. An adult female with an infant (estimated to be under 1 year old) and another female immigrated into the Mushamuka Group in a different month of 1994 and disappeared together in 1995 (Case 5). They probably joined the same group because they returned to the Mushamuka Group together in 1997, just after the death of Mushamuka (Case 7). The former female was accompanied with the grown male infant, and the latter female carried a newborn female infant, and another adult female with an infant was associated with them. Three other cases of multifemale immigration occurred in the Mubalala Group (Cases 3, 6, and 8). None of these were accompanied by immatures, and it

Table 4.3. *Patterns of female movement at Kahuzi: multifemale movement*

Case	Year	Movement[a]	Group	Individual[b]	Situation
1	1989	Transfer[c]	From Mushamuka to Nindja	AF + I, SA, SA, J	Associated with Nindja, who emigrated from his natal group to form a new group.
2	1989	Transfer	From Maeshe to Nindja	AF + I, AF, AF, J, J	Transferred into the Nindja Group 5 months after Nindja formed his new group.
3	1989	Immigration	To Mubalala	AF, AF, AF	Might have come from an unhabituated group.
4	1992	Emigration	From Mushamuka	AF, AF, AF, SA	SA was one of AF's daughters.
5	1995	Emigration	From Mushamuka	AF + I, AF	Both females previously joined the Mushamuka Group from unknown groups.
6	1995	Immigration	To Mubalala	AF, AF	Might have come from an unhabituated group.
7	1997	Immigration	To Mushamuka	AF + I, AF + I, AF + I	No mature silverback was present in the group. The two females in case 5 were involved.
8	1997	Immigration	To Mubalala	AF, AF, AF	Might have come from an unhabituated group.
9	1997	Immigration	To Maeshe	AF + J, AF + I, AF + I	Might have come from an unhabituated group 1 year after the new male took over the Maeshe Group.

[a] Transfer, a shift movement between habituated groups with continuous stay for more than 1 month; Immigration, an appearance of new individual in the habituated group with continuous stay; Emigration, a disappearance of a known individual from an habituated group without confirmation of death.

[b] AF, adult female; SA, subadult; J, juvenile; I, suckling infant.

[c] Occurred as a group fission.

was assumed that they came from unhabituated groups. The final case was three females with suckling infants who immigrated into the Maeshe Group 11 months after Lambchop became the new leader (Case 9).

Female dispersal with no accompanying individuals was confirmed in 11 cases (Table 4.4: 1 transfer, 6 immigrations, and 4 emigrations). A female immigrated with a suckling infant in one case (to the Mushamuka Group). Both single female dispersal and multifemale dispersal occurred more frequently when the leading silverback was present (alive: 9 and 8 cases, respectively) than when he was absent (by death: 2 and 1 cases, respectively). A female's dispersal with a juvenile or an infant occurred in 9 cases when the leader alive, while it occurred in only one case in which three females immigrated to the group without any mature male (Case 7). In the Virungas, female gorillas are known to usually transfer alone, and to disperse with other females when their leading silverback dies (Harcourt, 1978; Stewart & Harcourt, 1987; Sicotte, this volume). In Kahuzi, by contrast, female gorillas often dispersed when the leading silverback was still alive and rarely dispersed together after the leading silverback died.

Another difference between female dispersal at the two sites is that at Kahuzi females did not disperse but maintained a cohesive group for a prolonged period after the death of their leading silverback. For instance, female gorillas maintained a group without any mature male for 29 months after Maeshe died, for at least 15 months after Mushamuka died, and for at least 9 months after Nindja died. In the Virungas, bisexual groups of mountain gorillas always disintegrate after the leading male dies, and the females join the neighboring groups or solitary males within a few months (Fossey, 1983; Watts, 1989).

Opportunity for infanticide
In the Virungas, infanticide is always expected in such cases of infant transfer with the mother. In one case, at least four sucking infants and one old infant (32 months old) were killed during the inter-group encounter following the death of the leading male (Watts, 1989). There are two cases in which female juveniles and infants survived in the new group. One was a 3-year-old female who was confiscated from poachers and introduced artificially to an all-male group. She survived for about 1 year with two silverbacks, two blackbacks, and two subadult males, and died of pneumonia (Fossey, 1983). Another case followed the death of a leading silverback, Nk. Four remaining juvenile and infant females were accepted and survived in the new group. But juvenile males of Group NK were not tolerated in the new group and finally joined an all-male group (Watts, 1989). An old infant who had followed his mother in her transfer was also

Table 4.4. *Patterns of female movement at Kahuzi: number of females who associated with each age and sex class at the time of movement*

| | Silverback male was | | | | | | | |
| | Present | | | | Absent (dead) | | | |
Associated with	Transfer	Immigration	Emigration		Transfer	Immigration	Emigration
None	1 (1)	6 (6)	2 (2)		0 (0)	0 (0)	2 (2)
Other female	6 (2)	9 (3)	8 (3)		0 (0)	3 (1)	0 (0)
Juvenile	3 (2)	1 (1)	0 (0)		0 (0)	0 (0)	0 (0)
Suckling infant	3 (3)	3 (2)	1 (1)		0 (0)	3 (1)	0 (0)

Number of cases is shown in parentheses. A simultaneous movement of more than two individuals was scored as one case.

attacked and wounded by the silverbacks in his new group (Sicotte, 2000, this volume).

In Kahuzi, ten females carried suckling infants when they transferred between groups (Table 4.4). Four females were accompanied by weaned juveniles. Because of the data from the Virungas, infanticide was expected in such cases, but no immature was killed or wounded in any of the cases.

In at least one case (the group fission; Case 1 in Table 4.3), the male was possibly related to the mother and the infant, and had been a mating partner of the infant's mother, which may explain the lack of aggression. However, in other cases, the new male was unlikely related to the infant. For instance, in Case 2 in Table 4.3, infant Chai was born in the Maeshe Group and was thus unrelated to Nindja. The two female juveniles who transferred into the Nindja Group in Case 2 might not have been related to him either. Nevertheless, they were not wounded or killed by Nindja. In March, 1990, another female transferred from the Maeshe Group into the Nindja Group with her 9-month-old female infant. Nindja was very tolerant toward the female and infant, and they did not show fear of Nindja.

Female immatures can be the future mates for a new male in the group into which they transfer. This may explain why Nindja did not kill female juveniles and infants in Kahuzi. However, Chai was a male infant. Explanation for the male-biased infanticide is not applicable in this case. Among 4 juveniles and 7 infants who transferred or immigrated into the habituated groups including at least a mature male, the sex of 8 individuals was known (3 juveniles and 2 infants were females, and 3 infants were males). All of them survived.

Infanticide did not follow the death of the leading silverback or takeover by an extra-group male. A few days after the death of Mushamuka, a 12-year-old male disappeared from the Mushamuka Group. Three females with infants immigrated into the Mushamuka Group about 2 weeks later. Although their immigration was not seen, they were all observed in the group within 2 days of each other and probably joined together (Case 7). They were accepted by two blackbacks and two resident adult females, and no wounds were observed on either immigrant female or their infants after joining.

A young silverback, named Lambchop, joined the Maeshe Group as the new leader in 1996. The group did not include dependent immatures. In 1997, a female with a juvenile and two females with infants were confirmed to immigrate into the Maeshe Group on the same day, and Lambchop was seen to be tolerant toward them. No infanticide occurred after their joining. These observations suggest that at Kahuzi the death of the leading male may not stimulate female transfer or cause infanticide by the new immigrant male. Multifemale immigration with infants into the Mushamuka Group after the death of Mushamuka suggests that a group without mature males can attract females from neighboring unhabituated groups. These aspects of group formation at Kahuzi are probably linked to the lack of infanticide and are in pronounced contrast with those in the Virungas, where female gorillas always join a group with mature males or a solitary male.

Response of females and immatures to the leading silverback's death and to a new male's joining

Due to the absence of infanticide, females in Kahuzi may not have an urgent need to seek a protector male after the death of their leading silverback. However, if this is the case, why do they associate with each other instead of traveling alone or forming temporary parties like chimpanzees? Independent travel may reduce the cost of within-group feeding competition, while it may also increase the risk of predation. Nesting patterns of females after the death of the leading silverback may possibly answer this question. We observed a total of 3547 nests at 285 nest sites built by a group of gorillas (the Ganyamulume Group habituated for research, not for tourism) consisting of a silverback and several females and immatures in 1994–96. The mean proportion of ground nests (including all adult and immature nests) for each nest site was 87.9% (range: 41.7–100). This indicates that the Kahuzi gorillas tend to build nests on the ground. However, after Maeshe died, the remaining females and immatures showed a decrease in the proportion of ground nests. We observed the nest sites during the period before the death of Maeshe (seven nest sites in 1991, 1992, and 1993), during the period after his death until Lambchop (the new leading male) joined (14 nest sites in 1993, 1994, and 1995), and during the period after Lambchop's joining (five nest sites in 1996). An adult's nest was identified by the size of feces remaining in each nest. The mean proportion of ground nests for adults was 68.6% ($n = 7$, range: 50%–100%) before Maeshe's death, 22.9% ($n = 14$, range: 0%–60%) after his death until Lambchop's joining, and 60.0% ($n = 5$, range: 25%–100%) after Lambchop's joining. The difference is significant between the former two periods (Mann–Whitney U test: $U = 3.0$, $p < 0.0001$) and between the latter two periods ($U = 8.0$, $p < 0.05$). The proportion was not different between the period before Maeshe's death and after Lambchop's joining. These observations clearly show that female gorillas may avoid sleeping on the ground without a silverback. Immatures responded more drastically to the death of Maeshe by increasing arboreal sleeping (Yamagiwa, 2001). Since female gorillas are vulnerable to large terrestrial predators, such as

Table 4.5. *Patterns of new group formation*

Pattern of group formation	Number of cases	
	Kahuzi	Virunga
Acquire females after a solitary life	6	2
Acquire females into an all-male group	0	1
Takeover by ousting the leading silverback	0	0
Immigration after death of the leading silverback	3	0
Group fission	2	1
Succession of natal group	1	3

Source: Kahuzi, this study; Virunga, Robbins (1995).

leopards, they tend to build nests in trees in the absence of a protector male. Although no leopards appear to exist in the montane forest of Kahuzi at present, local people had spotted them until recently (in the 1970s). In the lowland sector of Kahuzi-Biega National Park, leopards are frequently seen everywhere. These nocturnal predators may have discouraged gorillas from sleeping on the ground in the absence of the protector male. Predation risk may prevent female gorillas from traveling alone and possibly stimulate them to associate with each other and to seek another protector male after the leading male's death.

Male emigration before maturity and immigration after the death of leading males

There is a marked contrast in social organization between the Kahuzi gorillas and the Virunga gorillas in the proportion of multimale groups and the presence of all-male groups in the population. About 40% of groups contain two or more silverbacks in the Virunga population (Stewart & Harcourt, 1987; Robbins, 1995, this volume), while multimale groups constitute less than 10% of the groups in the Kahuzi population. Matured and maturing males occasionally form all-male groups in Virunga (Yamagiwa, 1987; Robbins, 1995), but such groups have never been observed in Kahuzi. These differences are possibly related to other aspects of group formation (Table 4.5).

From the long-term record on the demography of the four habituated groups, it seems that male gorillas tend to emigrate from their natal groups around puberty. Among 14 males who had been followed from birth in the habituated groups, all males emigrated from their natal groups before reaching 15 years old. Excluding the case in which males left their natal groups after the death of the leading male or after the new male's joining (even subadult males emigrated), the mean age at emigration is 12.6 years old (n = 6, range: 9.6–14.4). After emigration, males tend to spend a solitary life for months or years and eventually acquire females to form their own groups. This process of new group formation is similar to that of the Virunga gorillas (Harcourt, 1978). Group takeover by an extra-group male who forces out the leading male or immigration of a matured male into a bisexual group in the presence of the leading male has never been observed in either region. In the possible case of group takeover in Kahuzi described earlier (MacKinnon, 1978), the extra-group male joined the group after the death of the leading male. However, when the leading silverback dies, the process of new group formation differs between Kahuzi and the Virungas.

The Maeshe Group did not disintegrate after the death of the leading silverback. The females and immatures ranged together without a silverback for 29 months. Then they formed an association with a new male from a neighboring group. The same process of male immigration was observed in another two cases at Kahuzi. In the Virungas, by contrast, females did not associate with each other in the absence of a silverback for a prolonged period but usually transferred into neighboring groups or to solitary males after the death of the leading silverbacks. Female dispersion directly into another group is regarded as a tactic to associate with a protector male to avoid the expected infanticide by extra-group males (Watts, 1989, 1996; Sicotte, this volume). Due to the absence of infanticide, the Kahuzi females could associate with each other and with their immatures for a long period until finally accepting a new male as the leading silverback of their group.

Another difference between Kahuzi and Virunga is the choice of maturing males in their natal groups. At the group fission of the Mushamuka Group, Nindja succeeded in taking females away from their natal groups and formed his own group with them before acquiring females from other groups. No agonistic interactions were recorded between Mushamuka and Nindja before formation of the new group. Mubalala formed his new group in 1986 and possibly made the same choice as Nindja. A maturing silverback also took females away from a semi-habituated group with an aged silverback in 1996. Although the process of group formation was not well known in Kahuzi, other sons of Mushamuka possibly formed their new groups through group fission when the group size decreased. From the daily monitoring, the trackers noticed that young silverbacks left the Mushamuka Group with several females in 1979, 1981, and 1982. But these observations remain anecdotal, because we did not name all females and could not identify the individuals leaving the group. One case of group fission has been reported in Virunga, but it occurred between putative half-brothers (Robbins, this volume). In Kahuzi, group fission occurred between the putative father and sons. These suggest that the tolerance between related males within a group may be different between Kahuzi and Virunga. In Kahuzi, maturing silverbacks tend to establish their own groups with members of their natal groups through group fission and to share home range with their fathers, while in the Virungas males tend to stay in their natal groups after maturity to gain reproductive mates.

Female transfer and reproductive features: comparison with the Virunga population

In Kahuzi, the mean age of the first emigration (or transfer) among eight subadult and adult females who had been followed from birth to the first emigration is calculated to be 9.6 years old ($n = 8$, range: 7.5–15.3). Females who dispersed as immatures with mothers are excluded from this calculation. The mean age at the first emigration for females who emigrated independently (10.5 years, $n = 4$) is older than that for females who emigrated with another female (7.4 years, $n = 4$). In the Virungas, the median age of eight females at first transfer was approximately 8 years old (Harcourt, 1978). This suggests a similar age at which females first transfer in both regions. However, if we consider that most females emigrate independently in the Virungas, it suggests that they tend to emigrate at an earlier age from their natal groups than females in Kahuzi.

Among 43 females who had immigrated into the four habituated groups, ten females (23%) emigrated again. Six out of these ten females stayed with the groups for more than 5 years.

Among 71 infants that were born between 1978 and 1998, the sex of 64 infants was confirmed. The sex ratio at birth is close to 1:1 (31 males vs. 33 females). Based on 46 infants that were followed during immaturity, infant mortality is higher in the first year after birth than in the later years (Table 4.6). This tendency is similar to that in the Virungas (Watts, 1991).

Among the infants whose sex was known, more males (4) died than females (1) in the first year. The mortality of all the infants is 26.1%, which is not significantly different from that (34.9%) of Virunga (Fisher's Exact Test, $p = 0.411$). Mortality during the first year is not significantly different from that in later years ($p = 0.119$), and the mortality of first-born infants is not significantly higher than that of subsequent infants ($p = 0.335$). These proportions and tendencies are also similar to those in the Virungas.

The first parturition occurred at 10.6 years old on average (Table 4.6). This is similar to the Virunga mountain gorillas (Watts, 1991). The interval between consecutive births of nine females when the former infant

Table 4.6. *Comparison of reproductive features between the Kahuzi and Virunga populations*

| | Kahuzi | | Virunga | |
	n		n	
Infant mortality (1)				
First year	46	19.6%	65	26.2%
Later years	46	6.5%	65	7.7%
Infant mortality (2)				
First-born	21	33.3%	14	42.9%
Later born	25	20.0%	45	17.8%
First parturition	6	10.6 yrs (9.1–12.1)	8	10.1 yrs (8.7–12.8)
Interbirth interval (viable)	9	4.6 yrs (3.4–6.6)	26	3.9 yrs (3.0–7.3)

Source: Kahuzi, this study; Virunga, Watts (1991).

survived is slightly longer than that of the Virunga mountain gorillas. The median female produces a viable offspring every 4.6 years (range: 3.4–6.6). Given a reproductive life of females of 25 years and 26% infant mortality, a female might produce three to five offspring that survive to adulthood. The maximum number of surviving offspring in the females that we monitored was three. In Kahuzi, no difference was found in the mean interval when the former infant was male (4.5 years, $n = 5$, range: 3.7–6.6) and when the infant was female (4.6 years, $n = 4$, range: 3.6–6.6). In the Virungas, the difference ($n = 11$, range 3.0–7.3 following the birth of sons vs. $n = 15$, range 3.0–5.1 following the birth of daughters) was also not significant (Watts, 1991). When the newborn infant died, the next parturition occurred 2.2 years ($n = 3$, range: 1.4–2.7) after the death. It is longer than the interval calculated in the Virungas (1.0 years, $n = 15$, range: 0.9–3.1), but the sample size is still too small to consider differences.

In Kahuzi, among 18 female gorillas who were followed from birth to the first parturition or the first transfer (excluding infants at transfer), five females (28%) gave birth in their natal groups and the others emigrated before parturition. Four of these five females eventually emigrated from their natal groups after the first parturition. In the Virungas, among 16 females who were followed from birth to the first parturition, seven females (44%) gave birth in their natal groups (Watts, 1991).

In Kahuzi, the mean age of females giving birth in their natal group at the first parturition is 9.8 years old (*n* = 3, range: 9.1–10.5). The mean ages of females leaving their natal groups without giving birth at the first transfer and at the first parturition are 8.3 years old (*n* = 8, range: 6.2–10.3) and 11.4 years old (*n* = 3, range: 10.3–12.1), respectively. Transfers may delay initial age of parturition at Kahuzi, as observed in the Virunga mountain gorillas (Harcourt *et al.*, 1981*a*; Watts, 1991).

Discussion

Features of social organization and reproduction differing between eastern lowland gorillas and mountain gorillas

Demographic changes in the four habituated groups suggest that the basic social structure of eastern lowland gorillas is similar to that of the Virunga mountain gorillas (Harcourt *et al.*, 1976; Fossey, 1982, 1984). The Kahuzi gorillas usually form a cohesive group that consists of one male and several females with their offspring. The maximum number of silverbacks observed in a group was three, and they were probably an aged father and his maturing sons. Both female and male eastern lowland gorillas tend to leave their natal groups before maturity, and only females transfer into reproductive groups.

The female gorillas at Kahuzi tend to leave their natal groups before having their first birth and secondary transfer occurs, as observed in the Virungas (Stewart & Harcourt, 1987; Watts, 1991). Many similarities were found in the reproductive characteristics of female gorillas between Kahuzi and the Virungas, such as delayed conception by transfer, the age at first parturition, and infant mortality (Harcourt *et al.*, 1981*a*; Fossey, 1982; Watts, 1991).

These results suggest that ecological differences between Kahuzi and the Virungas, such as food distribution and availability, may not be influential for producing marked differences in the basic social structure of gorillas. Home-range shift and range overlap may decrease the ecological cost of female transfer for folivorous mountain gorillas (Watts, 1996). Our results indicate that this is also the case for the Kahuzi gorillas who are seasonal frugivores, having a longer day range and a larger home range than the Virunga mountain gorillas. Among the more frugivorous western lowland gorillas at Lopé, females also tend to transfer between groups (Tutin, 1996). These observations suggest that the ecological cost of female transfer may be negligible despite the differences in ecological features among subspecies of gorillas.

However, marked differences between Kahuzi and the Virunga populations were found in association among females when they shifted between groups or when the leading silverback died. In the Virungas, most of the

110   *Juichi Yamagiwa & John Kahekwa*

females transferred alone, irrespective of natal or secondary transfer (Harcourt, 1978; Fossey, 1983; Stewart & Harcourt, 1987). When the leading silverback died, females subsequently transferred with their infants into neighboring groups or to solitary males within a few months, but only to suffer from infanticide by the new male, with a few exceptions (Fossey, 1984; Watts, 1989; Sicotte, this volume). By contrast, the Kahuzi females often transferred with another female and carried infants when the leading silverbacks were still alive. Females continued to have prolonged association amongst themselves without any matured male after the death of the leading male for months and even years.

These differences in female association may be linked to the patterns of new group formation. In the Virungas, males often remain in their natal groups after maturity and share mating partners with their putative fathers (Robbins, 1995; Sicotte, this volume). Males that adopt this strategy even appear to be more successful reproductively than males that form a new group (Robbins, 1995; Watts, 2000). In Kahuzi, by contrast, males rarely stay with their putative fathers after maturity and occasionally take females out of their natal groups to form their own groups. Male immigration also occurs into widow groups after the death of the leading male.

Whether infanticide occurs or not in a population is likely to influence female transfer decisions. In the Virungas, the possibility of infanticide probably prevents females from transferring with infants (Watts, 1989, 1994*a*). Female mountain gorillas seek the best protector male at transfer in order to avoid infanticide. This tendency is also suggested by the positive correlation of female group size and number of males per group and the tendency of females to transfer to large groups with two or more males (Watts, 1990; Robbins, 1995). Females tend to transfer into the group in which stronger protection is expected. By contrast, since infanticide is not expected in Kahuzi, females do not need to compete for the protector male as security against extra-group males. They can transfer together with infants to other groups or accept an extra-group male as a new leader instead of transfer or disperse alone.

Why doesn't infanticide occur in Kahuzi?

Infanticide was expected but not observed in 14 cases in Kahuzi. Four explanations are applicable for the absence of infanticide in such cases. First, the male to which a female transferred with infants was related to one or both of them. In Case 1 (Table 4.3), Nindja was the half-brother of a female carrying an infant and might be the possible father of the infant. Killing the infant might have reduced Nindja's reproductive success and was not expected. These facts suggest that infanticide may not follow group fission because the new leader is usually related to the females and infants. However, this explanation does not fit for Case 2 in which the new male was unlikely to have been related to an infant and two juveniles that transferred with females.

Second, if the infant that accompanies its mother at transfer is female, then the infant could be a potential mate for the male in the future. In chimpanzees of Gombe and Mahale, male infants were more frequently killed than female infants, and this tendency was prominent when the killer was male (Takahata, 1985; Goodall, 1986; Hiraiwa-Hasegawa, 1987). If infanticide by male gorillas aims to kill a potential competitor for mates, female immatures may not be the direct target of infanticide. This may explain the fact that the female juveniles and infants were not killed in Cases 1 and 2. However, Chai, a male infant, was not attacked by Nindja when he transferred with his mother in Case 2. At least two other male infants survived after immigration into new groups with their mothers. Moreover, no sex difference was observed for victims of infanticide in the Virungas (Fossey, 1984; Watts, 1989). The second interpretation thus does not seem applicable to this case, although the sample size is small.

The third explanation is based on the lack of potential killers. In the Virungas, most of infanticidal males were solitary males (Fossey, 1984; Watts, 1989). Solitary males are strongly motivated to attract females to establish their own groups. They tend to follow reproductive groups more persistently and encounter with them more intensively than males who have their own groups (Caro, 1976; Yamagiwa, 1986; Sicotte, 1993; Watts, 1994a, 1998). Thus, unmated males constitute a greater infanticidal threat than do breeding males of neighboring groups. The same tendency is suggested for monogamous gibbons (Palombit, 1999). Considering this tendency, it is possible that the lower number of solitary or unmated males may decrease the opportunity of infanticide. In other words, the occurrence of infanticide in the Virungas may be linked to the large number of solitary males. From 1973 to 1986, population censuses were made four times in the Virungas and the number of solitary males fluctuated between 5 and 15. Infanticide was not observed from 1979 until 1982 when the number of solitary males was the lowest (Aveling & Aveling, 1987; Watts, 1989). The number of solitary males in the Kahuzi population was relatively low in each census, although the size of population was similar to that in the Virungas (Table 4.1). However, the number of solitary males may be somewhat inaccurate because their nests were more difficult to find during censuses. The results of censuses thus may not represent the difference in the actual number of solitary males between the Virungas and Kahuzi. At least five solitary males temporarily visited the widow groups after the deaths of the leading males in Kahuzi. In spite of these potential opportunities, infanticide was not observed in these groups. Moreover, in the Virungas the majority of infanticidal males were not solitary males but were extra-group males with own groups (Watts, 1989). The small number of solitary males in Kahuzi may explain a decrease of infanticide but may not fully explain the absence of infanticide.

The last explanation is that infanticide is a common feature of gorillas and some unknown factors cause it to occur in the Virungas, while it has been suppressed as a potential reproductive tactic of males in Kahuzi. In the Virungas, Sicotte (2000) reported a case in which fearfulness expressed by a male infant who transferred with his mother seemed to induce intensive male aggression. In Kahuzi, both females and immatures showed no fear of the new males at immigration or at the male's joining as new leader in Kahuzi. The absence of fear in immatures possibly induced tolerance in the unfamiliar males at transfers in Kahuzi. It may also reflect a lack of experience with infanticide or intensive attacks by unfamiliar males in their past history. Although observations are still insufficient to identify the causing factors of infanticide, it is important to note the strong relationships between infanticide, prolonged male–male associations within a group, and the tendency for females to transfer alone in the Virungas. On the contrary, the absence of infanticide promotes male emigration from their natal groups and prolonged female–female association through transfer between groups and after the death of the leading male in Kahuzi. Infanticide may thus take an important role in changing social organization of gorillas over relatively short periods.

Factors influencing social structure of gorillas

Watts (1996) argues that the "male protection" model predicts two consequences following the death of a breeding male in a one-male group: high vulnerability to infanticide and the group's fragmentation with females seeking to avoid infanticide. Data on the Virunga mountain gorillas support this interpretation (Fossey, 1984; Watts, 1989). However, our findings on prolonged female associations after the death of the leading male and the fact that infants survived after being in new groups with unfamiliar males suggest that the female gorillas in Kahuzi may seek a protector male not to avoid infanticide but to avoid predators. The decrease in the proportion of ground nests built by females after the death of the leading male supports this idea. It is likely that gorillas inhabiting the montane forest of Kahuzi have been exposed to predation by leopards until recently and are now still vigilant against them. Protection against predators may also be an important factor influencing the social organization of gorillas

(Wrangham, 1979; Stewart & Harcourt, 1987).

In spite of their large body size and terrestrial features, vulnerability to large predators may prevent females and immature gorillas from independent travels and may force them into a cohesive group in the two populations. In the Virungas, the pressure of infanticide further affects female transfer decisions and group formation. Females seem motivated to transfer into a group in which better protection is available against infanticide by extra-group males. This results in females transferring alone and in their preference for transferring into multimale groups. Males can attain a higher reproductive success by remaining in their natal groups (Watts, 2000). This may promote association among related males after maturity and the formation of multimale groups. This leads to the question, however, of why these multimale groups do not become the rule, and why they are not more stable (Robbins, this volume). Female gorillas show inconspicuous sign of estrus in a short period around the day of ovulation (Harcourt *et al.*, 1981*a*; Fossey, 1982). Even in a group including two or more silverbacks, females do not show promiscuous mating but tend to have a prolonged consortship with the particular male (usually with the leader male). These tendencies of exclusive mating may prevent males from forming permanent associations like chimpanzees.

By contrast, in Kahuzi the absence of infanticide may not induce females to transfer alone. Although females need protection against predators, they need not form a reliable relationship with a particular male individually to avoid infanticide. This situation may promote flexibility in female choice at transfer or at the death of the leading male (Figure 4.3). Females can transfer between groups in various ways (alone, with other females, and with immatures). Associations among females without any mature male may form for a prolonged period after the death of the leading male, and group takeover by an unfamiliar male subsequently occurs. Since multimale groups may not confer an advantage to females in the form of reduced risk of infanticide, females may not choose multimale groups to join at transfer. Thus, maturing males do not need to remain in their natal groups. Although males can find their mates both inside and outside their natal group, most males emigrate from their natal groups. As females occasionally associate with each other at transfer, males can establish large groups in a short period by taking females from their natal groups, by luring females from neighboring groups, or by takeover of a widow group after the death of its leading male.

Although intensive studies with habituation of groups have yet to be conducted on western lowland gorillas, no evidence of infanticide has been reported. The absence of infanticide may contribute to the low frequency

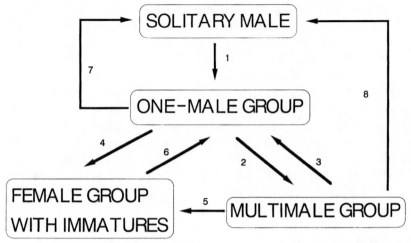

Figure 4.3. Transition between social units in eastern lowland gorillas in Kahuzi.
(1) Solitary male acquires females to form one-male group. (2) Maturing males
remain in natal group and the group becomes multimale temporarily. (3) Death
of a silverback in a multimale group may cause it return to the one-male group,
or a multimale group fission results in both groups returning to the one-male
group. (4, 5) Death of a silverback in a one-male group or a multimale group
may cause a prolonged association among females with immatures.
(6) Immigration of a mature male into the female group results in establishment
of one-male group. (7, 8) Males may emigrate to become solitary from both
one-male and multimale groups.

of multimale groups in western lowland gorillas as in eastern lowland
gorillas in Kahuzi (Remis, 1994; Tutin, 1996; Doran & McNeilage, 1998;
Robbins, this volume). The absence of infanticide may also influence
female dispersal and social organization of western lowland gorillas. Tutin
(1996) reported a case in which a maturing silverback presumably took a
young female away from his father's group at his emigration in the Lopé
Reserve, Gabon. A gradual process of group fission, as observed in Case 1
of this study, was reported in Bai Hokou, Central African Republic
(Remis, 1994). Prolonged subgrouping of western lowland gorillas reported in Bai Hokou and Ndoki (Mitani, 1992; Remis, 1997; Goldsmith,
1999) has never been observed in either Kahuzi or the Virungas. Doran &
McNeilage (1998) suggested that responses to increased within-group
competition among frugivorous western lowland gorillas might include
increased female philopatry and more differentiated female relationships.
The absence of infanticide may promote female flexible choice at dispersal
with various ecological factors in lowland forests. It is hoped that future

long-term studies on the demography of gorillas in Kahuzi and in other lowland forests will be made for comparison with those of mountain gorillas in the Virungas and will clarify this issue.

Summary

The habituated population of eastern lowland gorillas of the Kahuzi-Biega National Park is compared to the mountain gorillas of Karisoke. The similarities between the two populations include the age at first parturition, the inter-birth interval, high infant mortality during the first year after birth, high infant mortality of primiparous females, natal transfer and reproduction in the group into which they transfer, and multiple transfers in their lifetime. There are, however, marked differences between the two populations that relate to the pattern of female transfer and new group formation and may be related to the presence or absence of infanticide. Female mountain gorillas tend to transfer alone, while female eastern lowland gorillas usually transfer with another female and their offspring. In the Virungas, all but one infant transferring with its mother was killed by the stranger silverbacks, while all infants in these cases survived in Kahuzi. In the Virungas, male mountain gorillas tend to remain in their natal groups after maturity to share mating partners with their putative fathers. By contrast, in Kahuzi male eastern lowland gorillas rarely stay with their putative fathers but occasionally take females out of their natal groups to form their own groups through group fission. In the Virungas, when the leading silverbacks die, females subsequently transfer into other groups or to solitary males, while in Kahuzi females tend to associate with each other for prolonged periods without associating with any silverback. In Kahuzi, females may not seek a protector male to avoid infanticide, and multifemale transfer may enable maturing males to establish their own groups in various ways. However, the low proportion of ground nests after the death of a leading male suggests that females also seek a protector male to avoid predation in Kahuzi. Protection against predators may be an important factor influencing the social organization of gorillas. Infanticide may be a potential reproductive tactic common to male gorillas across subspecies. Although its causing factors are still unknown, the occurrence of infanticide may be linked with an increase in male–male association within groups (multimale groups and all-male groups) and an increase in female transfer alone in the Virungas. The absence of infanticide in Kahuzi may enlarge female choice of dispersal, promote female–female association, and influence the way that maturing males participate in new group formation.

Box 4.1. Slaughter of gorillas in the Kahuzi-Biega National Park in 1999

Unfortunately, data collection on the long-term demographic patterns of the four habituated groups in the Kahuzi-Biega National Park that enabled the analysis of the present study is no longer possible due to the large-scale poaching of gorillas that occurred in 1999. Four out of the five habituated groups have been decimated, and it is suspected that more than half of the population in the highland sector of the park has been killed recently. Poaching activities in the lowland sector (the rest of the 6000 km^2) is anticipated to be even more severe. The eastern lowland gorillas are now in critical danger of extinction.

Until the end of July 1998, four groups (Mushamuka, Maeshe, Nindja, and Mubalala) of gorillas had been monitored by the park on a daily basis for tourism. Another group (Ganyamulume) had been monitored by us for the purpose of research. A total of 94 gorillas were confirmed to survive in these five groups. However, after the outbreak of the civil war in August 1998, the park guards were disarmed and could not enter the park, due to the establishment of the new government. Although monitoring on the Ganyamulume Group resumed 3 months later, no direct observations had been made and little information had been available on the four tourist groups until the end of March 1999 when the park guards resumed monitoring of the groups and regular patrols without arms.

In February 1999, the park staff found a large number of gorilla bones scattered within the range of the Mubalala Group. It is suspected that most members of the Mubalala Group were killed by poachers for the bushmeat trade. The Maeshe Group was found as a large group (23 gorillas) in the beginning of April 1999, when the park staff started to monitor this group again. Several females had immigrated and some babies had been born in this group since August 1998. However, the group moved to the former range of the Mubalala Group where poaching activities were high. Gunfire was frequently heard. In September, the park staff arrested a group of poachers in possession of numerous fragments of gorilla fur, skull, and bone. It is possible that most members of the Maeshe Group were slaughtered by these poachers for bushmeat.

The Mushamuka Group had probably disintegrated before April 1999. A small group consisting of a silverback/blackback, three females, a subadult male, a juvenile, and an infant was found in the former range of Mushamuka Group. The maturing silverback, named Kaboko, was confirmed to have been born in the Mushamuka Group in 1987. He was given a new name, Mugaruka (the name of the present chief in the village nearby), and the park staff have continued to monitor this group.

On April 11, the sound of gunfire was frequently heard from within the

(continues opposite)

range of Nindja Group and since then, the group has not been found. Some time later, a number of dead gorilla bodies were seen being carried by poachers to villages nearby. It is likely that most members of the group were shot and killed by the poachers.

The Ganyamulume Group has escaped the slaughter, although a young silverback and three females were shot dead in August 1999 and January 2000, respectively. A solitary male has begun associating with the group, and all females and immatures have remained as a unit and moved together.

As of April 2000, the park staff keep 20% of the highland sector safe for gorillas with frequent patrols. The rest of the sector is not monitored, and the lowland sector is completely out of control of the park. From the tragedies of the former habituated groups, it is estimated that poachers have killed more than 60 gorillas. If the possible killing of gorillas in unmonitored areas of the park (in the worse situation) is taken into account, more than half of the gorilla population in the Kahuzi-Biega National Park was lost in 1999.

Starvation and the spread of guns among the local people during the war are the main cause of this situation. The peaceful life history of eastern lowland gorillas and a unique social organization in Kahuzi analyzed in this study have been lost and will never be observed again. It is a great loss for the understanding of gorillas. The first effective solution is, needless to say, to attain peace by means of positive negotiations among all political forces. However, urgent measures for conservation should be initiated even during the war. The efforts of conservation will hopefully attract international attention to the tragedies of gorillas at Kahuzi and will improve attitudes of local people towards the most precious life of eastern lowland gorillas as a national and world heritage.

Acknowledgements

This paper was originally prepared for the conference "Mountain Gorillas of Karisoke: Research Synthesis and Potentials" held January 18–19, 1999 at the Max Planck Institute for Evolutionary Anthropology in Leipzig, Germany. We thank Pascale Sicotte, Martha Robbins, Kelly Stewart, and Christophe Boesch for their extraordinary organizational efforts. This study was financed by the Monbusho (Ministry of Education, Science, Sports and Culture, Japan) International Scientific Research Program (No. 08041146) in cooperation with CRSN (Centre de Recherches en Sciences Naturelles) and ICCN (Institut Congolais pour Conservation de la Nature). We thank Dr. S. Bashwira, Dr. B. Baluku, Mr. M.O. Mankoto, Mr. L. Mushenzi, and Mr. K. Basabose for their administrative help and hospitality. We are also greatly indebted to all guides, guards, and field assistants at the Kahuzi-Biega National Park for their technical help and hospitality throughout the fieldwork.

References

Aveling, C & Harcourt, A H (1983) A census of the Virunga gorillas. *Oryx*, **18**, 8–13.
Aveling, R & Aveling, C (1987) Report from the Zaire Gorilla Conservation Project. *Primate Conservation*, **8**, 162–4.
Caro, T M (1976) Observations on the ranging behaviour and daily activity of lone silverback mountain gorillas (*Gorilla gorilla beringei*). *Animal Behaviour*, **24**, 889–97.
Casimir, M J (1975) Feeding ecology and nutrition of an eastern gorilla group in the Mt. Kahuzi region (République du Zaïre). *Folia Primatologica*, **24**, 81–136.
Casimir, M J (1979) An analysis of gorilla nesting sites of the Mt. Kahuzi region (Zaire). *Folia Primatologica*, **32**, 290–308.
Casimir, M J & Butenandt, R (1973) Migration and core area shifting in relation to some ecological factors in a montane gorilla group in the Mt. Kahuzi region. *Zeitschrift für Tierpsychologie*, **33**, 514–22.
Doran, D & McNeilage, A (1998) Gorilla ecology and behavior. *Evolutionary Anthropology*, **6**, 120–31.
Emlen, J T & Schaller, G B (1960) Distribution and status of the mountain gorilla (*Gorilla gorilla beringei*) – 1959. *Zoologica*, **45**, 41–52.
Fossey, D (1982) Reproduction among free-living mountain gorillas. *American Journal of Primatology*, Supplement, **1**, 97–104.
Fossey, D (1983) *Gorillas in the Mist*. Boston MA: Houghton Mifflin.
Fossey, D (1984) Infanticide in mountain gorillas (*Gorilla gorilla beringei*) with comparative notes on chimpanzees. In *Infanticide: Comparative and Evolutionary Perspectives*, ed. G Hausfater & S B Hrdy, pp. 217–36. Hawthorne NY: Aldine Press.
Fossey, D & Harcourt, A H (1977) Feeding ecology of free-ranging mountain gorilla (*Gorilla gorilla beringei*). In *Primate Ecology*, ed. T H Clutton-Brock, pp. 415–47. New York: Academic Press.
Garner, K J & Ryder, O A (1996) Mitochondrial DNA diversity in gorillas. *Molecular and Phylogenetic Evolution*, **6**, 39–48.
Goldsmith, M L (1999) Ecological constraints on the foraging effort of Western Gorillas (*Gorilla gorilla gorilla*) at Bai Hokou, Central African Republic. *International Journal of Primatology*, **20**, 1–23.
Goodall, A G (1977) Feeding and ranging behaviour of a mountain gorilla group (*Gorilla gorilla beringei*) in the Tshibinda–Kahuzi region (Zaïre). In *Primate Ecology*, ed. T H Clutton-Brock, pp. 450–79. New York: Academic Press.
Goodall, J (1986) *Chimpanzees of Gombe: Patterns of Behavior*. Cambridge MA:

Belknap Press of Harvard University Press.

Hall, J S, White, L J T, Inogwabini, B-I, Omari, I, Morland, H S, Williamson, E A, Saltonstall, K, Walsh, P, Sikubwabo, C, Ndumbo, B, Kaleme, P K, Vedder, A & Freema, K (1998) Survey of Grauer's gorillas (*Gorilla gorilla graueri*) and eastern chimpanzees (*Pan troglodytes schweinfurthii*) in the Kahuzi-Biega National Park lowland sector and adjacent forest in eastern Democratic Republic of Congo. *International Journal of Primatology*, **19**, 207–35.

Harcourt, A H (1978) Strategies of emigration and transfer by primates, with particular reference to gorillas. *Zeitschrift für Tierpsychologie*, **48**, 401–20.

Harcourt, A H (1996) Is the gorilla a threatened species? How should we judge? *Biological Conservation*, **75**, 165–76.

Harcourt, A H, Stewart, K J & Fossey, D (1976) Male emigration and female transfer in wild mountain gorillas. *Nature*, **263**, 226–7.

Harcourt, A H, Stewart, K J & Fossey, D (1981*a*) Gorilla reproduction in the wild. In *Reproductive Biology of the Great Apes*, ed. C Graham, pp.265–79. New York: Academic Press.

Harcourt, A H, Fossey, D & Sabater Pi, J (1981*b*) Demography of *Gorilla gorilla*. *Journal of Zoology, London*, **195**, 215–33.

Hiraiwa-Hasegawa, M (1987) Infanticide in primates and a possible case of male-biased infanticide in chimpanzees. In *Animal Societies: Theories and Facts*, ed. Y Ito, J L Brown & J Kikkawa, pp. 125–39. Tokyo: University of Tokyo Press.

Hrdy, S B (1979) Infanticide among animals: a review, classification, and examination of the implications for the reproductive strategies of females. *Ethology and Sociobiology*, **1**, 13–40.

Hrdy, S B, Janson, C H & van Schaik, C P (1995) Infanticide: Let's not throw out the baby with the bath water. *Evolutionary Anthropology*, **3**, 151–4.

Kuroda, S, Nishihara, T, Suzuki, S & Oko, R A (1996) Sympatric chimpanzees and gorillas in the Ndoki Forest, Congo. In *Great Ape Societies*, ed. W C McGrew, L F Marchant & T Nishida, pp. 71–81. Cambridge: Cambridge University Press.

MacKinnon, J (1978) *The Apes within Us*. London: Collins.

Mankoto, M O (1988) La gestion du Parc National du Kahuzi-Biega (Zaïre). *Cahiers d'Ethologie Appliquée*, **8**, 447–50.

Mankoto, M O, Yamagiwa, J, Steinhauer, B B, Mwanza, N, Maruhashi, T & Yumoto, T (1994) Conservation of eastern lowland gorilla in the Kahuzi-Biega National Park, Zaire. In *Current Primatology*, Vol. 1, *Ecology and Evolution*, ed. B Thierry, J R Anderson, J J Roeder & N Herrenschmidt, pp. 113–22. Strasbourg: Université Louis Pasteur.

Mitani, M (1992) Preliminary results of the studies on wild western lowland gorillas and other sympatric diurnal primates in the Ndoki Forest, northern Congo. In *Topics in Primatology*, vol. 2, *Behavior, Ecology and Conservation*, ed. N Itoigawa, Y Sugiyama, G P Sackett & R K R Thompson, pp. 215–24. Tokyo: University of Tokyo Press.

Murnyak, D F (1981) Censusing the gorillas in Kahuzi-Biega National Park. *Biological Conservation*, **21**, 163–76.

Nadler, R D (1980) Reproductive physiology and behaviour of gorillas. *Journal of Reproduction and Fertility,* Supplement **28**, 79–89.

Nadler, R D (1982) Laboratory research on sexual behavior and reproduction of gorillas and orangutans. *American Journal of Primatology,* Supplement **1**, 57–66.

Palombit, R A (1999) Infanticide and the evolution of pair bonds in nonhuman primates. *Evolutionary Anthropology*, 7, 117–29.

Remis, M J (1994) Feeding ecology and positional behavior of western lowland gorillas (*Gorilla gorilla gorilla*) in the Central African Republic. PhD thesis, Yale University.

Remis, M J (1997) Ranging and grouping patterns of a western lowland gorilla group at Bai Hokou, Central African Republic. *American Journal of Primatology*, **43**, 111–33.

Robbins, M M (1995) A demographic analysis of male life history and social structure of mountain gorillas. *Behaviour*, **132**, 21–47.

Ruvolo, M, Pan, D, Zehr, S, Golberg, T, Disotell, T R & Dornum, M von (1994) Gene trees and hominoid phylogeny. *Proceedings of the National Academy of Sciences USA*, **91**, 8900–4.

Sarmiento, E E, Butynski, T M & Kalina, J (1996) Gorillas of Bwindi Impenetrable Forest and the Virunga Volcanoes: taxonomic implications of morphological and ecological differences. *American Journal of Primatology*, **40**, 1–21.

Sicotte, P (1993) Inter-group interactions and female transfer in mountain gorillas: influence of group composition on male behavior. *American Journal of Primatology*, **30**, 21–36.

Sicotte, P (1995) Interpositions in conflicts between males in bimale groups of mountain gorillas. *Folia Primatologica*, **65**, 14–24.

Sicotte, P (2000) A case study of mother–son transfer in mountain gorillas. *Primates*, **41**, 95–103.

Sievert, J, Karesh, W B & Sunde, V (1991) Reproductive intervals in captive female western lowland gorillas with a comparison to wild mountain gorillas. *American Journal of Primatology*, **24**, 227–34.

Sterck, E H M, Watts, D P & van Schaik, C P (1997) The evolution of female social relationships in nonhuman primates. *Behavioral Ecology and Sociobiology*, **41**, 291–309.

Stewart, K J & Harcourt, A H (1987) Variation in female relationships. In *Primate Societies*, ed. B B Smuts, D L Cheney, R M Seyfarth, R W Wrangham & T T Struhsaker, pp. 155–64. Chicago: University of Chicago Press.

Takahata, Y (1985) Adult male chimpanzees kill and eat a male newborn infant: newly observed intragroup infanticide and cannibalism in Mahale National Park, Tanzania. *Folia Primatologica*, **44**, 161–70.

Treves, A & Chapman, C A (1996) Conspecific threat, predation avoidance, and resource defence: implications for grouping in langurs. *Behavioral Ecology and Sociobiology*, **39**, 45–53.

Tutin, C E G (1996) Ranging and social structure of lowland gorillas in the Lopé Reserve, Gabon. In *Great Ape Societies*, ed. W C McGrew, L F Marchant & T Nishida, pp. 58–70. Cambridge: Cambridge University Press.

Tutin, C E G & Fernandez, M (1992) Insect-eating by sympatric lowland gorillas (*Gorilla g. gorilla*) and chimpanzees (*Pan t. troglodytes*) in the Lopé Reserve, Gabon. *American Journal of Primatology*, **28**, 29–40.

Tutin, C E G & Fernandez, M (1993) Composition of the diet of chimpanzees and comparisons with that of sympatric lowland gorillas in the Lopé Reserve, Gabon. *American Journal of Primatology*, **30**, 195–211.

van Schaik, C P (1996) Social evolution in primates: the role of ecological factors and male behaviour. *Proceedings of the British Academy*, **88**, 9–31.

van Schaik, C P & Dunbar, R I M (1990) The evolution of monogamy in large primates: a new hypothesis and some crucial tests. *Behaviour*, **115**, 30–62.

Watts, D P (1984) Composition and variability of mountain gorilla diets in the central Virungas. *American Journal of Primatology*, **7**, 325–56.

Watts, D P (1989) Infanticide in mountain gorillas: new cases and a reconsideration of the evidence. *Ethology*, **81**, 1–18.

Watts, D P (1990) Mountain gorilla life histories, reproductive competition, and sociosexual behavior and some implications for captive husbandry. *Zoo Biology*, **9**, 185–200.

Watts, D P (1991) Mountain gorilla reproduction and sexual behavior. *American Journal of Primatology*, **24**, 211–25.

Watts, D P (1992) Social relationship of immigrant and resident female mountain gorillas. I. Male–female relationships. *American Journal of Primatology*, **28**, 159–81.

Watts, D P (1994*a*) The influence of male mating tactics on habitat use in mountain gorillas (*Gorilla gorilla beringei*). *Primates*, **35**, 35–47.

Watts, D P (1994*b*) Agonistic relationships of female mountain gorillas. *Behavioral Ecology and Sociobiology*, **34**, 347–58.

Watts, D P (1996) Comparative socio-ecology of gorillas. In *Great Ape Societies*, ed. W C McGrew, L F Marchant & T Nishida, pp. 16–28. Cambridge: Cambridge University Press.

Watts, D P (1997) Agonistic interventions in wild mountain gorilla groups. *Behaviour*, **134**, 23–57.

Watts, DP (1998) Long-term habitat use by mountain gorillas (*Gorilla gorilla beringei*). I. Consistency, variation, and home range size and stability. *International Journal of Primatology*, **19**, 651–80.

Watts, D P (2000) Causes and consequences of variation in male mountain gorilla life histories and group membership. In *Primate Males*, ed. P M Kappeler, pp. 169–79. Cambridge: Cambridge University Press.

Wrangham, R W (1979) On the evolution of ape social systems. *Social Science Information*, **18**, 334–68.

Wrangham, R W (1987) Evolution of social structure. In *Primate Societies*, ed. B B Smuts, D L Cheney, R M Seyfarth, R W Wrangham & T T Struhsaker, pp. 282–97. Chicago: University of Chicago Press.

Yamagiwa, J (1983) Diachronic changes in two eastern lowland gorilla groups (*Gorilla gorilla graueri*) in the Mt. Kahuzi region, Zaire. *Primates*, **24**, 174–83.

Yamagiwa, J (1986) Activity rhythm and the ranging of a solitary male mountain gorilla (*Gorilla gorilla beringei*). *Primates*, **27**, 273–82.

Yamagiwa, J (1999) Socioecological factors influencing population structure of gorillas and chimpanzees. *Primates*, **40**, 87–104.

Yamagiwa, J (2001) Factors influencing the formation of ground nests by eastern lowland gorillas in Kahuzi-Biega National Park: some evolutionary implications of nesting behavior. *Journal of Human Evolution*. (in press)

Yamagiwa, J, Mwanza, N, Yumoto, T & Maruhashi, T (1992) Travel distances and food habits of eastern lowland gorillas: a comparative analysis. In *Topics in Primatology*, vol. 2, *Behavior, Ecology and Conservation*, ed. N Itoigawa, Y Sugiyama, G P Sackett & R K R Thompson, pp. 267–81. Tokyo: University of Tokyo Press.

Yamagiwa, J, Mwanza, N, Spangenberg, A, Maruhashi, T, Yumoto, T, Fischer, A & Steinhauer, B B (1993) A census of the eastern lowland gorillas *Gorilla gorilla graueri* in Kahuzi-Biega National Park with reference to mountain gorillas *G. g. beringei* in the Virunga Region, Zaire. *Biological Conservation*, **64**, 83–9.

Yamagiwa, J, Mwanza, N, Yumoto, Y & Maruhashi, T (1994) Seasonal change in the composition of the diet of eastern lowland gorillas. *Primates*, **35**, 1–14.

Yamagiwa, J, Maruhashi, T, Yumoto, T & Mwanza, N (1996) Dietary and ranging overlap in sympatric gorillas and chimpanzees in Kahuzi-Biega National Park, Zaire. In *Great Ape Societies*, ed. W C McGrew, L F Marchant & T Nishida, pp. 82–98. Cambridge: Cambridge University Press.

5 Subspecific variation in gorilla behavior: the influence of ecological and social factors

DIANE M. DORAN & ALASTAIR McNEILAGE

Umuco feeding on wild celery in Beetsme's Group in the Virunga Volcanoes.
(Photo by Alastair J. McNeilage.)

Introduction

Primates exhibit considerable diversity in their social systems (Smuts *et al.*, 1987), which is thought to have evolved through an interaction of several factors including ecological variables, social factors, phylogeny, and demographic and life history variables. Attempts to understand how this variation in social organization evolves have focused primarily on (1) ecological variables, particularly predation pressure and the abundance and distribution of food (Alexander, 1974; Wrangham, 1979, 1980, 1987; van Schaik, 1983, 1989, 1996; Sterck *et al.*, 1997), (2) social factors, primarily sexual selection and the potential risk of infanticide (Wrangham, 1979; Watts, 1989; van Schaik, 1996), and (3) phylogenetic inertia (DiFiore & Rendall, 1994). Gorillas provide a unique opportunity to reevaluate proposed models of ecological and social influences on social organization in African apes. Western lowland and mountain gorillas seem to differ dramatically in their habitats, resource availability, and foraging strategies. To what degree these differences are associated with differences in social organization is an intriguing question.

The genus *Gorilla* occurs in two widely separated forest habitats, one in western central Africa and one in eastern central Africa. Three subspecies of gorillas are generally recognized, western lowland (*G. gorilla gorilla*), eastern lowland (*G. g. graueri*), and mountain gorillas (*G. g. beringei*) (but see Doran & McNeilage, 1998 for review of taxonomic debate: Morell, 1994; Ruvolo *et al.*, 1994; Butynski & Sarmiento, 1995). The three subspecies have not been studied equally. Most of our knowledge of gorilla behavior is based on the pioneering work of Dian Fossey (1974) and subsequent researchers on the well-studied mountain gorillas of Karisoke, Rwanda. At Karisoke, several groups have been directly observed through time, providing extensive information on almost all aspects of mountain gorilla behavioral ecology (for reviews see Watts, 1996; Doran & McNeilage, 1998, this volume). This small population of 300 mountain gorillas occurs in a habitat quite different from that of gorillas elsewhere. Although there have been numerous studies of gorillas at other sites (mountain gorillas in Bwindi, Uganda; eastern lowland gorillas in Kahuzi-Biega, Democratic Republic of Congo; and western lowland gorillas in Congo–Brazzaville, Cameroon, Gabon, Equatorial Guinea, and Nigeria) there are few published data, and none on social behavior, on *habituated* gorillas from any other site than Karisoke (but see Yamagiwa *et al.*, 1996; Bermejo, 1997). Until recently, it has been generally assumed that the behavior of mountain gorillas is representative of all gorillas. However, with studies of additional gorilla populations from other geographical areas, there are striking hints of diversity in gorilla behavior. The aim of

124

this chapter is to assess the variability in gorilla behavior across sites from an evolutionary perspective. To this end we will (1) review how social and ecological factors are proposed to influence the evolution of sociality in primates, (2) discuss mountain gorilla behavior in light of these factors, (3) outline how these factors differ in western lowland gorillas, (4) make specific predictions about how these differences (changes in resource availability and potential risk of infanticide) may influence western lowland gorilla behavior relative to that of mountain gorillas, and (5) assess whether currently available data support these predictions. Original data are presented from a new study site, the Mondika Research Center, Central African Republic, in addition to reviewing the growing body of data from several other sites. Finally, important areas for future research will be highlighted.

What factors influence social behavior?
Ecological factors
Since female reproductive success is strongly limited by access to food, ecological explanations of sociality have focused on explaining how female distribution and relationships change in response to differing ecological circumstances (Trivers, 1972; Emlen & Oring, 1977; Bradbury & Vehrencamp, 1977). Van Schaik (1989) hypothesized that female primates form spatial associations to reduce the risk of predation. This group formation leads to increased feeding competition (if resources are limiting), with the type and degree of feeding competition directly influencing female social relationships. Specifically, when resources are either evenly and abundantly distributed, or sparsely distributed, they are not monopolizable by any individuals, regardless of size or rank, and scramble competition results. In this case, there is no selection for aggression (since it will not result in greater resource acquisition), no predictable dyadic relationships between individuals, no clear linear dominance hierarchy among females, and no alliance formation. Furthermore, since under scramble competition, group size is the major factor influencing a female's access to resources (and reproductive success), females should move freely between groups and particularly to smaller groups (if other factors allow) to reduce competition. Thus, these females should not be strongly philopatric. Females that fit this description are described as "non-female-bonded" primates (Wrangham, 1980). However, when resources are limited and can be monopolized (i.e. patchy in distribution, where a patch is large enough for some but not all members of a group), then contest competition arises within a group, and there is variance in individual resource acquisition. In this case, access to resources is based directly on power or rank, and there is selection

for aggression, resulting in female dominance relationships (and a rank hierarchy) and alliance formation with other females, particularly with kin. Female philopatry is likely to be favored and maintained under such conditions. Females that fit this description are "female bonded primates" (Wrangham, 1980). These two basic patterns are influenced by competition between groups as well, so that when contest competition is high between groups, resident females should exhibit more tolerant relationships (van Schaik, 1989; Sterck *et al.*, 1997). In general, van Schaik's (1989) model has been successful, although incomplete, in describing or predicting differences in female social relationships between primate taxa (Mitchell *et al.*, 1991; Barton *et al.*, 1996; Koenig *et al.*, 1998: but see Boinski, 1999; Kappeler, 1999; Strier, 1999).

Social factors
Attempts to understand the role of ecological factors on sociality have focused on female–female relationships, and have largely ignored the role of male–female associations (van Schaik, 1996). However, diurnal primates are characterized by permanent male–female association, unlike the majority of mammals (reviewed in van Schaik, 1996). Male–female relations are ultimately shaped by sexual selection and particularly male–male competition. Infanticide has been documented in many primate species, and it has been argued to be an adaptive strategy if the infanticidal male (1) kills unrelated, unweaned infants, which results in the mother going into estrus (and being fertilizable) sooner than she would have been otherwise, and (2) is in a position to sire the next offspring (Hrdy, 1979; Hrdy *et al.*, 1995). Van Schaik (1996, 1999) has argued that infanticide is a particularly viable reproductive strategy in many species of primates, as a result of primates' (relative to other mammals) very slow life histories, and in particular, long periods of lactational amenorrhea. As a result, female primates may evolve differing counter-strategies to infanticide, including, among others, the tendency to associate with males to decrease infanticide risk, in species that would be otherwise unlikely to group (van Schaik 1996, 1999). Van Schaik (1996) hypothesized that the tendency to group, as a defense against infanticide, could occur if (1) the cost to female grouping was low, (2) there is variation in male quality (so several females would be independently attracted to the same male), and (3) female transfer was common.

Phylogenetic factors
There is clearly some phylogenetic limit to the degree of plasticity a species can exhibit in differing ecological conditions; species are constrained by morphology, physiology, body size, and life history patterns (Brooks &

McLennan, 1991; Harvey & Pagel, 1991; Fleagle, 1999). In 1994, DiFiore & Rendall evaluated the phylogenetic contribution to primate social systems through a comparative study of primate social organization. They concluded that social organization might be strongly conserved in some lineages, particularly in Old World monkeys, in the face of considerable ecological variability, largely as a result of the retention of some key trait, such as female philopatry, which they argue, is highly conserved after it appears in a lineage.

However, two recent comparative studies of Old World primates indicate that differences in the distribution of resources at two sites had a major (and predictable) although variable impact on female social relationships (Barton *et al.*, 1996; Koenig *et al.*, 1998). Barton *et al.* (1996) found that although olive baboons spent only 30% of their feeding time feeding on clumped items (in trees), the resulting contest competition was correlated with a linear dominance hierarchy among females, strong female–female affiliation, coalition formation, and female philopatry, none of which was characteristic of the chacma baboon, which spent all feeding time on dispersed items (and was free of predator pressure). In a comparative study of folivorous Hanuman langurs, differences in the distribution of key resources at two sites also led to some, but not all, predictable differences in female social relationships (Koenig *et al.*, 1998). Thus differences in resource distribution, even in folivorous primates, which are traditionally thought of as non-female-bonded, can result in more female-bonded groups.

These findings raise the question of exactly to what extent phylogeny constrains social behavior. In van Schaik's (1996) discussion of the topic, he acknowledges that in some cases, species show similar behavior in widely different habitats and situations (for example captive versus wild), so that social behavior seems phylogenetically constrained and ecological differences seems to have little impact on behavior. However, the studies reported above indicate that there is some flexibility in a trait (female philopatry) that has been argued to be highly conserved once evolved, both in recently divergent species (chacma and olive baboons) and even within one species (Hanuman langurs). In addition, although the amount of plasticity in any given taxa has not been extensively addressed, it is clear that there can be large-scale differences in social organization due to subtle changes in ecology (reviewed in Boinski, 1999; see also Mitchell *et al.*, 1991; Kinzey & Cunningham, 1994; Barton *et al.*, 1996; Koenig *et al.*, 1998; Strier, 1999). Understanding the influence of phylogeny on social behavior requires further study of primate taxa in a wide variety of ecological and social conditions.

How do these factors influence African ape sociality?
Exactly how changes in resource distribution and availability should influence ape sociality is unclear. Because of their large body size (relative to monkeys), apes are much less susceptible to predation. This eliminates the major ecological benefit proposed for grouping, and in fact, orangutans and chimpanzees have a much less cohesive society than other primates, often spending large periods of time alone. However, unlike other apes, mountain gorillas live in cohesive groups. It has been proposed that social factors may have played an especially important role in the evolution of grouping patterns in apes, and that for mountain gorillas in particular, avoidance of infanticide may be the primary explanation for group cohesion (Wrangham, 1979; Watts, 1990; van Schaik, 1996). Gorillas provide a unique opportunity to further assess the influences of food distribution and infanticide on behavior in apes – independent of predation and phylogeny. Western lowland and mountain gorillas seem to differ in their habitats and foraging strategies. Following from ecological models (van Schaik, 1989) is the prediction that concurrent change in social behavior would be expected. In addition to potential changes in group composition and life history variables, variation in habitat and foraging strategies may have a direct impact on the frequency of situations in which infanticide may occur.

Mountain gorilla behavior
Mountain gorillas of Karisoke live in high-altitude montane forests of the Virunga Volcanoes, Rwanda. They are herbivorous, feeding primarily on leaves, stems, shoots, and piths of terrestrial herbs (Fossey & Harcourt, 1977; Vedder, 1984; Watts, 1984, 1985, 1996). Seasonality has little influence on the diet (Watts, 1996). Although mountain gorillas are selective feeders, choosing vegetation that is relatively high in protein and low in fiber and tannins, these "high-quality" terrestrial herbs, or terrestrial herbaceous vegetation, are abundant and widely distributed, which results in little within-group feeding competition; group size has little influence on an average day range of 500 m per day (Fossey & Harcourt, 1977; Vedder, 1984; Watts, 1984, 1985, 1996).

Mountain gorillas live in stable and cohesive age-graded groups (*sensu* Eisenberg *et al.*, 1972), the majority of which have one adult male, although up to four males may be present (Schaller, 1963; Fossey, 1974; Harcourt, 1979; Robbins, 1995; Watts, 1996). Multimale groups are generally the result of a male maturing and remaining in his putative father's group (Robbins, 1995). The most recent census indicated that groups ranged in size from two to 34 gorillas per group (mean = 9.15; n = 32) and that 28% of mixed-sex groups had more than one fully adult silverback male (Sholley, 1991). Within groups, adult social bonds are strongest between females and silverbacks; most females are unrelated and do not associate regularly with each other. Although some females in gorilla groups can be ranked on the basis of decided non-aggressive approach–retreat interactions, they do not form clear agonistic linear dominance hierarchies, formally signal subordinate status to each other (although they do to males), or enlist familial support to consistently influence agonistic relationships (Harcourt, 1979; Stewart & Harcourt, 1987; Watts, 1994). As such female–female relationships can be defined as "egalitarian".

Both sexes may disperse at maturity, with female, rather than male, natal dispersal more common [natal dispersal occurs in 72% of known females (n = 29; Watts, 1996) and 36% of maturing males in heterosexual groups (n = 11; Robbins, 1995)]. Females disperse during inter-group encounters and transfer directly into other groups; dispersing males remain solitary or join all-male groups (Harcourt, 1978). No male takeovers of groups from the outside have been recorded. Infanticide accounts for a minimum of 37% of infant mortality (Fossey, 1984; Watts, 1989). It occurs most frequently after the death of a silverback in a one-male group, but may also occur during inter-group encounters, when an outside male kills an infant when the presumed sire is present (Fossey, 1984; Watts, 1989). Gorilla infants born in one-male groups are more likely to die from infanticide than those born in multimale groups (Robbins, 1995), a pattern that has been noted in some but not all primate species (see Borries & Koenig, 2001, for review). Demographic constraints, such as a long time to male maturation combined with an average male tenure length that is short in comparison, may limit the number and duration of groups in the population with a multimale structure (Robbins, 1995). Mountain gorilla inter-unit encounters (encounters between two groups or between a group and a lone male) occur rarely, less than once per month on average (Sicotte, 1993; Watts, 1994) and most frequently involve aggressive encounters (Harcourt, 1978). Thus, opportunities for both female transfer and infanticide in mountain gorillas are rare. Mountain gorilla social organization is consistent with the predictions of van Schaik's models of female sociality and group formation to avoid infanticide.

How do ecological factors differ in western lowland gorillas?

What kind of data are available on western lowland gorillas?

Habituation of western lowland gorillas has proven difficult. Data on western lowland gorilla behavior are derived from studies with three distinct types of aims and methods. First, there is currently one site (Lossi, Democratic Republic of Congo) where western lowland gorillas are reported to be habituated (for ecotourism) to observers. Published results include group size and day range (Bermejo, 1997). Second, there are two study sites in the Democratic Republic of Congo (Mbeli Bai and Maya Maya: Olejniczak, 1996, 1997; Magliocca & Querouil, 1997; Magliocca, 1999; Magliocca *et al.*, 1999; Parnell, 1999) where researchers observe gorillas from a platform at very large open swampy clearings referred to as "bais". At these sites, gorillas become accustomed to the presence of the researchers, who can observe the gorillas as they feed and socialize. However, researchers are unable to follow the gorillas when they leave the area, due to the physical nature of the swamps. These studies are rich resources for demographic and life history information, because many different groups can be observed over time. Finally, there are three study sites (in Gabon: Lopé; in Central African Republic: Bai Hokou and Mondika, this study) where the process of habituation is currently on-going. This process entails tracking a gorilla group daily from nest to nest and contacting them as frequently as possible. During the habituation process, indirect data from trail signs, fecal samples, and nest sites can be used, in addition to limited direct observation, to provide information on ranging, grouping patterns, and diet. Currently no data on western lowland gorilla within-group interactions of known individuals are reported from any site.

Ecological factors

Western lowland gorillas live in lowland tropical forest, where the abundance and distribution of available food differs considerably from that of mountain gorilla habitat. There are three major differences in food availability between western lowland gorillas and mountain gorillas. First, preferred high-quality terrestrial herbs (Kuroda *et al.*, 1996) are much less abundant and are more sparsely distributed in lowland forests than in the Virungas (Watts, 1984; Rogers & Williamson, 1987; Malenky *et al.*, 1994; White *et al.*, 1995; Goldsmith, 1996; Fay, 1997; McNeilage & Doran, personal observation). Second, superabundant patches of either aquatic herbs or Marantaceae forest (super patches) are present at some, but not all, western lowland gorilla sites. Finally, fruit is widely available, unlike at Karisoke, and western lowland gorillas, unlike mountain gorillas, incorporate a significant amount of fruit in their diet (reviewed in Tutin, 1996 and Doran & McNeilage, 1998). Commonly used fruit trees are both seasonal in fruit production and rare in the habitat (unpublished data). Predictions can be made for how each of these factors should alter western lowland gorilla behavior, relative both to mountain gorillas and to western lowland gorillas at sites with differing resource availability and distribution. Specific predictions will be based on conditions at Mondika, although implications for other western lowland gorilla sites will be discussed.

Predictions for behavioral differences in western lowland gorillas
Reduced density of preferred herbs
Overall density of terrestrial herbaceous vegetation should constrain group size, through within-group scramble competition (Malenky & Stiles, 1991; Malenky *et al.*, 1994). Since predation pressure is reduced, group spread (or average inter-individual distances) can increase during foraging up to a certain point (Kuroda *et al.*, 1996; Doran & McNeilage, 1998). It is probable that the inherent upper limit to gorilla group spread is determined by a combination of factors (e.g. the point at which individuals can no longer communicate effectively with each other, or when females, particularly females in estrus or with infants susceptible to infanticide, cannot receive the necessary protection from males). Thus, the maximum amount of potential group spread that western lowland gorillas can undergo is relatively fixed or inherent in gorillas (or at least in gorillas who communicate the same way) but different densities and distribution of herbs at different sites should limit gorilla density within that group spread. Therefore, maximum possible group spread and herb densities should impose limits on group size, if herbs are limiting. Since herb densities are reduced at all western lowland gorilla sites relative to mountain gorilla sites and at Mondika relative to some other western lowland gorilla sites (density in stems per m^2: Mondika = 0.78; Karisoke = 8.81,Watts, 1984; Lopé = 1.87, White *et al.*, 1995) it is predicted that there is a constraint on the upper limit of group size at Mondika, relative to mountain gorillas and to western lowland gorilla at sites of greater herb density. In addition, day range length and home range size should be greater for western lowland gorillas relative to mountain gorillas, and variable across western lowland gorilla sites depending on herb density and distribution. Within a western lowland gorilla site, if there is seasonal variation in the relative importance of terrestrial herbaceous vegetation in the diet, then average day range should be reduced when feeding primarily on such vegetation as compared to when feeding on fruit or aquatic herbs (see below). In addition, the relative inter-unit encounter rate should increase slightly for western lowland gorillas based on their longer day

travel distances. Furthermore, within a site, inter-unit encounter rates should be lower when gorillas are feeding primarily on terrestrial herbaceous vegetation as compared to aquatic herbs or fruit, if predictions are based on ecological factors alone. Since these herbs are not a contestable resource, consumption should not result in any differences in female social behavior at Mondika relative to Karisoke.

Use of swamps

These patches can be either very large open swampy clearings referred to as "bais" (e.g. Mbeli Bai = 20–30 ha, C. Olejniczak, personal communication), or long and narrow swamps bordering rivers (Mondika = roughly 3 km long × 500 m wide, personal observation) where gorillas feed on aquatic herbs while wading in water. Long-term observation at "bais" indicates that, although many different groups use these resources, they are used infrequently by any one group (average visits by a group: Maya Maya: = 3 days in 8 months, Magliocca *et al.*, 1999; Mbeli Bai = <2% of total time, Doran & McNeilage, 1998 calculated from Olejniczak, 1996), and thus they probably have little effect on gorilla sociality. However, at Mondika, gorillas use the swamps often (swamp use on 28% of days, *n* = 208) and remain in them for variable amounts of time, ranging from minutes to days. Swamps, although large, are restricted in distribution, and should therefore, on occasion, influence day range, with day ranges longer when traveling to and from them, versus on days when they are not used. Since many groups are drawn to these patches (Mbeli Bai: 14 groups and 10 lone silverbacks, Olejniczak, 1996, 1997; Maya Maya: 36 groups and 18 solitaries, Magliocca *et al.*, 1999), there may be considerable overlap of gorilla home ranges at all sites where they are present. It is possible that rather than continuous access along the length of the swamp, gorillas enter and exit the swamp from a few sites where there are tree falls or other "natural bridges". If swamp use is predictable based on availability of other resources, and if the swamp is accessed from relatively few sites, then the frequency of inter-unit encounters may be higher for western lowland gorillas at Mondika compared to mountain gorillas or for western lowland gorillas at sites where these resources are not present. Although these patches are large, if the number of feeding sites are limiting (because of depth of swamp) then contest competition could arise. If feeding sites are not limiting, then scramble competition should prevail.

Use of fruit

The fruits that western lowland gorillas feed on vary considerably in their mean patch size, density, distribution and phenological patterns, and should therefore have differing impacts on behavior. Fruits from small trees [of diameter at breast height (DBH) < *c.* 20 cm], shrubs, and terrestrial herbaceous vegetation offer only one or two gorilla feeding sites and are probably fed on opportunistically as gorillas are spread out foraging on various foods. As such, access to these resources should influence gorilla grouping and ranging patterns in a manner similar to that for terrestrial herbaceous vegetation (discussed above). Fruits from large fruit trees (DBH > *c.* 80 cm) probably do not limit party size, if and when the fruit crop is sufficiently large, because they provide enough feeding sites for all individuals in a group. These resources, if preferred, should influence ranging patterns, and if rare enough in the habitat should draw several gorilla groups to them when they are ripe, which would increase the rate of inter-unit encounters. Medium to large fruit trees (DBH = 20–80 cm which includes the majority of species) are large enough to provide feeding sites for several (perhaps three to five on average), but not all gorillas in a group (see Table 5.1 for group sizes). If these resources are strongly preferred, and groups remain cohesive, then increased reliance on fruit should lead to increased within-group contest competition, with predictable consequences including longer day ranges, larger home ranges, and smaller groups, as well as more differentiated female relationships. An alternate hypothesis is that groups subdivide or form separate foraging parties; if average distance between fruit patches is greater than that covered by gorilla short-range communication, then a portion of the group may splinter off (at some potential cost) to gain access to fruit in a separate patch. As long as subgroups continue to feed on this resource, it may be advantageous to continue to follow separate foraging paths for several days. In this case, subgrouping should be a seasonal phenomenon, occurring most frequently during the time of fruit (of medium-sized trees) abundance.

Considerable differences in the availability of these three resources at different western lowland gorilla sites should result in variation in the relative importance of these resources, and their subsequent influence on behavior. Comparative studies across sites will be important to assess the relative impact of these resources on gorilla behavior.

Are these predictions correct? What do we know to date about western lowland gorilla behavior?
Western lowland gorilla diet
The exact degree of frugivory in western lowland gorillas is unclear, because of the absence of studies based on direct observations of feeding. However, a pattern is emerging from the sites where dietary studies have

Table 5.1. *Gorilla group size: is there evidence for subgrouping and/or supergrouping across sites?*

Site	Type of study[a]	Group size	n	Subgrouping and supergrouping?
WESTERN LOWLAND GORILLAS				
Gabon				
Lopé[1]	P	Median = 10 (4–16)	8 groups	No subgrouping; large group spread (> 500 m)
Central African Republic				
Ndakan[2]				
Transect		6.0 (2–24)[b]	229 nest sites	No subgroups
Trail follow		8.5 (3–20)[b]	29 nest sites	
Group follow		12.2 (8–14)	12 day follow	
Bai Hokou[3]	P	12–15[b]	1 group	4 confirmed occasions of subgroups sleeping 950 m apart (in group with 2 SB)
Bai Hokou[4]	T	9.4[b] (1–19)	232 nest sites	Subgroups (a.m. and p.m. nest count differ by 2) on 31% of days
Mondika[5]				
All fresh nest sites	P	6–7 (1–20)[b]	362 nest sites	25% of consecutive nest follows had nest counts that differed by more than 4 nests
Consecutive tracking		7.4 + 3.4 (1–16)[b]	91 nest sites	Verified subgroups on 2 days
Congo				
Mbeli Bai[6]	B	7.4 (based on direct observation)	13 groups	Subunits of group with 2 SB seen on several occasions, enter and leave by different directions
Lossi[7]	H	14 (largest group = 32)	7 groups	Fusion of 2 groups, nest site of 40 (*Dialium*)
Maya Maya[8,9]	B	11.2 (2–22)	31 groups	More than one group seen mingling without aggression, solitary females seen
MOUNTAIN GORILLAS				
Karisoke[10]	H	9.15 (2–34)	32 groups	Cohesive groups
Bwindi Impenetrable[11]		10 (2–23)	28 groups	No evidence of subgrouping
EASTERN LOWLAND GORILLAS				
Kahuzi-Biega[12]		13.5 (5–20)	4 groups	Cohesive groups

[a] H = habituated, P = process of habituation is on-going, T = variant of habituation process where gorilla groups are tracked but not contacted, and B = observed at "bais".
[b] Group size estimates are based on the number of weaned individuals in a group (nest counts).
Sources: [1]Tutin, 1996; [2]Fay, 1997; [3]Remis, 1997a; [4]Goldsmith, 1996; [5]this study; [6]Olejniczak, 1996; [7]Bermejo,1997; [8]Magliocca, 1999; [9]Magliocca & Querouil, 1997; [10]Watts, 1996; [11]McNeilage et al., 1998; [12]Yamagiwa et al., 1996.

been conducted. Western lowland gorillas eat a wide variety of fruit (more than 70 species eaten at each site); they incorporate some fruit in the diet year-round (more than 90% of all fecal samples contain fruit); and they are selective in their fruit choice, opting for ripe, succulent, sweet fruit that is low in proteins and fat, while avoiding unripe fruits (Rogers *et al.*, 1990; Williamson *et al.*, 1990; Kuroda, 1992; Remis, 1994, 1997b; Nishihara, 1995; Goldsmith, 1996, 1999; Kuroda *et al.*, 1996; Tutin, 1996). There is much more seasonality in the diet of western lowland gorillas than in that of mountain gorillas (Rogers *et al.*, 1988; Mitani *et al.*, 1992; Yamagiwa, 1992; Tutin & Fernandez, 1993; Remis, 1994, 1997b; Nishihara, 1995; Goldsmith, 1996, 1999; Kuroda *et al.*, 1996; Tutin, 1996). During periods of resource abundance, fruit seems to make up the major portion of the diet. During periods of resource scarcity when succulent fruits are unavailable, there is an increase in consumption of herbs (at least low-quality), bark, woody pith and non-preferred fruits (e.g. *Duboscia*). Such fruits are higher in fiber and tannin content than preferred fruits and, although present throughout the year, are ignored when ripe succulent fruits are available (Rogers *et al.*, 1988; Tutin *et al.*, 1991).

In addition to fruits, western lowland gorillas regularly consume herbs and piths both on dry land and in swamps or "bais". Currently, little is known about the importance of swamp vegetation to western lowland gorillas. However, on dry land, piths and leaves are important to gorillas year round: at Mondika, herbs/leaves were present in 99.7% of fecal samples, and on average, 61% of each sample by volume is fiber (unpublished data). Kuroda *et al.* (1996) have recently suggested that high-quality herbs (defined as rich in minerals and proteins), such as *Haumania* and *Hydrocharis,* are ingested throughout the year, whereas lower-quality herbs such as *Palisota* and *Aframomum* are incorporated in greater quantities when fruit is unavailable. Our data from Mondika support this hypothesis. Greatest diversity in herb and/or leaf consumption occurs during months with lowest fruit diversity. However, the pattern of use differs between high-quality herbs (*Haumania,* and to a lesser extent, *Megaphrynium*) and the other three most commonly used herbs. Feeding remains of *H. danckelmaniana* are present on 79 ± 13% of days (range = 55–94); it is fed on consistently throughout the year, whereas other common herb species (*Aframomum* sp. and *Palisota ambigua*) are used as fallback foods, as predicted by Kuroda *et al.* (1996). In addition to fruits and herbs, gorillas consume leaves of trees and shrubs, bark, and small quantities of insects.

Group size and composition

Overall social structure appears to be roughly similar between western lowland gorilla and mountain gorillas. Available data indicate that average western lowland gorilla group size is not uniformly smaller than that of mountain gorillas, in spite of reduced herb density and increased frugivory (Table 5.1). Average group size is variable across sites; group size is largest at Lossi, a site where both herb and gorilla densities are reported to be high, and smallest at Mondika and Ndakan. Data on group composition based on direct observation of groups at "bais" indicate that multimale (versus one-male) groups occur less frequently in western lowland gorillas relative to mountain gorillas (percentage of groups that are multimale: Mbeli Bai = 8%, *n* = 13, Olejniczak, 1996; Maya Maya = 0%, *n* = 31, Magliocca *et al.*, 1999). There are no reports of all-male western lowland gorilla groups.

Dispersal patterns

Little is known about western lowland gorilla dispersal patterns. There are two reports of female transfer between groups (Tutin, 1996; Parnell, 1999). In addition, there are several instances of females disappearing from groups at an age appropriate for both natal and secondary transfer (Tutin, 1996). Tutin (1996) also reports the case of emigration of two males (14–15 years) from a group of known composition. Thus it appears that in western lowland gorillas, like mountain gorillas, both males and females may disperse from groups. Since there are more multimale (versus one-male) groups in western lowland gorillas than mountain gorillas, it would seem likely that rates of male dispersal would be higher in western lowland gorillas than mountain gorillas.

Group cohesion

At all sites, group spread is reported to be larger in western lowland gorillas relative to mountain gorillas. Tutin (1996) reports group spread to be greater than 500 m at times at Lopé; at other sites with lower herb densities, group spread may be larger. From the one site where western lowland gorillas are habituated to observers, subadults were observed wandering several hundred meters from their group (M. Bermejo, personal communication, reported in Magliocca *et al.*, 1999). In addition, adult females have been observed several hundred meters from a male on occasions at four different sites (Mondika: personal observation; Ndoki: M. Mitani, personal communication; Bai Hokou: A. Blom, personal communication; Maya Maya, Magliocca *et al.*, 1999). These long inter-individual distances may result in some subspecific differences in communications; for example, we have observed a western lowland gorilla female hand-clapping, indicating her position to a silverback, when she was alarmed and he was at a considerable distance, a behavior that is not seen in mountain gorillas. Changes in types and frequencies of vocalizations, associated with changes in resource distribution and use have been documented in other species of primates (Boinski & Mitchell, 1992) and will be an interesting area of future research. The occurrence of subgrouping and/or supergrouping (where two groups unite and travel together for several days) has been reported at some, but not all sites (Table 5.1). Subunits of known larger groups have entered and exited "bais" without the rest of the group, although it is unclear how long these subgroups remained apart (Olejniczak, 1996). At Lossi, two groups (with a total of 40 individuals) nested together, making a supergroup near a preferred fruit tree. More data derived from direct follows of gorillas are needed to examine grouping and spacing patterns.

Day range, home range size and implications for inter-unit encounters

Average day range of western lowland gorillas, although variable across sites, is two to three times greater than that of mountain gorillas (Table 5.2). Western lowland gorillas are reported to range significantly farther when feeding on succulent fruit, than when feeding primarily on herbs (Table 5.3). Although there is considerable intersite variation in western lowland gorilla annual home range size ($7–23\,km^2$), it is, on average, at least double that reported for mountain gorillas (Table 5.4). Although both mountain gorillas and western lowland gorillas have extensive overlap in their home ranges, western lowland gorilla groups, unlike mountain gorillas are reported to use the same parts of the home range at the same time. Several different gorilla groups use the same "bais": 14 groups and 10 lone silverbacks frequent Mbeli Bai (Olejniczak, 1996, 1997) and 36 groups and 18 solitaries are reported to use Maya Maya (Magliocca *et al.*, 1999). Olejniczak (1996) reports that more than one group is present during 10% of observations. In addition to congregating at "bais", there is also a tendency, although less well documented, for different gorilla groups to be in proximity at individual fruit trees, which are both rare in the environment and seasonal in their fruiting patterns (Tutin, 1996; personal observation). Genetic work (at Mondika) confirms the presence of several groups (a minimum of five) with overlapping home ranges that actually appear to track and follow each other on some occasions (Bradley, unpublished data) This more frequent home range overlap and simultaneous use of rare resources results in a much greater potential for inter-group en-

Table 5.2. *Variability in gorilla day ranges*

Site	Study type[a]	Mean (m)	Range (m)	n (full day ranges)
WESTERN LOWLAND GORILLAS				
Gabon				
Lopé[1]	P	1105 ± 553	220–2790	80 (5 gorilla groups)
Central African Republic				
Bai Hokou[2]	P	2300	1000–3200	8
Bai Hokou[3]	T	2590 ± 1010	342–5237	95
Mondika[4] (all groups)	P	1553 ± 841	200–4040	94
Mondika[4] (one group)	P	1904 ± 411	1485–2651	n = 6 monthly means
Congo				
Lossi[5]		1853 ± 807	300–5500	63 (one group of 20)
MOUNTAIN GORILLAS				
Rwanda				
Karisoke[6]	H	500		
Bwindi[7]	T	918 ± 480	148–2036	81 (10 groups)
EASTERN LOWLAND GORILLAS				
Democratic Republic of Congo				
Kahuzi-Biega[8]		600–1100	200–1800	36
Kahuzi-Biega[9]		813	154–2280	31

[a] H = habituated, P = process of habituation is on-going, T = variant of habituation process where gorilla groups are tracked but not contacted.
Sources: [1]Tutin, 1996; [2]Remis, 1997a; [3]Goldsmith, 1996; [4]this study; [5]Bermejo, 1997; [6]Watts, 1996; [7]Achoka, 1993; [8]Casimir, 1975; [9]Yamagiwa *et al.*, 1996.

counters for western lowland gorillas compared to mountain gorillas. Currently few data are available on rates of western lowland gorilla inter-group encounters.

Type of inter-unit encounters

Olejniczak (1996) reports variable reactions during inter-group encounters at Mbeli Bai: the response of the same two groups varies over the course of a few days, with members of different groups sometimes seen feeding in proximity, and then a few days later appearing to avoid each other. Reports from several sites indicate occasional peaceful intermingling of groups (Tutin, 1996; Magliocca & Querouil, 1997). Bermejo (1997)

Table 5.3. *Seasonal variation in gorilla day range*

Site	Study type [a]	Good fruit month or rainy season[b]	Poor fruit month or dry season[c]
WESTERN LOWLAND GORILLAS			
Gabon			
Lopé[1]	P	1266 m ($n = 32$)	749 m ($n = 48$)
Central African Republic			
Bai Hokou[2]	T	3100 m	2110 m
Mondika[3]	P	1648 m ($n = 77$)	1118 m ($n = 17$)
MOUNTAIN GORILLAS			
Rwanda			
Karisoke[4]	H	No seasonal difference	No seasonal difference

[a]H = habituated, P = process of habituation is on-going, T = variant of habituation process where gorilla groups are tracked but not contacted.
[b]Good fruit months or rainy season months are months in which succulent fruit makes up a major portion of the diet.
[c]Poor fruit months or dry season months are when herbs make up the major portion of the diet.
[1]Tutin, 1996; [2]Goldsmith, 1996; [3]this study; [4]Watts, 1996.

documented two large gorilla groups nesting together near a *Dialium* fruit tree. Tutin (1996) has documented (1) possible contest over females (lone males pursuing one-male groups for up to several days), (2) direct competition over fruit (63% of interactions, $n = 22$), and (3) tolerance between certain groups that nest in proximity and spend long periods of time within auditory range. Although peaceful intermingling of groups has been documented in mountain gorillas, it is a rare occurrence (accounting for less than 7% of 58 encounters; Sicotte, 1993), which occurs between familiar groups (those that have recently fissioned) or individuals who were previously members of the same group. Although quantitative data are not yet available on western lowland gorilla encounters, the fact that peaceful intermingling has been noted at so many sites suggests that it may be a more common phenomenon in western lowland gorillas compared to mountain gorillas. Thus, western lowland gorilla inter-unit interactions may occur more frequently and be more variable in nature than those of mountain gorillas. This flexibility in behavior during inter-group encounters is somewhat reminiscent of bonobo (*Pan paniscus*) inter-group encounters, which are also variable and include avoidance between groups, aggressive conflicts which may result in serious injuries, or peaceful intermingling of groups, depending upon which groups are interacting (Kano, 1992).

Table 5.4. *Gorilla home range size*

Site	Study type[a]	Home range (km²)	Home range overlap	Inter-group interactions
WESTERN LOWLAND GORILLAS				
Gabon				
Lopé[1]	P	10 (core area) 22 (3 years)	100%	$n = 40$; peaceful and aggressive encounters
Central African Republic				
Bai Hokou[2]	P	23		
Mondika[3]	P	≈20	100%	
Congo				
Mbeli Bai[4]	B		At least 17 different groups, 6 lone males (total 134 gorillas use Mbeli Bai)	Variable responses to other groups in saline
Lossi[5]	H	7–14	Extensive	
Maya Maya[6]	B		37 groups use saline, several observed at same time	More than one group seen mingling without aggression, solitary females observed
MOUNTAIN GORILLAS				
Rwanda				
Karisoke[7,8,9]	H	4–11 (mean annual home range) 1–4 (mean annual core area)	Overlap is present, but generally groups do not use the same area at the same time	Occur <1/month on average. Majority of encounters are aggressive. Peaceful intermingling rare.
EASTERN LOWLAND GORILLAS				
Kahuzi-Biega[10]		31 (15 month study)	Overlap is present but groups tried to keep a minimum distance of 2 km between them	Rare – another group was heard only once during 15 month study

[a]H = habituated, P = process of habituation is on-going, T = variant of habituation process where gorilla groups are tracked but not contacted, B = direct observation at "bai".

[1]Tutin, 1996; [2]Remis, 1997a; [3]Goldsmith, 1996; [3]this study; [4]Olejniczak, 1996; [5]Bermejo, 1997; [6]Magliocca, 1999; [7]Watts, 1998; [8]Hartcourt, 1978; [9]Sicotte, 1993; [10]Casimir, 1975.

Discussion

Although data based on direct observation of western lowland gorillas are limited, current evidence suggests that differences in resource availability and distribution lead to dietary changes that influence gorilla sociality. Current data indicate that western lowland gorillas, relative to mountain gorillas, have fewer multimale (versus one-male) groups, increased group spread, and probably have reduced group cohesion (more flexible grouping patterns with subgrouping and/or supergrouping) and more frequent and varied types of inter-group encounters. This mosaic of behavior makes it difficult to predict the changes one would expect to find in female sociality. The increased within-group contest competition which would be predicted for western lowland gorillas (relative to mountain gorillas) based on western lowland gorillas' increased consumption of patchier resources (fruits of limited DBH), is somewhat mitigated by the western lowland gorillas' increased group spread and the ability to feed on other foods. However, not all dietary items are of equal quality; if some preferred resources occur in patches that are large enough to accommodate some, but not all, members of a gorilla group, then one would predict higher frequencies of feeding aggression and more differentiated female relationships in western lowland gorillas compared to mountain gorillas. It is not currently possible to assess whether this is, in fact, the case but it is an interesting area of further inquiry. Furthermore, there may be inter-site differences in the degree to which contestable dietary items are used by western lowland gorillas due to differences in herb density, the size, density, and distribution of preferred fruit species, and the availability of swamps and "bais". As ongoing work at several gorilla sites continues, it will be possible to gain a clearer understanding of the ecological influences on gorilla behavior.

It has been proposed that social factors may have played an especially important role in the evolution of grouping patterns in apes, and that for mountain gorillas in particular, avoidance of infanticide may be the primary explanation for group cohesion (Wrangham, 1979; Watts, 1990; van Schaik, 1996). However, differences in western lowland gorilla sociality (relative to that of mountain gorillas) may lead to increased opportunity and decreased risk for western lowland gorilla males attempting to commit infanticide. First, increased inter-group encounter rate (with both groups and lone males) should result in increased opportunities for females to assess male quality and to transfer between groups, and for males to engage in direct male–male competition, including opportunities for males to commit infanticide. Second, increased seasonal variation in resource availability, reduced herb density and increased frugivory may result in

142   *Diane M. Doran & Alastair McNeilage*

longer inter-birth intervals for western lowland gorillas (relative to mountain gorillas), resulting in longer periods of infancy, and thus, vulnerability to infanticide (Janson & van Schaik, 1993; Leigh, 1994; van Schaik *et al.*, 1999; see also Leigh & Shea, 1995 and Doran, 1997 for discussion). Third, the cost (to non-group males) of female harassment and infanticidal attempts may be considerably reduced in western lowland gorillas relative to mountain gorillas, since the group male (and presumed sire) may be at a considerable distance to any given female as a result of reduced group cohesion (or increased group spread) and thus will be less able to defend her against infanticidal males. However, in spite of this greater potential for infanticide, current data suggest that western lowland gorillas are found more frequently in one-male groups than mountain gorillas, although long-term demographic records of mountain gorillas indicate that infants born in one-male groups face a greater risk of infanticide than those in multimale groups (Robbins, 1995). In addition, although evidence is scarce, reports of tolerance between, at least some, western lowland gorilla groups, raises interesting questions about the nature of male–male competition. These factors raise the possibility that either that the costs of continuous grouping for western lowland gorillas may be too great, or the benefits of female infanticide avoidance strategies too small, to act as strongly a selective force in gorilla behavior as previously suggested. If true, this again raises the question of why gorillas remain in groups.

Summary
A review of the subspecific variation in gorilla behavior is made to consider whether changes in resource availability and the potential risk of infanticide influence western lowland gorilla behavior, relative to that of mountain gorillas, in a predictable manner. Decreased herb density, increased fruit availability, and the presence of swamps or "bais" are associated predictably with longer day ranges and larger home ranges in western lowland gorillas relative to mountain gorillas. Western lowland gorilla group size is not predictably smaller than that of mountain gorillas, but rather, reduced herb densities and increased frugivory result in increased group spread, with individuals sometimes foraging hundreds of meters apart from each other. In addition, although mountain gorilla and western lowland gorilla groups may be similar in size, there is evidence that western lowland gorillas may be more flexible in their grouping patterns in response to changing resource availability. Increased frugivory, and in some cases the use of swamps and "bais", results in more contemporaneous overlap of home ranges between western lowland gorillas (relative to mountain gorillas), which may result in more frequent encounters with

other groups and lone males. The nature of these encounters is not yet well documented, but the hints are puzzling. These factors raise the question whether female infanticide avoidance acts as strongly a selective force in gorilla behavior as previously suggested.

Acknowledgements
We gratefully acknowledge the Ministries of Eaux et Forêts and Recherches Scientifiques in Central African Republic and the Ministry of l'Enseignement Primaire, Secondaire et Supérieur Chargé de la Recherche Scientifique in the Democratic Republic of Congo for permission to carry out research at the Mondika Research Center. In addition we would like to thank M. Gatoua Urbain, the Directeur National of the Dzanga Ndoki National Park, Allard Blom and Lisa Steele of the World Wide Fund for Nature and Bryan Curran of the Wildlife Conservation Society for their continued support and logistical assistance. Thanks to Chloe Cippoletta, Patrick Mehlman, Andreas Koenig, and Carola Borries for comments on the manuscript and fruitful discussion. We thank Martha Robbins, Pascale Sicotte, and Kelly Stewart for the invitation to participate in the Conference and for their many helpful comments on the chapter. Financial support is from the National Science Foundation (SBR-9422438 and SBR-9729126), National Geographic Society, and the L.S.B. Leakey Foundation. This research would not be possible without the efforts of many people in the field, including Dr. Carolyn Bocian, Dr. Patrick Mehlman, Dr. Alecia Lilly, Natasha Shah, Brenda Bradley, Piet Demmers, Mitch Keiver, Terry Brncic, Tripp Holman, Monica Wakefield, David Greer, Angelique Todd, Cleve Hicks, and Amy Parker. Finding gorillas would be impossible without the skilled tracking and botanical lessons of many field assistants, and especially Ndoki, Mopetu, Mangombe, Mokonjo, Ndima, and Mamandele.

References

Achoka, I (1993) Home range, group size and group composition of mountain gorilla (*Gorilla gorilla berengei*) in the Bwindi-Impenetrable National Park, South-western Uganda. MSc thesis, Makerere University.
Alexander, R D (1974) The evolution of social behavior. *Annual Review of Ecology and Systematics*, **5**, 325–83.
Barton, R A, Byrne, R W & Whiten, A (1996) Ecology, feeding competition and social structure in baboons. *Behavioral Ecology and Sociobiology*, **38**, 321–9.
Bermejo, M (1997) Study of Western Lowland Gorillas in the Lossi Forest of North Congo and a pilot gorilla tourism plan. *Gorilla Conservation News*, **11**, 6–7.

Boinski, S (1999) The social organizations of squirrel monkeys: implications for ecological models of social organization. *Evolutionary Anthropology*, **8**, 101–12.

Boinski, S & Mitchell, C L (1992) Ecological and social factors affecting the vocal behavior of squirrel monkeys. *Ethology*, **92**, 316–30.

Borries, C & Koenig, A (2001) Infanticide in Hanuman langurs: social organization, male migration and weaning age. In *Infanticide by Males and its Implications*, ed. C P van Schaik & C H Janson. Cambridge: Cambridge University Press. (in press)

Bradbury, J W & Vehrencamp, S L (1977) Social organisation and foraging in emballonurid bats. *Behavioral Ecology and Sociobiology*, **2**, 1–17.

Brooks, D R & McLennan, D H (1991) *Phylogeny, Ecology and Behavior: A Research Program in Comparative Biology*. Chicago: University of Chicago Press.

Butynski, T M & Sarmiento, E E (1995) Gorilla census on Mt. Tshiaberimu: preliminary report. *Gorilla Journal*, **10**, 11–12.

Casimir, M J (1975) Feeding ecology and nutrition of an eastern gorilla group in the Mt. Kahuzi region (Republic of Zaire). *Folia Primatologica*, **24**, 81–136.

DiFiore, A & Rendall, D (1994) Evolution of social organization: a reappraisal for primates by using phylogenetic methods. *Proceedings of the National Academy of Sciences, USA*, **91**, 9941–5.

Doran, D M (1997) Ontogeny of locomotion in mountain gorillas and chimpanzees. *Journal of Human Evolution*, **32**, 323–44.

Doran, D M & McNeilage, A (1998) Gorilla ecology and behavior. *Evolutionary Anthropology*, **6**, 120–31.

Eisenberg, J F, Muckenhirn, N A & Rudran, R (1972) The relation between ecology and social structure in primates. *Science*, **176**, 863–74.

Emlen, S T & Oring, L W (1977) Ecology, sexual selection, and the evolution of mating systems. *Science*, **197**, 215–23.

Fay, J M (1997) The Ecology, Social Organization, Populations, Habitat and History of the Western Lowland Gorilla (*Gorilla gorilla gorilla*). PhD thesis, Washington University.

Fleagle, J G (1999) *Primate Adaptation and Evolution*. San Diego CA: Academic Press.

Fossey, D (1974) Observations on the home range of one group of mountain gorillas (*Gorilla g. beringei*). *Animal Behaviour*, **22**, 568–81.

Fossey, D (1984) Infanticide in mountain gorillas (*Gorilla gorilla beringei*) with comparative notes on chimpanzees. In *Infanticide: Comparative and Evolutionary Prespectives*, ed. G Hausfater & S B Hrdy, pp. 217–36. New York: Aldine Press.

Fossey, D & Harcourt, A H (1977) Feeding ecology of free-ranging mountain gorillas. In *Primate Ecology*, ed. T H Clutton-Brock, pp. 415–49. New York: Academic Press.

Goldsmith, M L (1996) Ecological Influences on the Ranging and Grouping Behavior of Western Lowland Gorillas at Bai Hoköu, Central African Republic. PhD thesis, State University of New York, Stony Brook.

Goldsmith, M L (1999) Ecological constraints on the foraging effort of western lowland gorillas (*Gorilla gorilla gorilla*) at Bai Hokou, Central African Republic. *International Journal of Primatology*, **20**, 1–23.

Harcourt, A H (1978) Strategies of emigration and transfer by primates, with particular reference to gorillas. *Zeitschrift für Tierpsychologie*, **48**, 401–20.

Harcourt, A H (1979) Social relationships among adult female mountain gorillas. *Animal Behaviour*, **27**, 252–64.

Harvey, P & Pagel, M (1991) *The Comparative Method in Evolutionary Biology*. Oxford: Oxford University Press.

Hrdy, S B (1979) Infanticide among animals: a review, classification, and examination of the implications for the reproductive strategies of females. *Ethology and Sociobiology*, **1**, 13–40.

Hrdy, S B, Janson, C H & van Schaik, C P (1995) Infanticide: Let's not throw out the baby with the bath water. *Evolutionary Anthropology*, **3**, 151–4.

Janson, C H & van Schaik, C P (1993) Ecological risk aversion in juvenile primates: slow and steady wins the race. In *Juvenile Primates: Life History, Development, and Behavior*, ed. M E Pereira & L A Fairbanks, pp. 57–74. New York: Oxford University Press.

Kano, T (1992) *The Last Ape: Pygmy Chimpanzee Behavior and Ecology*. Stanford CA: Stanford University Press.

Kappeler, P M (1999) Lemur social structure and convergence in primate socioecology. In *Comparative Primate Socioecology*, ed. P C Lee, pp. 273–99. Cambridge: Cambridge University Press.

Kinzey, W G & Cunningham, E P (1994) Variability in platyrrhine social organization. *American Journal of Primatology*, **34**, 185–98.

Koenig, A, Beise, J, Chalise, M K & Ganzhorn, J U (1998) When females should contest for food: testing hypotheses about resource density, distribution, size and quality with Hanuman langurs (*Presbytis entellus*). *Behavioral Ecology and Sociobiology*, **42**, 225–37.

Kuroda, S (1992) Ecological interspecies relationships between gorillas and chimpanzees in the Ndoki-Nouabale Reserve, Northern Congo. In *Topics in Primatology*, vol. 2, *Behavior, Ecology and Conservation*, ed. N Itoigawa, Y Sugiyama, G Sackett & R K R Thompson, pp. 385–94. Tokyo: University of Tokyo Press.

Kuroda, S, Nishihara, T, Suzuki, S & Oko, R A (1996) Sympatric chimpanzees and gorillas in the Ndoki Forest, Congo. In *Great Ape Societies*, ed. W C McGrew, L F Marchant & T Nishida, pp. 71–81. Cambridge: Cambridge University Press.

Leigh, S R (1994) Ontogenetic correlates of diet in anthropoid primates. *American Journal of Physical Anthropology*, **94**, 499–522.

Leigh, S R & Shea, B T (1995) Ontogeny and the evolution of adult body size dimorphism in apes. *American Journal of Primatology*, **36**, 37–60.

Magliocca, F (1999) 1998 update on the gorillas of the Odzala National Park, Popular Republic of Congo. *Gorilla Conservation News*, **13**, 3.

Magliocca, F & Querouil, S (1997) Preliminary report on the use of the Maya-

Maya North Saline (Odzala National Park, Congo) by lowland gorillas. *Gorilla Conservation News*, **11**, 5.

Magliocca, F, Querouil, S & Gautier-Hion, A (1999) Population structure and group composition of western lowland gorillas in north-western Republic of Congo. *American Journal of Primatology*, **48**, 1–14.

Malenky, R K & Stiles, E W (1991) Distribution of terrestrial herbaceous vegetation and its consumption by *Pan paniscus* in the Lomako Forest, Zaïre. *American Journal of Primatology*, **23**, 153–69.

Malenky, R K, Kuroda, S, Vineberg, E O & Wrangham, R W (1994) The significance of terrestrial herbaceous foods for bonobos, chimpanzees and gorillas. In *Chimpanzee Cultures*, ed. R W Wrangham, W C McGrew, F B M de Waal & P G Heltne, pp. 59–75. Cambridge MA: Harvard University Press.

McNeilage, A, Plumptre, A J, Vedder, A & Brock-Doyle, A (1998) *Bwindi Impenetrable National Park, Uganda Gorilla and Large Mammal Census, 1997*. Wildlife Conservation Society, Working Paper no. 14.

Mitani, M, Moutsambote, J & Oko, R (1992) Feeding behaviors of the western lowland gorilla in the Ndoki Forest, the Ndoki-Nouabale planning Reserve, in the Congo: Why can they live in high density in this forest? In *Rapport Annuel 1991–1992*, pp. 9–21. Kyoto: University of Kyoto Press.

Mitchell, C L, Boinski, S & van Schaik, C P (1991) Competitive regimes and female bonding in two species of squirrel monkeys (*Saimiri oerstedi* and *S. sciureus*). *Behavioral Ecology and Sociobiology*, **28**, 55–60.

Morell, V (1994) Will primate genetics split one gorilla into two? *Science*, **265**, 1661.

Nishihara, T (1995) Feeding ecology of western lowland gorillas in the Nouabale-Ndoki National Park. *Primates*, **36**, 151–68.

Nishihara, T (1996) Insect-eating by western lowland gorillas. In *Proceedings of the XVth Congress of the International Primatological Society*, Madison WI, Abstract no. 664, p. 53.

Olejniczak, C (1996) Update on the Mbeli Bai gorilla study, Nouabalé-Ndoki National Park, northern Congo. *Gorilla Conservation News*, **10**, 5–8.

Olejniczak, C (1997) 1996 update on the Mbeli Bai gorilla study, Nouabalé-Ndoki National Park, northern Congo. *Gorilla Conservation News*, **11**, 7–10.

Parnell, R (1999) Gorilla exposé. *Natural History*, **108**, 38–43.

Remis, M J (1994) Feeding Ecology and Positional Behavior of Western Gorillas (*Gorilla gorilla gorilla*) in the Central African Republic. PhD thesis. Yale University.

Remis, M J (1997*a*) Ranging and grouping patterns of a western lowland gorilla group at Bai Hokou, Central African Republic. *American Journal of Primatology*, **43**, 111–33.

Remis, M J (1997*b*) Gorillas as seasonal frugivores. *American Journal of Primatology*, **43**, 87–109.

Robbins, M M (1995) A demographic analysis of male life history and social structure of mountain gorillas. *Behaviour*, **132**, 21–47.

Rogers, M E & Williamson, E A (1987) Density of herbaceous plants eaten by gorillas in Gabon: some preliminary data. *Biotropica*, **19**, 278–81.

Rogers, M E, Williamson, E A, Tutin, C E & Fernandez, M (1988) Effects of the dry season on gorilla diet in Gabon. *Primate Report*, **22**, 25–33.

Rogers, M E, Maisels, F, Williamson, E A, Fernandez, M & Tutin, C E (1990) Gorilla diet in the Lopé Reserve, Gabon: a nutritional analysis. *Oecologia*, **84**, 326–39.

Ruvolo, M, Pan, D, Zehr, S, Golberg, T, Disotell, T R & Dornum, M von (1994) Gene trees and hominoid phylogeny. *Proceedings of the National Academy of Sciences, USA*, **91**, 8900–4.

Schaller, G (1963) *The Mountain Gorilla: Ecology and Behavior*. Chicago: University of Chicago Press.

Sholley, C R (1991) Conserving gorillas in the midst of guerrillas. *American Association of Zoological Parks and Aquariums, Annual Conference Proceedings*, 30–37.

Sicotte, P (1993) Inter-group encounters and female transfer in mountain gorillas: influence of group composition on male behavior. *American Journal of Primatology*, **30**, 21–36.

Smuts, B B, Cheney, D L, Seyfarth, R M, Wrangham, R W & Struhsaker, T T (1987) *Primate Societies*. Chicago: University of Chicago Press.

Sterck, E H M, Watts, D P & van Schaik, C P (1997) The evolution of female social relationships in nonhuman primates. *Behavioral Ecology and Socioecology*, **41**, 291–309.

Stewart, K J & Harcourt, A H (1987) Gorillas: variation in female relationships. In *Primate Societies*, ed. B B Smuts, D L Cheney, R M Seyfarth, R W Wrangham & T T Struhsaker, pp. 165–77. Chicago: University of Chicago Press.

Strier, K B (1999) Why is female bonding so rare? Comparative sociality of neotropical primates. In *Comparative Primate Socioecology*, ed. P C Lee, pp. 300–19. Cambridge: Cambridge University Press.

Trivers, R L (1972) Parental investment and sexual selection. In *Sexual Selection and the Descent of Man*, ed. B Campbell, pp. 136–79. Chicago: Aldine Press.

Tutin, C E G (1996) Ranging and social structure of lowland gorillas in the Lopé Reserve, Gabon. In *Great Ape Societies*, ed. W C McGrew, L F Marchant & T Nishida, pp. 58–70. Cambridge: Cambridge University Press.

Tutin, C E G & Fernandez, M (1985) Foods consumed by sympatric populations of *Gorilla g. gorilla* and *Pan t. troglodytes* in Gabon: some preliminary data. *International Journal of Primatology*, **6**, 27–43.

Tutin, C E G & Fernandez, M (1993) Composition of the diet of chimpanzees and comparisons with that of sympatric lowland gorillas in the Lopé Reserve, Gabon. *International Journal of Primatology*, **30**, 195–211.

Tutin, C E G, Fernandez, M, Rogers, M E, Williamson, E A & McGrew, W C (1991) Foraging profiles of sympatric lowland gorillas and chimpanzees in the Lopé Reserve, Gabon. *Philosophical Transactions of the Royal Society, Series B*, 179–86.

van Schaik, C P (1983) Why are diurnal primates living in groups? *Behaviour*, **87**, 120–44.

van Schaik, C P (1989) The ecology of social relationships among female primates.

In *Comparative Socioecology*, ed. V Standen & R Foley, pp. 195–218. Oxford: Blackwell Scientific Publications.

van Schaik, C P (1996) Social evolution in primates: the role of ecological factors and male behavior. *Proceedings of the British Academy*, **88**, 9–31.

van Schaik, C P, van Noordwijk, M A & Nunn, C L (1999) Sex and social evolution in primates. In *Comparative Primate Socioecology*, ed. P C Lee, pp. 204–40. Cambridge: Cambridge University Press.

Vedder, A L (1984) Movement patterns of a group of free-ranging mountain gorillas and their relation to food availability. *American Journal of Primatology*, **7**, 73–88.

Watts, D P (1984) Composition and variability of mountain gorilla diets in the Central Virungas. *American Journal of Primatology*, **7**, 323–56.

Watts, D P (1985) Relations between group size and composition and feeding competition in mountain gorilla groups. *Animal Behaviour*, **33**, 72–85.

Watts, D P (1989) Infanticide in mountain gorillas: new cases and a reconsideration of the evidence. *Ethology*, **81**, 1–18.

Watts, D P (1990) Ecology of gorillas and its relationship to female transfer in mountain gorillas. *International Journal of Primatology*, **11**, 21–45.

Watts, D P (1994) Agonistic relationships of female mountain gorillas. *Behavioral Ecology and Sociobiology*, **34**, 347–58.

Watts, D P (1996) Comparative socio-ecology of gorillas. In *Great Ape Societies*, ed. W C McGrew, L F Marchant & T Nishida, pp. 16–28. Cambridge: Cambridge University Press.

Watts, D P (1998) Long-term habitat use by mountain gorillas (*Gorilla gorilla beringei*). I. Consistency, variation, and home range size and stability. *International Journal of Primatology*, **19**, 651–80.

White, L J T, Rogers, M E, Tutin, C E G, Williamson, E A & Fernandez, M (1995) Herbaceous vegetation in different forest types in the Lopé Reserve, Gabon: implications for keystone food availability. *African Journal of Ecology*, **33**, 124–41.

Williamson, E A, Tutin, C E G, Rogers, M E & Fernandez, M (1990) Composition of the diet of the lowland gorillas at Lopé in Gabon. *American Journal of Primatology*, **21**, 265–77.

Wrangham, R W (1979) On the evolution of ape social systems. *Social Sciences Information*, **18**, 334–68.

Wrangham, R W (1980) An ecological model of female bonded primate groups. *Behaviour*, **75**, 262–300.

Wrangham, R W (1987) Evolution of social structure. In *Primate Societies*, ed. B B Smuts, D L Cheney, R M Seyfarth, R W Wrangham & T T Struhsaker, pp. 282–97. Chicago: University of Chicago Press.

Yamagiwa, J (1992) Population structure and dietic diversity of western lowland gorillas in the Ndoki Forest, Northern Congo. In *Rapport Annuel 1991–1992*, pp. 22–7. Kyoto: University of Kyoto.

Yamagiwa, J, Mwanza, N, Spangenberg, A, Maruhashi, T, Yumoto, T, Fisher, A, Steinhauser, B B & Refisch, J (1992) Population density and ranging pattern of chimpanzees in Kahuzi-Biega National Park, Zaire: a comparison with a sympatric population of gorillas. *African Study Monographs*, **13**, 217–30.

Yamagiwa, J, Maruhashi, T, Yumoto, T & Mwanza, N (1996) Dietary and ranging overlap in sympatric gorillas and chimpanzees in Kahuzi-Biega National Park, Zaïre. In *Great Ape Societies*, ed. W C McGrew, L F Marchant & T Nishida, pp. 82–98. Cambridge: Cambridge University Press.

Part II
Within-group social behavior

6 *Development of infant independence from the mother in wild mountain gorillas*

ALISON FLETCHER

Ubibwenge and Jenny during a feeding session. (Photo by Alison Fletcher.)

Introduction

Non-human primates exist within a wide array of complex social systems in varied ecological contexts, with the social structure of a group resulting from a network of continually changing social relationships. The life cycle of primates can be divided into four developmental stages: infancy, juvenescence, adolescence, and adulthood, each of which is characterized by an emphasis on different behavioral patterns and social relationships. Since mortality is higher in infancy than in any other life stage (Eisenberg, 1981), an infant's caretaker plays a key role in its survival. During early infancy, the mother is the primary socializing agent and, partly as a result of this, the mother–infant dyad has been the focus of many primate studies examining various aspects of their developing relationship, resulting in models for the investigation of other dyadic relationships (e.g. Hinde, 1974, 1979).

Most of the studies detailing mother–infant relationships have been carried out on a small number of species in captive conditions (e.g. chimpanzees, *Pan troglodytes*: Horvat & Kraemer, 1981; Indian langurs, *Presbytis entellus*: Dolhinow & Krusko, 1984; Japanese macaques, *Macaca fuscata*: Eaton *et al.*, 1985; rhesus macaques, *M. mulatta*: Simpson & Simpson, 1986; patas monkeys, *Erythrocebus patas*: Chism, 1986; owl monkeys, *Aotus trivirgatus*: Dixon & Fleming, 1981). Several field studies on a number of species have also contributed significantly to our knowledge (e.g. chimpanzees: Lawick-Goodall, 1968; orangutans, *Pongo pygmaeus*: Horr, 1977; olive baboons, *Papio anubis*: Nash, 1978; yellow baboons, *P. cynocephalus*: Altmann, 1980; rhesus macaques: Berman, 1984; owl monkeys and titi monkeys, *Callicebus moloch*: Wright, 1984; Japanese macaques: Tanaka, 1989; ringtailed lemurs, *Lemur catta*: Gould, 1990; ruffed lemurs, *Varecia variegata*: Morland, 1990; howling monkeys, *Alouatta palliata*: Clarke, 1990). For gorillas, published accounts detailing the progression of mother–infant relationships over the period of infancy are few, with accounts of mountain gorillas being mostly qualitative and lacking investigation of sex differences (Schaller, 1963; Fossey, 1979). For captive western lowland gorillas, accounts detail the development of infant independence in three individuals (Hoff *et al.*, 1981) and development of maternal transport (Hoff *et al.*, 1983). Such a paucity of data is partly due to the long period of infancy in gorillas and the small size of groups. For eastern lowland gorillas and wild western lowland gorillas there are no published accounts detailing the mother–offspring relationship. Beyond infancy, few studies have dealt with the topic of social development in detail (Pereira & Altmann, 1985), particularly in the great apes (Watts & Pusey, 1993).

154

Gorillas reside in polygynous, cohesive groups containing several females, their offspring, and one or more adult males. Thus, infants mature in a usually stable social environment in which adult females and males remain together, sometimes for several years (Harcourt *et al.*, 1981; Watts, 1991; Robbins, 1999). Periphalization and emigration of young adult males to become solitary or join a "bachelor" group is common (Stewart & Harcourt, 1987), although they may remain in their natal group to reproduce. Like males, females may also reside as adults in their natal group, but most emigrate as young adults, transferring directly either into another breeding group or to a lone adult male (Harcourt *et al.*, 1976).

As a result of female transfer, relatedness and affinitive behavior among females are usually low (Harcourt, 1979*a*) although "friendships" may occasionally develop between unrelated females (Watts, 1994, this volume). While there is generally a lack of extensive matrilines, related females residing together favor one another over non-relatives (Watts, 1994) with alliances and help in contests occurring more frequently between kin than non-kin (Stewart & Harcourt, 1987; Harcourt & Stewart, 1989). Although interactions are usually infrequent, linear hierarchies do exist between females based on non-aggressive approach–retreat interactions, but not on aggression (Watts, 1985, this volume).

Bonds between dominant adult males and adult females are strong, with females spending more time near the mature male than near a resident blackback (Harcourt, 1979*b*). Females are primarily responsible for proximity with the dominant male (Harcourt, 1979*c*), although during estrus, males play a greater role in maintaining proximity with proceptive females (Sicotte, 1994). Furthermore, females with young infants spend a greater proportion of time near the silverback than females without, with an initial increase in proximity being associated with the birth of the new offspring (Harcourt, 1979*b*). Such behavior may have evolved because infanticide is a major cause of infant death (Fossey, 1984; Watts, 1989).

Females give birth to their first infant, on average, at the age of 10 years (Harcourt *et al.*, 1980; Watts, 1991) with the overall sex ratio of births being 1:1 (Watts, 1991). Inter-birth intervals are long, most recently stated as 3.92 years and apparently not varying in relation to the sex of the previous offspring (Watts, 1991) or with age of the mother (Stewart, 1988; Watts, 1991). The highest number of surviving offspring for a female is reported to be at least eight (Watts, 1991). Females with young infants have a period of about 3 years of lactational amenorrhea (Harcourt *et al.*, 1981; Fossey, 1982) before resumption of the estrus cycle, with the frequency of suckling having a major influence (Stewart, 1988).

The course of development in mother–infant relationships is fairly consistent across species (e.g. Altmann, 1980) but the factors that influence the relationship vary. Various studies have found that sex of infant, group size and composition, parity and reproductive status of the mother may contribute to variation in mother–offspring relationships. Sex differences have often been explained in terms of differential maternal investment in males and females. Males might elicit more investment because they are more expensive to raise (Gomendio *et al.*, 1990) and, in polygynous species like the gorilla, because male reproductive success is more variable than for females (Clutton-Brock, 1988).

Demography of groups is also likely to influence the developing mother–infant relationship. For example, the presence of older siblings or peers may lead to an early first break in the continuous mother–infant body contact that typifies early infancy (Fossey, 1979; Simpson, 1983; Berman, 1984). Availability of peers for social interaction may also affect the course of spatial independence from the mother. Fossey (1979) reported that siblings were found most consistently near the mother–infant dyad throughout infancy, whereas Stewart (1981) found that infants tended to spend more time near preferred play partners (other infants). Although these results may seem contradictory, Fletcher (1994), taking into account group activity patterns, found that the closest associates during feeding were juveniles (likely to be siblings) but that during rest periods infants were preferred associates. A larger group is more likely to contain more siblings and more peers and thus might differentially affect the speed at which an infant attains spatial independence when compared to a smaller group. In addition, the benefit of having adult kin as potential allies within a group (Stewart & Harcourt, 1987; Harcourt & Stewart, 1989) could have an effect on both the infant's confidence and the mother's restriction of its movements, and thus could influence the timing of its progression further and further from the mother.

Parity, and therefore previous experience with infants, has been shown to have an effect on the mother–infant relationship in some primates. For example, in rhesus macaques, primiparous mothers are less restrictive than multiparous mothers (Berman, 1984). In captivity, some studies have shown that parity interacts with infant sex in complex ways to affect mother–infant interactions (Hooley & Simpson, 1981). Resumption of estrus has also been shown to affect mother–infant relationships in chimpanzees, specifically that of grooming (Nishida, 1988).

Nutritional weaning is the most commonly cited example of potential parent–offspring conflict and has been widely investigated in primates (e.g. Clark, 1977; Altmann, 1980; Nicolson, 1987; Worlein *et al.*, 1988; Gomendio, 1991). Whimpering increases around the time of weaning, a period when infants appear generally depressed and often distressed at their mothers' high frequency of rejections (Clark, 1977; Altmann, 1980; Worlein *et al.*, 1988) and general lack of interest (Altmann, 1980; Collinge, 1987). The influence of infant sex on weaning has not been investigated in gorillas.

The aim of this chapter is to examine the ontogeny of the mother–infant relationship, from complete dependence to weaning. Data are presented on changes in (1) physical contact and mother–offspring proximity during the course of infancy, (2) relative contribution of mother and infant in the maintenance of contact and proximity, (3) grooming of the infant, and (4) conflict between mother and offspring addressed by the examination of suckling, whimpering, and agonistic behaviour, and related to the period of weaning. The influence of infant sex and group demography on the interactions is considered, with parity of the mother being discussed as a possible factor in the changing relationship between mother and infant. Although dominance rank of the mother has been seen to affect mother–infant relationships in some species – see reviews by Berman (1984) and Nicolson (1987) – it was not taken into account in the present study due to its subtle nature in gorillas (Watts, 1985).

Methods
Study individuals
The data presented here on infants are from a larger study of 26 infant, juvenile, and adolescent mountain gorillas in two habituated groups (Table 6.1), the subjects for which comprised about one-fifth of the total Virunga Volcano gorilla population. Data were collected, from a developmental perspective, on all infants in the two groups, Group 5 and Group BM. Group 5 was the larger of the two groups with several more adults and nearly twice as many immatures as Group BM (Table 6.2). Subjects included 11 infants present at the start and five born during the study. Infancy was considered to be from birth through the 3rd year, individuals of 3 years through the 6th year were juveniles, and immatures between 6 and 8 years were classed as adolescents (Fossey, 1979). Individuals were included in analyses of weaning until they were last seen to suckle, even if this was beyond 36 months. Parity of the mother and sex of the infant are confounded since all females had multiparous mothers whereas only 5 of 10 males did so. Sex differences that were found, therefore, should be interpreted with caution.

Table 6.1. *Demography of study groups at the beginning and end of the study period*

		Group composition							Demographic changes			
		Infant	Juvenile	Adolescent	Adult female	Blackback	Silverback	Total	Births	Deaths	Immigration	Emigration
Group 5	Start	7	6	0	11	2	2	28	–	–	–	–
	End	7	4	7	11	–	4	33	5	1	2	1
Group BM	Start	4	1	1	7	–	2	15	–	–	–	–
	End	5	2	2	7	–	2	18	4	1	–	–

Table 6.2. *Details of all infant subjects including sex, age, mother and parity of mother.*

Identity	Group	Sex	Age at start of study[a]	Mother	Parity of mother[b]	Notes
Nh	5	F	Born March, 1992	Wa	m	
Ki	BM	M	Born September, 1991	Mw	p	
Ur	5	M	Born June, 1991	Kb	p	
Pk	BM	F	Born April, 1991	Pa	m	
Ts	5	M	Born March, 1991	Py	m	
In	BM	M	0:2	Je	p	
Bw	BM	M	0:5	Gi	p	
Kp	5	F	0:6	Pd	m	
Mk	5	F	1:0	Pu	m	
Um	BM	F	1:0	Tu	m	
Sa	5	M	1:3	Mg	p	Died October 13, 1991
Ty	5	F	1:5	Ef	m	
Jo	BM	M	2:6	Fu	m	
Is	5	M	2:7	Si	m	
Gw	5	M	2:10	Li	m	
Ug	5	M	2:10	Wa	m	

[a] Age is listed as years:months.
[b] Parity of mother is that when the subject was born; m = multiparous; p = primiparous.

Sampling methods

Data were collected by two methods, both using focal sampling techniques (Altmann, 1974), during September 1990, March through November 1991, and January through December 1992. Scheduling of samples was rotated in order to observe each subject at mid-month, based on age, so that data were comparable between subjects. Data were collected from as many individuals as was logistically possible each month. Individual focal samples were at least 5 hours per day whenever possible, beginning between 0730 and 0900 hrs and ending at 1400 hrs. Continuous sampling of behavioral patterns was facilitated by a hand-held computer which enabled the recording of behavioral patterns and activity patterns (Fletcher, 1994). Activity patterns, which in mountain gorillas have a tendency to be synchronized (Harcourt, 1978), were classed as feeding, resting, or traveling, noted for both the focal subject and the group as a whole. Instantaneous point sampling at 5–minute intervals recorded spatial distribution of individuals using a dictaphone and stopwatch. Periods of time where the focal individual was obscured, either by other gorillas nearby or by mass movement of the group, were withdrawn from analyses.

Definitions of behavioral terms
Spatial proximity
Several categories were used: the focal subject was judged to be either in contact; within 1 m (but not in contact); between 1 m and 2 m; between 2 m and 5 m; or beyond 5 m of its mother. Where vegetation was dense, 5 m was often the maximum range within which the observer, and therefore probably the gorillas, could visually determine the presence of another individual. A measure of < 5 m between mother and infant was used to represent some degree of close association. Data were analyzed taking into account group activity patterns. Time spent by infants (from instantaneous sample points) in any of the following categories: < 1 m, < 2 m, and < 5 m was represented as a proportion of time out of contact with the mother.

The relative role of partners in maintenance of contact was evaluated by comparing contact initiation and termination by partners to produce a responsibility index for contact (Hinde & White, 1974), henceforth referred to as HRIc. Since breaking of contact was rare until the age of 9 months, no indices were calculated below this age. Data are presented as medians by age.

The approaches and departures between mother and infant in the distance categories mentioned above were also measured on a continuous basis. A variation on the responsibility index used for physical contact determined the role of mothers and infants in their maintenance of spatial proximity (Hinde & Atkinson, 1970). In this context, the percentage of approaches minus the percentage of departures due to the infant formed the proximity index HRIp.

Grooming: mother manipulated the hair of her infant, or vice versa, using fingers and lips.

Weaning and associated behavioral patterns
Suckling: continued contact between the infant's mouth and the mother's nipple, even though it was not possible to be sure if the infant was receiving milk. Suckling was difficult to evaluate in very young infants due to their small size and the close protection provided by mothers. However, once aged 3 to 4 months, observations of infant suckling were no longer problematical. Data are presented as the median frequency and duration per hour and median bout lengths. Responsibility for the termination of a suckling bout was noted when it was clearly attributable to mother or infant. Mothers actively prevented or stopped their infants suckling in several ways: (1) movement of arms to cover breast(s); (2) re-orientation of ventrum away from the infant; (3) lying down, ventrum to substrate; (4)

eating with arms close to chest; (5) grooming the infant.

Tantrum: infant screaming, rolling on ground, pulling/shaking nearby vegetation.

Resumption of estrus by mother: persistent following of mother by adult male(s) and subsequent attempt at copulation.

Whimper: a high-pitched whining vocalization used almost invariably by infants when they were distressed.

Agonism directed from mother to infant: usually occurred in one of three forms:

(1) pig grunts – a series of short, rough guttural grunts (Fossey, 1972; 'cough grunts' Harcourt *et al.*, 1993);
(2) threat – brisk movement of mother's head towards the infant, with open mouth;
(3) bite – gentle but firm nip administered usually to the infant's arm or leg.

Threat and bite are subsequently classed together as threat/bite.

Data analysis

In total, 665 hours of infant observation were amassed. Social behavioral patterns were calculated as median frequency per hour, median duration per hour, and median proportion of total observation time, at each age, for each focal individual. To illustrate the ontogeny of a behavioral pattern data were grouped in two ways:

1. By month: the median value was taken from the range of values provided by all individuals at any one age and used to illustrate the detailed progression of a behavioral pattern.
2. By "age-range": the interval used to describe the changing behavior was 3 months. For each individual, the median value of scores derived from all observation months during the 3-month period was calculated. Subsequently the median value of these individual medians was used as a composite data point for each 3-month category.

Spatial proximity data were analyzed as the proportion of the total sample points that a mother was at a particular distance from her infant. This was determined by calculating the median for each subject, from observation days within a 3-month period, and subsequently taking the median value of these medians. Data are presented as the median rate of sample points per half-hour, giving a maximum of six.

Data were analyzed using nonparametric statistics (Siegel & Castellan, 1988). Spearman's rank correlation coefficient (r_s) was used to investigate age-related changes in behavioral patterns; "age-range" was the independent variable against which sex differences or group differences were investigated using Wilcoxon matched-pairs signed ranks (Z); Friedman two-way analysis of variance by ranks (χ^2) was employed to compare between three different proximity indices; and Mann–Whitney U was used to compare progression beyond 5 m between males and females. All tests were two-tailed.

Results
Contact between mother and infant
Duration of contact

The age at which mother–infant pairs were first observed to break body contact varied little between dyads. This usually occurred in the 4th or 5th month and separation lasted only a few seconds. In general, infants spent their first 6 months of life almost constantly in contact with their mother, irrespective of group activity (Figure 6.1).

During group feeding periods, the rate of mother–infant contact decreased rapidly beyond 6 months, with contact reaching a fairly constant level of less than 2.5 minutes per half-hour by the age of 2 years. Above the age of 33 months individuals spent a negligible amount of feeding time in contact with their mothers. This very significant decrease with age of contact during feeding periods ($r_s = -0.94$, $n = 12$, $p < 0.0001$) contrasted with the less significant decrease in contact time during rest periods ($r_s = 0.23$, $n = 12$, $p < 0.05$), the latter being attributable mainly to a decrease in contact levels during the first year of life. Beyond this, contact remained between 8–20 minutes per half-hour.

Sex differences   During feeding periods, females aged between 6 and 12 months showed slightly higher levels of contact behavior than males of the same age. Beyond this, female–mother contact decreased sharply with age whilst male–mother contact decreased rather more gradually and exhibited a consistently higher rate than females. This sex difference did not quite reach statistical significance over infancy ($Z = -1.84$, $n = 11$, $p = 0.07$). During rest periods, females exhibited a nonsignificant trend of a higher rate of contact than males ($Z = -1.36$, $n = 11$, $p = 0.17$). Females spent more time in contact with their mothers when resting than feeding ($Z = -2.09$, $n = 12$, $p < 0.05$) whereas for males there was no difference.

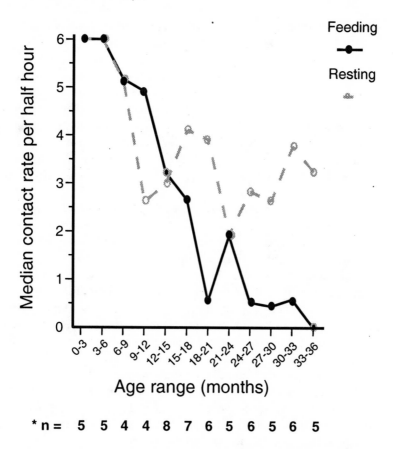

Figure 6.1. Mother–infant contact during group feeding and resting periods. Rate of contact is given as the number of 5-minute scans in which infants were in contact with their mothers, out of a maximum of six in any half-hour. Sample size, $n = 15^*$.

Group differences During feeding periods, mother–infant contact differed between the two study groups. Infants in Group 5 – the larger of the two groups – spent significantly less time in contact with their mothers than did infants in Group BM ($Z = -2.43$, $n = 12$, $p < 0.01$). By the age of 24 months, individuals in Group 5 spent less than 3% of feeding time in contact although this level was not reached until 33 months in Group BM.

There were no significant inter-group differences in time spent in contact during rest periods ($Z = -1.17$, $n = 12$, $p > 0.1$). In both groups most mother-infant contact time occurred during resting (Group 5, $Z = -2.09$, $n = 12, p < 0.05$; Group BM, $Z = -1.32, n = 12, p > 0.1$).

Role of mother and infant in maintenance of mutual body contact
At only one stage (9.5 months) was the HRIc negative, indicating that the mother was more responsible for contact at this age. There was a positive correlation between the HRIc and the age of individuals indicating that the infant was increasingly responsible for the maintenance of mother–infant contact ($r_s = 0.69$, $n = 24$, $p < 0.001$). The number of clear makes and breaks of contact were not sufficient to consider sex or group differences.

Spatial proximity – ontogeny of association with the mother
General transition from continual physical contact to a reduced level of association
As expected, infants proceeded gradually further from their mothers with increasing age (Figure 6.2). Movement beyond 5 m was not often observed until 12 months and did not start to increase until 18–21 months, after which it occurred frequently. By the age of 30 months the infant spent only about half its time associating with its mother. Individually there was some variation in the age at which each infant ventured to increasing distances from its mother. Although observations were not evenly distributed over age for all individuals, some individuals made a transition from contact to > 5 m over a very short period (e.g. Pk broke contact and reached > 5 m at 5 and 10 months, respectively; Ki at 7 and 11 months, respectively) while that of others was more gradual. There was no difference in the age at which males and females definitely moved > 5 m from their mothers (Mann–Whitney $U = 7.0, n = 8, p > 0.1$).

Mother-infant "association" – within 5 metres
There was a sharp, age-related drop in the time an infant spent within 5 m of the mother, from 100% as a newborn to between 25% (feeding) and 40% (resting) by the age of 36 months (feeding: $r_s = -0.92, n = 11, p < 0.001$; resting: $r_s = -0.93, n = 11, p < 0.001$). There was a tendency for infants to spend more time within 5 m when feeding than when resting although this did not reach statistical significance ($Z = -1.75, n = 11, p < 0.10$). There was, however, some variation in mother–infant association.

Sex differences   There was no difference between male and female infants in the time they spent, overall, associating with their mothers either during feeding periods ($Z = -0.02, n = 11, p > 0.1$) or resting periods ($Z = -0.67, n = 19, p > 0.1$). However, the pattern of change in association

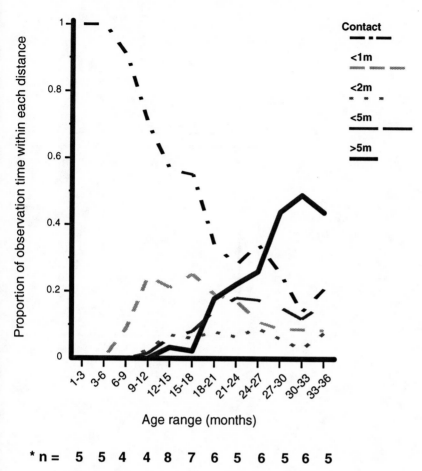

Figure 6.2. Change in spatial association between mother and infant during infancy. Sample size, *n* = 15*.

during rest periods differed between the sexes. Resting males spent 80% of their first 18 months of life associating with their mothers whereas females had dropped well below this level by 12 months. The sharp drop to less than 50% by males beyond 18 months was not matched by females who spent relatively more time within 5 m of their mothers until they reached the age of 27 months. At this point mother–female association decreased sharply.

Group differences Throughout infancy, individuals in Group BM spent more time feeding in association with their mothers than individuals in Group 5 ($Z = -2.19$, $n = 10$, $p < 0.05$). This was particularly

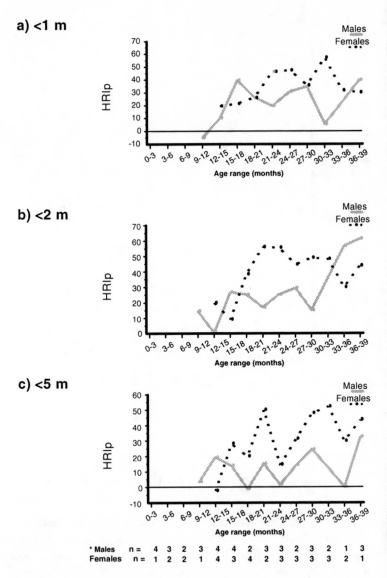

| | * Males | n = | 4 | 3 | 2 | 3 | 4 | 4 | 2 | 3 | 3 | 2 | 3 | 2 | 1 | 3 |
| | Females | n = | 1 | 2 | 2 | 1 | 4 | 3 | 4 | 2 | 3 | 3 | 3 | 3 | 2 | 1 |

Figure 6.3. Comparison of the roles of male and female infants in maintaining association with their mothers. Positive HRIp indicates responsibility for mother–infant proximity is primarily due to infant, negative HRIp indicates responsibility primarily due to mother. The higher the HRIp, the greater the role of the infant relative to that of the mother. Sample size, *n* = 15*.

Table 6.3. *Mother–infant spatial proximity: Spearman's rank correlation coefficients of association between age and HRIp for males and females*

Distance between individuals	Sex	r_s	n	Significance level
<1 m	male	0.47	9	n.s
	female	0.78	9	$p<0.01$
<2 m	male	0.45	8	n.s.
	female	0.76	8	$p<0.02$
<5 m	male	0.21	7	n.s.
	female	0.57	7	n.s.

marked in the period between 18 and 36 months where Group BM spent on average 18% more time in mother-associated feeding. Overall during rest periods, there were no significant differences between the two groups ($Z = -0.17$, $n = 10$, $p > 0.1$).

Relative responsibility of mother and infant in maintaining association

In all three approach-departure distance categories there was an increase in HRIp with age and thus also in the role of the infant in maintenance of proximity. There was a strong positive correlation with age for infant responsibility at <1 m and at <2 m ($r_s = 0.93$, $n = 9$, $p < 0.001$, and $r_s = 0.87$, $n = 9$, $p < 0.001$, respectively). Infants moved less frequently between <5 m and >5 m resulting in fewer data and a less obvious trend of increasing infant responsibility. However, a Friedman analysis of variance resulted in no significant difference between the indices at different distances ($\chi^2 = 1.56$, $n = 9$, d.f. 2, $p > 0.1$). Lack of data prevented the consideration of differences due to group activity.

Sex differences Female infants exhibited a trend of higher HRIps than males for all categories (Figure 6.3), although none of the comparisons reached statistical significance when tested using a Wilcoxon signed ranks test (<1 m, $Z = -1.01$, $n = 7$, $p > 0.1$; <2 m, $Z = -1.40$, $n = 7$, $p > 0.1$; <5 m, $Z = -1.52$, $n = 7$, n.s.). Females also had a stronger positive correlation of HRIp and age than did males for all distance categories (Table 6.3), indicating that female infants achieved responsibility for maintenance of proximity at a faster rate than males.

Grooming by the mother

Infants were rarely seen to solicit grooming from their mothers; the mother usually initiated bouts and frequently restrained the infant while she

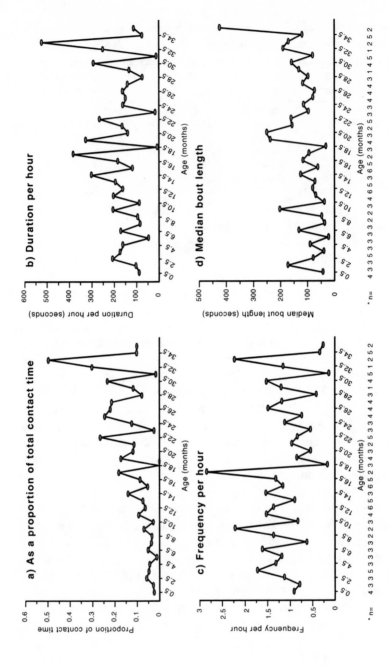

Figure 6.4. Grooming received from the mother (medians). Duration is given as a proportion of contact time (a), and per hour of total observation (b). Frequency is per hour of observation (c). Bout length is given in seconds (d). Sample size, *n* = 15*.

groomed. Termination of bouts was often unclear – sometimes a small movement by the infant apparently caused the mother to cease grooming, or the mother simply stopped of her own accord. Infants also groomed mothers but there were great individual differences in the frequency of this behavior.

The proportion of contact time during which individuals were groomed varied, increasing with age overall during infancy (Figure 6.4a, $r_s = 0.60$, $n = 36$, $p < 0.001$). Whilst grooming frequency decreased (Figure 6.4c, $r_s = -0.28$, $n = 36$, $p < 0.05$), bout length increased (Figure 6.4d, $r_s = 0.47$, $n = 36$, $p < 0.002$), resulting in a fairly constant duration per hour (Figure 6.4b).

Sex differences There was no significant difference between males and females in the time they were groomed as a proportion of mother–infant contact time ($Z = -1.19$, $n = 31$) or in the frequency with which they were groomed per hour of observation time ($Z = -0.27$, $n = 31$). However, the duration of grooming per hour and median bout length were considerably different with females being groomed for longer periods and for a longer total duration than males (bout length: $Z = -3.57$, $n = 31$, $p < 0.001$; duration per hour: $Z = -3.12$, $n = 31$, $p < 0.002$). In addition, the frequency and duration of grooming received exhibited a significant, although gradual, decrease over infancy for females (frequency: $r_s = -0.38$, $n = 31$, $p < 0.025$; duration: $r_s = -0.32$, $n = 31$, $p < 0.05$), whereas grooming received by males increased during the first 18 months, and then remained more variable (overall frequency: $r_s = 0.13$, $n = 31$, $p > 0.1$; duration: $r_s = 0.18$, $n = 31$, $p > 0.1$).

Resumption of the mother's estrus cycle In the five individuals whose mothers resumed cycling during the study there was no apparent change in grooming by the mother at this time, even when the mother was proceptive and when she was pregnant. For two of the three subjects whose siblings were born (and survived) during the study period, grooming received from the mother declined after the birth of the new offspring. Samples were too small to allow statistical testing.

Weaning and associated behavioral patterns
Suckling
Weaning is often defined as being completed beyond the infant's last suckling from the nipple (Clark, 1977). In the present study, six individuals were observed to suckle for a final time. Of these, four males (Gw, Is, Jo and Ug) were above the age of 40 months and the remaining two females

170   *Alison Fletcher*

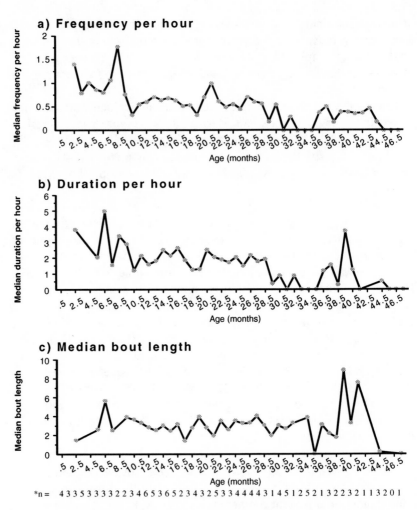

Figure 6.5. Suckling by individuals throughout infancy and weaning. Bouts and durations presented as minutes. Sample size, $n = 15^*$.

(Mk and Um) were 29.5 and 30.5 months respectively. Another female (Ty) was still seen to be suckling at the age of 40 months when the study ended. The oldest age at which suckling was seen in this study was 44 months. All mothers were multiparous.

Suckling frequency and duration decreased significantly with age (Figure 6.5a, $r_s = -0.80$, $n = 46$, $p < 0.001$; Figure 6.5b, $r_s = -0.70$, $n = 42$, $p < 0.001$). Between the ages of 28 and 36 months, suckling frequency and duration dipped to lower levels than previously and recovered briefly afterwards. Although the length of suckling bouts remained fairly constant over the entire period (Figure 6.5c), those individuals who were suckled past the age of 40 months tended to have particularly long bouts.

Prevention and termination of suckling by the mother
Mothers can curtail suckling either by preventing an infant from gaining access to the nipple or by terminating a bout before the infant has finished. Mothers prevented infants from suckling most frequently when infants were aged between 14 and 33 months. Preventions were rather idiosyncratic and not a good indicator of weaning stage. Some infants were denied often, for example Bw, whose attempts were thwarted 16 of 48 times while aged between 14.5 and 32.5 months, whilst others were not prevented from suckling: all of Ty's 18 attempts between 27.5 and 40 months of age were successful.

Suckling decreased in duration after the mother resumed her estrus cycle. Although preventions did occur around this time, infants would usually not attempt to suckle without testing the mother first, for example by whimpering or snuggling close to the breast. Infants were usually distressed when refused access to the nipple and whimpered (which occasionally escalated in intensity to a scream), before usually huddling close to the mother. Only one infant, Um, was seen to throw tantrums, which sometimes occurred several times a day when her mother was proceptive.

Termination of suckling bouts varied in frequency, displayed no pattern with the age of the subject, and was sometimes associated with pig-grunting at and/or nipping of the infant. The qualitative impression gained was one of inter-mother differences in tolerance levels. The mother was never observed to initiate a suckling bout although some mothers did lift the infant when it was searching for the nipple.

Whimpering by the infant
Accompanied by a pouted expression, whimpering was observed in this study to occur throughout infancy in a wide variety of contexts. These included occasions when the infant apparently wanted to be carried, groomed, or suckled, or generally sought contact. In addition, whimpering sometimes occurred if an infant suddenly realized that it was "lost" whereupon it wandered around whimpering pitifully. In this situation other adult individuals, particularly silverbacks, were often observed to belch grunt ("double-grunt", Harcourt *et al.*, 1993) in reassurance. On several occasions, one mother (Mw), who was nearby, but out of sight of, her infant (Ki) and fully aware that he was looking for her (as judged by her movements), ignored his whimpering completely and left him to move

even further away from her.

Whimpering was observed as early as the first month after birth, when it was probably due to discomfort; whimpers were short and resulted in the mother moving her infant to another position. Infants began to whimper more regularly between 3 and 7 months and continued through the weaning period. When the mother resumed her estrus cycle, the effect on the infant's social behaviour was considerable. The infant followed the mother almost constantly, played rarely, and whimpered very frequently, appearing especially distressed during copulations. Beyond weaning, especially after the birth of a sibling, whimpering decreased to a very low level.

Agonism directed from mother to infant

Pig-grunts   Pig-grunting directed from mother to infant was rarely seen before 10 months of age except between Mw and Ki. There was no apparent relationship between the frequency of pig-grunts and age or intense weaning conflict. Pig-grunting by the mother was positively associated with whimpering of the infant ($r_s = 0.54$, $n = 40$, $p < 0.001$) and occurred both before and after infant whimpering. Three of the 15 mothers in this analysis were never heard to pig-grunt at their offspring.

Threat/bite

There was no relationship between threat/bite by the mother and age of the infant although there were slight peaks in frequency between 14 and 22 months, and again between 29 and 35 months. Some mothers did threat/bite more frequently when they were proceptive, during which time there was already heightened tension in the group.

Discussion

A summary of the main results is given in Table 6.4. Over the period of infancy, contact between mother and offspring declined, spatial proximity gradually increased, and both of these age-related changes were due mainly to the infant. There were no statistically significant sex differences but there was variation according to which group the infants were in. During feeding periods, infants in Group BM spent more time in contact with and in association (< 5 m) with their mothers than did infants in Group 5. An increasing bout length and decreasing frequency of grooming led to a fairly constant duration which, as a proportion of contact time, increased over infancy. Females received longer bouts and consequently more grooming than males. Suckling decreased in frequency and duration while bout lengths remained fairly constant throughout infancy. Maternal agonism and infant whimpering showed no relation to the decrease in suckling.

Table 6.4. *Summary of results, indicating overall age trend, sex differences, and group differences during infancy*

	Age trend[a]	Sex differences[b]	Group differences[c]
Contact			
Feeding	↓	n.s.	BM > Gp 5
Resting	↓	n.s.	n.s.
HRIc	↑	—[d]	—
Proximity			
Feeding	↓	n.s.	BM > Gp 5
Resting	↓	n.s.	n.s.
HRIp	↑	—	n.s.
Grooming			
% contact time	↑	n.s.	—
Frequency	↓	n.s.	—
Duration	← →	F > M	—
Bout length	↑	F > M	—
Suckling			
Frequency	↓	—	—
Duration	↓	—	—
Bout length	← →	—	—

[a] ↑ significant increase with age; ↓ significant decrease with age; ← → no overall significant change with age.
[b] n.s., no significant difference; F, females; M, males.
[c] BM, Beetsme's Group; 5, Group 5.
[d] where no entry is made, this comparison was not considered through lack of available data.

However, there was some evidence to suggest an increase in both agonism and whimpering when copulations resumed, and a decrease after the birth of another infant.

Mother–infant contact
Many studies have examined the roles of mother and infant in their maintenance of mutual proximity (e.g. Hinde & Atkinson, 1970; Nash, 1978; Altmann, 1980), with the common conclusion that responsibility changes from being largely due to the mother to being increasingly due to the infant as it becomes older. Data from mountain gorillas support this conclusion and are clearly demonstrative of a weakening dependence on the part of the infant. First break of mother–infant contact at the age of 4 to 5 months was consistent with that observed by Fossey (1979) and Stewart (1981). The decrease in contact observed by Fossey (1979) was also similar up to the age of 1 year. However, results from the present study

have highlighted the importance of accounting for different activity patterns in measures of proximity. Fossey recorded the rate of contact as being 30% of total time by the age of 3 years but in this study infants spent a median 50% of resting time in contact at the end of infancy, but no longer fed in contact.

Close proximity to the mother at an early age is probably necessary for the infant to acquire complex feeding skills (see Byrne, this volume). The extremely low contact rate while feeding in the final year of infancy may have been related to the infant's processing of its own solid food (Fossey, 1979), thus increasing the spatial area needed by the mother–infant pair. The lack of such a decline during resting periods indicates the continuing strength of the mother–infant bond, and the importance of grooming which occurred mainly during these periods.

Hoff *et al.* (1981) reported that in captive lowland gorillas mother–infant contact was first broken at 11 weeks, more than 1 month before the earliest break of contact seen in wild mountain gorillas. Although the presence of older siblings can result in the earlier attainment of independence of young infants (e.g. Fossey, 1979; Simpson, 1983; Berman, 1984), the infants in Hoff *et al.*'s (1981) study had primiparous mothers so siblings could not have accounted for their earlier development. In the present study, infants in Group 5 spent less time in contact with their mothers overall than infants in Group BM. This result may be linked to the fact that more infants in Group 5 had multiparous mothers and, therefore, older siblings. It is more likely that the behavioral differences in captivity were related to the less complex environmental conditions in which the lowland gorillas were observed, an effect that has also been documented for chimpanzees (Nicolson, 1977). The result of captive conditions is that the relatively short time spent feeding is unlikely to restrict resting time or social time, even during lactation (cf. Altmann, 1980; Dunbar & Dunbar, 1988).

Close association between mother and infant

A juvenile's relationship with its mother is expected to be considerably less intense than that of an infant (Pereira & Altmann, 1985), so when approaching juvenescence an individual would be expected to spend much of the day away from its mother (>5 m). The proportioning of infants' time at increasing distances from the mother was therefore not surprising. Association was highest during feeding periods, when the mother–infant pair was less likely to be close to other individuals, while rest periods became a time for social exploration. Decreasing association beyond

2 years was accentuated by the dominant silverback's increasing role in the socialization of older infants and juveniles (Fletcher, 1994; Stewart, this volume).

Fossey's (1979) observation that males associated more with their mothers up to 2 years but rapidly reduced such association beyond this age was not seen in the present study. However, females tended to have higher proximity indices than males, suggesting that they had to "work harder" to stay near their mothers. This might be related to differences in maternal investment in males and females (Clutton-Brock, 1988) or to differences between primiparous and multiparous mothers. A mother's behavior towards her infant has been associated with her experience (Kemps *et al.*, 1989) and therefore it could be that less experienced primiparous mothers were more restrictive of their infants. Since all female infants had multiparous mothers, results must be interpreted with care. Whilst not covered in this chapter, there was also a sex difference in the age at which infants were carried dorsally; males were carried dorsally later than females (Fletcher, 1994, unpublished data). This later occurrence could have been due to increased protectiveness in the mother or inexperience of dorsal carrying. Alternatively, earlier dorsal riding seen in female infants could have indicated a lower level of maternal investment or more experience of dorsal carrying. The fact that sex and parity were confounded prevents any conclusions at this point. Mother–infant contact differed in relation to varying activity periods but the relevance of this to levels of maternal investment is unclear.

Grooming of the infant
Similarly to the case in chimpanzees (Nishida, 1988), grooming by the mother remained fairly constant throughout infancy. It remained so also during subsequent pregnancy, conflicting with Stewart's (1981) report of a decrease in grooming with age (although a decrease was seen for females in this study). However, in support of Fossey's (1979) observations, grooming was habitually initiated by the mother and terminated by the infant, with the mother accepting such termination of grooming up to the age of 6 months, but rebuffing and persisting with older offspring. An increase with age in the proportion of contact time that involved grooming suggests that this was an increasingly important reason for contact between mother and infant. The lack of a sex difference in the proportion of mother–infant contact time during which individuals were groomed might indicate that contact time *per se* is not the important factor affecting the difference in grooming received. Rather, grooming was more likely to occur during rest periods and it was during this time that female infants were in contact with

their mothers for longer periods than males. After the birth of a sibling grooming decreased substantiating Stewart's (1981) observation under similar circumstances, and suggesting redistribution of the mother's resources.

Since mothers should invest more in the sex to whose reproductive success they can make the most difference (Clutton-Brock, 1988) female primates might be expected to invest more in whichever sex is philopatric. This argument has been used to explain the fact that, in some cercopithecines, for example vervets, mothers groom females more than males (Fairbanks & McGuire, 1985). In this study, longer grooming bouts for daughters meant that females, overall, received more grooming from their mothers than did males. However, while the duration of grooming for females was high at the start of infancy and decreased quickly in the first year, males received a low, but increasing level of grooming throughout the first year. Such a sex difference might indicate the reduction of maternal investment at an earlier age in females. Again, it is difficult to tease apart the effects of infant sex and parity of the mother since all but one of the males aged below 2 years who contributed to these data had primiparous mothers, whilst all the females were born to multiparous mothers. Thus the lack of previous experience with very young infants in primiparous mothers may have been reflected in such low levels of grooming.

Weaning

The development of nutritional independence observed in the present study suggests two possible strategies of weaning in mountain gorillas. Most commonly, weaning in gorillas has been considered to be a rather gradual, gentle process (Stewart, 1981), substantiated here by infants who were still suckling at 40 months and by the lack of apparent increase in whimpering, which is seen in baboons and rhesus macaques (Altmann, 1980; Collinge, 1987), or in pig-grunting by mothers during the weaning period. Only one infant (Um) was seen to throw tantrums, a behavior which is sometimes (Fossey, 1979), although not commonly (Stewart, 1981), seen in gorillas, as is the case for chimpanzees in the wild (Clark, 1977). In chimpanzees, suckling continues for 4–5 years but weaning begins by the second year of life, with rejections intensifying with time. The more cohesive nature of gorilla groups, with many distractions present in the form of constant availability of a number of play partners and the dominant silverback (Stewart, this volume), may contribute to the less intense conflict seen between mother and infant.

However, weaning for two female infants was notably earlier and considerably different from that seen in the majority. For Um weaning involved severe, intense conflict with her mother arising early in infancy, with a similar, though slightly less intense, conflict observed between Mk and her mother. These infants were weaned considerably earlier than others. Distress appeared to be particularly intense on days when the mother was copulating with an adult male. Such distress has been observed in other primates at the resumption of estrus, as a result of increased rejections by the mother (Clark, 1977; Gore, 1986; Collinge, 1987; Gomendio, 1991). It would seem then, that whilst weaning in gorillas is more usually gradual and without overt conflict or infant distress, a reproductively distinct schedule of relatively early estrus cycling and pregnancy may result in a more intense weaning period.

Future research

Ecological constraints acting on the nutritional quality of food (Hauser & Fairbanks, 1988) and arboreal feeding (Johnson & Southwick, 1987) have been seen to affect mother–infant relationships in vervet monkeys and rhesus macaques, respectively. The abundance and even distribution of food, and the fact that the two study groups ranged in similar habitats, precluded such investigation in the present study. A comparison of mother–infant relationships between regions within the Virunga Volcanoes with differences in food distribution (McNeilage, 1995, this volume), and also between different subspecies of *Gorilla* due to differences in feeding ecology, may highlight differential development due to varying environmental constraints placed on the mother–infant dyad. Future research on social development would also benefit greatly from more studies of groups with differing demographies, to enable the separation of the effects of infant sex and mother's parity on social development.

Summary

It is predicted by parental investment theory that mothers should invest more in their male offspring than female offspring in polygynous species (Clutton-Brock, 1988). Several features of the developing mother–offspring relationship in gorillas seemed to support this view. Mothers played a lesser part in the maintenance of proximity with female infants and began to decrease the amount of grooming they gave to females as early as the second half of infancy in comparison to the gradual increase which males received. In addition, the two individuals who were weaned earliest and with most resulting distress were females. However, one of the major limitations of the present study was the demography of the population, with primiparous mothers all having male offspring and females offspring all having multiparous mothers. Conclusions relating to relative investment in male and female offspring are thus confounded by the fact that primiparous mothers are likely to be more restrictive than experienced multiparous mothers. At this stage it is not possible to inextricably divide the two factors but such results are worth noting in anticipation of further study in this area. Further work is also needed on group differences.

Acknowledgements

Financial support came from a scholarship awarded by the University of Bristol. I thank the former Governments of Rwanda (Office Rwandais du Tourisme et des Parcs Nationaux) and Zaire (Institut Zairois pour la Conservation de la Nature) and The Digit Fund for permission to work at Karisoke. I am grateful to Stephen Harris for his supervision of this study, and especially to the trackers and staff of the Karisoke Research Center, in particular Alphonse Nemeye, Kana Munyanganga, Faustin Barabwiriza, and Antoin Banyangandora. I also thank reviewers for their helpful comments on earlier drafts of the manuscript.

References

Altmann, J (1974) Observational study of behaviour: sampling methods. *Behaviour*, **49**, 227–65.
Altmann, J (1980) *Baboon Mothers and Infants*. Cambridge MA: Harvard University Press.
Berman, C M (1984) Variation in mother–infant relationships: traditional and nontraditional factors. In *Female Primates: Studies by Female Primatologists*, ed. M F Small, pp. 17–36. New York: Alan R Liss.
Chism, J (1986) Development and mother–infant relations among captive patas monkeys. *American Journal of Primatology*, **7**, 49–81.
Clark, C B (1977) A preliminary report on weaning among chimpanzees of the Gombe National Park, Tanzania. In *Primate Bio-Social Development: Biological, Social and Ecological Determinants*, ed. S Chevalier Skolnikoff & F E Poirier, pp. 235–60. New York: Garland.
Clarke, M R (1990) Behavioral development and socialization of infants in a free-ranging group of howling monkeys (*Alouatta palliata*). *Folia Primatologica*, **54**, 1–15.
Clutton-Brock, T H (ed.) (1988) *Reproductive Success*. Chicago: University of Chicago Press.
Collinge, N E (1987) Weaning variability in semi-free-ranging Japanese macaques (*Macaca fuscata*). *Folia Primatologica*, **48**, 137–50.
Dixon, A F & Fleming, D (1981) Parental behaviour and infant development in owl monkeys (*Aotus trivirgatus griseimembra*). *Journal of Zoology, London*, **194**, 25–39.

Dolhinow, P & Krusko, N (1984) Langur monkey females and infants: the female's point of view. In *Female Primates: Studies by Women Primatologists*, ed. M F Small, pp. 37–57. New York: Alan R. Liss.

Dunbar, R I M & Dunbar, P (1988) Maternal time budgets of gelada baboons. *Animal Behaviour*, **36**, 970–80.

Eaton, G G, Johnson, D F, Glick, B B & Worlein, J M (1985) Development in Japanese macaques (*Macaca fuscata*): sexually dimorphic behavior during the first year of life. *Primates*, **26**, 238–48.

Eisenberg, J F (1981) *The Mammalian Radiations*. Chicago: University of Chicago Press.

Fairbanks, L A & McGuire, M T (1985) Relationships of vervet mothers with sons and daughters through three years of age. *Animal Behaviour*, **33**, 40–50.

Fletcher, A W (1994) Social development of immature mountain gorillas (*Gorilla gorilla beringei*). PhD thesis, University of Bristol.

Fossey, D (1972) Vocalisations of the mountain gorilla (*Gorilla gorilla beringei*). *Animal Behaviour*, **20**, 36–53.

Fossey, D (1979) Development of the mountain gorilla (*Gorilla gorilla beringei*): the first 36 months. *Perspectives in Human Evolution*, **5**, 139–84.

Fossey, D (1982) Reproduction amongst free living mountain gorillas. *American Journal of Primatology*, Supplement **1**, 97–104.

Fossey, D (1984) Infanticide in mountain gorillas (*Gorilla gorilla beringei*) with comparative notes on chimpanzees. In *Infanticide: Comparative and Evolutionary Perspectives*, ed. G Hausfater & S B Hrdy, pp. 217–35. New York: Aldine Press.

Gomendio, M (1991) Parent/offspring conflict and maternal investment in rhesus macaques. *Animal Behaviour*, **42**, 993–1005.

Gomendio, M, Clutton-Brock, T H, Albon, S D, Guinness, F E & Simpson, M J A (1990) Mammalian sex ratios and variation in costs of rearing sons and daughters. *Nature*, **343**, 261–63.

Gore, M A (1986) Mother–offspring conflict and interference at mother's mating in *Macaca fascicularis*. *Primates*, **27**, 205–14.

Gould, L (1990) The social development of free ranging infant *Lemur catta* at Berenty Reserve, Madagascar. *International Journal of Primatology*, **11**, 297–318.

Harcourt, A H (1978) Activity periods and patterns of social interaction: a neglected problem. *Behaviour*, **66**, 121–35.

Harcourt, A H (1979a) Social relationships among adult female mountain gorillas. *Animal Behaviour*, **27**, 251–64.

Harcourt, A H (1979b) Social relationships between adult male and female mountain gorillas in the wild. *Animal Behaviour*, **27**, 325–42.

Harcourt, A H (1979c) Contrasts between male relationships in wild gorilla groups. *Behavioral Ecology and Sociobiology*, **5**, 39–49.

Harcourt, A H & Stewart, K J (1989) Functions of alliances in contests within wild gorilla groups. *Behaviour*, **109**, 176–90.

Harcourt, A H, Stewart, K J & Fossey, D (1976) Male emigration and female transfer in wild mountain gorillas. *Nature*, **263**, 226–27.

Harcourt, A H, Stewart, K J & Watts, D P (1980) Reproduction in wild gorillas and some comparisons with chimpanzees. *Journal of Reproduction and Fertility*, Supplement **28**, 59–70.

Harcourt, A H, Stewart, K J & Fossey, D (1981) Gorilla reproduction in the wild. In *Reproductive Biology of the Great Apes: Comparative and Biomedical Perspectives*, ed. C E Graham, pp. 265–78. New York: Academic Press.

Harcourt, A H, Stewart, K J & Hauser, M (1993) Functions of wild gorilla 'close' calls. I. Repertoire, context and interspecific comparison. *Behaviour*, **124**, 89–122.

Hauser, M D & Fairbanks, L A (1988) Mother–offspring conflict in vervet monkeys: variation in response to ecological conditions. *Animal Behaviour*, **36**, 802–13.

Hinde, R A (1974) *Biological Bases of Human Social Behaviour*. New York: McGraw-Hill.

Hinde, R A (1979) *Towards Understanding Relationships*. London: Academic Press.

Hinde, R A & Atkinson, S (1970) Assessing the roles of social partners in maintaining mutual proximity, as exemplified by mother–infant relations in rhesus monkeys. *Animal Behaviour*, **18**, 169–76.

Hinde, R A & White, L E (1974) Dynamics of a relationship: rhesus mother–infant ventro-ventral contact. *Journal of Comparative Physiological Psychology*, **86**, 8–23.

Hoff, M P, Nadler, R D & Maple, T L (1981) Development of infant independence in a captive group of lowland gorillas. *Developmental Psychobiology*, **14**, 251–65.

Hoff, M P, Nadler, R D & Maple, T L (1983) Maternal transport and infant motor development in a captive group of lowland gorillas. *Primates*, **24**, 77–85.

Hooley, J M & Simpson, M J A (1981) A comparison of primiparous and multiparous mother–infant dyads in *Macaca mulatta*. *Primates*, **22**, 379–92.

Horr, D A (1977) Orang-utan maturation: growing up in a female world. In *Primate Bio-Social Development: Biological, Social, and Ecological Determinants*, ed. S Chevalier Skolnikoff & F E Poirier, pp. 289–321. New York: Garland.

Horvat, J R & Kraemer, H C (1981) Infant socialization and maternal influence in chimpanzees. *Folia Primatologica*, **36**, 99–110.

Johnson, R L & Southwick, C H (1987) Ecological constraints on the development of independence in Rhesus. *American Journal of Primatology*, **13**, 103–18.

Kemps, A, Timmermans, P & Vossen, J (1989) Effects of mother's rearing condition and multiple motherhood on the early development of mother–infant interactions in Java macaques (*Macaca fascicularis*). *Behaviour*, **111**, 61–76.

Lawick-Goodall, J van (1968) The behaviour of free-living chimpanzees in the Gombe Stream Reserve. *Animal Behaviour Monographs*, **1**, 165–311.

McNeilage, A J (1995). Mountain gorillas in the Virunga Volcanoes: ecology and carrying capacity. PhD thesis, University of Bristol.

Morland, H S (1990) Parental behavior and infant development in ruffed lemurs (*Varecia variegata*) in a Northeast Madagascar rain forest. *American Journal of Primatology*, **20**, 253–65.

Nash, L T (1978) The development of the mother–infant relationship in wild baboons (*Papio anubis*). *Animal Behaviour*, **26**, 746–59.

Nicolson, N A (1977) A comparison of early behavioral development in wild and captive chimpanzees. In *Primate Bio-Social Development: Biological, Social and Ecological Determinants*, ed. S Chevalier Skolnikoff & F E Poirier, pp. 529–60. New York: Garland.

Nicolson, N A (1987) Infants, mothers, and other females. In *Primate Societies*, ed. B B Smuts, D L Cheney, R M Seyfarth, R W Wrangham & T T Struhsaker, pp. 330–42. Chicago: University of Chicago Press.

Nishida, T (1988) Development of social grooming between mother and offspring in wild chimpanzees. *Folia Primatologica*, **50**, 109–23.

Pereira, M E & Altmann, J (1985) Development of social behavior in free-living nonhuman primates. In *Nonhuman Primate Models for Human Growth and Development*, ed. E S Watts, pp. 217–309. New York: Alan R Liss.

Robbins, M M (1999) Male mating patterns in wild multimale mountain gorilla groups. *Animal Behaviour*, **57**, 999–1004.

Schaller, G B (1963) *The Mountain Gorilla: Ecology and Behavior*. Chicago: University of Chicago Press.

Sicotte, P (1994) Effect of male competition on male–female relationships in bi-male groups of mountain gorillas. *Ethology*, **97**, 47–64.

Siegel, S & Castellan, N J Jr (1988) *Nonparametric Statistics for the Behavioral Sciences*. New York: McGraw-Hill.

Simpson, M J A (1983) Effect of the sex of an infant on the mother–infant relationship and the mother's subsequent reproduction. In *Primate Social Relationships: An Integrated Approach*, ed. R A Hinde, pp. 53–57. Oxford: Blackwell Scientific Publications.

Simpson, M J A & Simpson, A E (1986) The emergence and maintenance of interdyadic differences in the mother–infant relationship of rhesus macaques: a correlational study. *International Journal of Primatology*, **7**, 377–97.

Stewart, K J (1981) Social development in wild mountain gorillas. PhD thesis, University of Cambridge.

Stewart, K J (1988) Suckling and lactational anoestrus in wild gorillas. *Journal of Reproduction and Fertility*, **83**, 627–34.

Stewart, K J & Harcourt, A H (1987) Gorillas: variation in female relationships. In *Primate Societies*, ed B B Smuts, D L Cheney, R M Seyfarth, R W Wrangham & T T Struhsaker, pp. 155–64. Chicago: University of Chicago Press.

Tanaka, I (1989) Variability in the development of mother–infant relationships among free-ranging Japanese macaques. *Primates*, **30**, 477–91.

Watts, D P (1985) Relations between group size and composition and feeding competition in mountain gorilla groups. *Animal Behaviour*, **33**, 72–85.

Watts, D P (1989) Infanticide in mountain gorillas: new cases and a reconsideration of the evidence. *Ethology*, **81**, 1–18.

Watts, D P (1991) Mountain gorilla reproduction and sexual behaviour. *American Journal of Primatology*, **23**, 211–26.

Watts, D P (1994) Social relationships of immigrant and resident female mountain gorillas. II. Relatedness, residence, and relationships between females. *American Journal of Primatology*, **32**, 13–30.

Watts, D P & Pusey, A E (1993) Behavior of juvenile and adolescent great apes. In *Juvenile Primates*, ed. M E Pereira & L A Fairbanks, pp. 148–67. New York: Oxford University Press.

Worlein, J M, Eaton, G G, Johnson, D F & Glick, B B (1988) Mating season effects on mother–infant conflict in Japanese macaques, *Macaca fuscata. Animal Behaviour*, **36**, 1472–81.

Wright, P C (1984) Biparental care in *Aotus trivirgatus* and *Callicebus moloch*. In *Female Primates: Studies by Female Primatologists*, ed. M F Small, pp. 59–75. New York: Alan R. Liss.

7 Social relationships of immature gorillas and silverbacks

KELLY J. STEWART

Infant and silverback (Jozi and Icarus) during a pause in feeding. (Photo by
Kelly J. Stewart.)

Introduction

The importance of males to the survival of infants is argued to be a major force in the evolution of male–female association in primates (van Schaik, 1996; Palombit et al., 1997; Sterck et al., 1997; Palombit, 1999). Across taxa, however, the extent and nature of males' interactions with immatures, especially infants, vary enormously, from intense, directed caregiving in some New World monkeys like marmoset, tamarins, and titis (Wright, 1984; Goldizen, 1987), to mere tolerance with occasional affiliation as in many Old World monkeys (Whitten, 1987; Maestripieri, 1998). Differences across and within species have been linked to various functions of males' interactions with infants: while in some cases they represent investment by males in related infants, usually offspring (e.g. Busse and Hamilton, 1981; Anderson, 1992), in others, males appear to be using infants as social tools, either in their competitive interactions with other males (Paul et al., 1996), or as means to gain mating access to infants' mothers (Smuts, 1985; Ferrari, 1992; van Schaik & Paul, 1996). To add to the complexity, several functions might apply in the same species. In baboons, for example, all three have been invoked to explain male–infant interactions (Packer, 1980; Smuts & Gubernick, 1992).

In considering the function of male–immature relationships researchers have examined the benefits that immature animals gain from their associations with males. It has been argued that, for most polygynous primates, the primary contribution that males make to their offspring's fitness is protection from predators and/or infanticide by unrelated adult males (Dunbar, 1984; van Schaik, 1996; Palombit, 1999; van Schaik & Janson, 2000). Many older infants and juveniles, however, who have matured beyond the need for such protection, still exhibit close bonds with adult males (Whitten, 1987). Why?

Because of their attractiveness to immature animals, a general socializing role has been suggested for adult males in the sense that they are a spatial focus where infants and juveniles can find peers with whom to interact (Kummer, 1968; Stein, 1984). Additionally, the adult male may serve as an attachment figure at times when maternal attention is withdrawn, thus functioning as a psychological buffer during the weaning period or in cases of orphaning (Itani, 1962; Kummer, 1967; Bowlby, 1969; Whitten, 1987). Weaned infants and juveniles might also gain feeding benefits from proximity to the adult male, because of his protection from intra-group agonism, and his relatively high tolerance of immatures' proximity (Busse & Hamilton, 1981; Stein, 1984). Finally, associations formed during immaturity could provide an important basis for a future

184

cooperative relationship with an effective ally (Fairbanks, 1993; Strier, 1993).

Assessment of the relative roles that males and immatures play in interactions between them has been used to assess the balance of benefits that males and immatures gain from interacting with each other. For example, the observation that older infants and juveniles usually play the greater role in initiating and maintaining affiliation with males supports the contention that they are benefiting most from the association (Keddy-Hector *et al.*, 1989; Fairbanks, 1993; Horrocks & Hunte, 1993). In situations in which males are using infants to manipulate the behavior of other adult males, or the mating choices of infants' mothers, responsibility for affiliative interactions often lies with the male (Keddy-Hector *et al.*, 1989). The balance of benefits can change in relation to group size and composition. For example, in large groups, subordinate immatures might "need' a male's help more than in smaller groups (Stein & Stacey, 1981). In baboons and Barbary macaques, as the number of adult males increases in a group, so does mating competition and, potentially, the "usefulness' of infants to adult males (Busse, 1985; Paul *et al.*, 1996).

Research on mountain gorillas has provided support for theories about the role of paternal protection in social evolution. Male gorillas defend their groups with aggressive responses to potential predators, human and non-human (Schaller, 1963). They also engage in aggressive displays and sometimes violent fights with extra-group silverbacks who kill unrelated infants to increase their access to fertile females (Harcourt, 1978*b*; Hrdy, 1979; Fossey, 1984; Watts, 1989). In the Virungas, infanticide by unrelated males accounts for at least 37% of infant mortality. The vast majority of killings occur when the mature male of a group has died, and infants and their mothers are left unaccompanied and therefore, unprotected (Watts, 1989). Females avoid infanticide by associating more or less permanently with the fathers of their infants, and this attraction is considered to be the basis of gorilla groups. Males' abilities at repelling intruders are closely linked to female dispersal and residence patterns, and are possibly a basis for female mate choice (Wrangham, 1979; Stewart & Harcourt, 1987; Sicotte, 1993; Sterck, 1997).

But apart from general group defense, and distinguishing between their own and unrelated infants, in what way do male gorillas behave affiliatively with their offspring? Do they confer any other benefits during the course of immaturity? Do immatures gain anything from associating with related adult males other than protection when very young? If so, then silverbacks' behavior towards group immatures might play a role in female mate choice, either by influencing the likelihood of her future emigration, or her mating preferences within multi-male groups (Stewart & Harcourt, 1987).

The general picture for male–immature relationships in both captive and wild gorillas shows a strong attraction in infants and young juveniles to the dominant/older male (Schaller, 1963; Fossey, 1979; Yamagiwa, 1983; Enciso *et al.*, 1999). Because adult males take little active interest in newborns, close association with the silverback begins in early infancy when mothers increase their time near the adult male soon after giving birth (Harcourt, 1979*c*). As infants mature, they assume the active role in proximity to the silverback, a pattern also seen in mother–infant association (Fossey, 1979; Fletcher, this volume). Some individuals go on to develop extensive grooming relationships with the silverbacks (Watts & Pusey, 1993). As for the adult males, apart from intervening in aggression involving immatures (Harcourt & Stewart, 1989; Watts & Pusey, 1993), they appear to be relatively passive, if tolerant, partners in their dealings with young individuals (Stewart & Harcourt, 1987; Enciso *et al.*, 1999). Although adult males have been described as cuddling, grooming, and carrying young infants (Fossey, 1979) prolonged friendly contact between silverbacks and infants under 2 years old is extremely rare. Adult males are most often ascribed the function of being a spatial focus for infants and juveniles: a silverback surrounded by playing immatures during a rest period, or followed by a group of young animals during feeding are common sights (Schaller, 1963; Fossey, 1979; Fletcher, 1994).

Most published information on interactions between adult males and young gorillas comes from observations of either single-male groups, or bi-male groups in which there were large rank differences between the silverbacks, i.e. the older male was clearly dominant and the social focus of females and their young, whereas the younger male was obviously subordinate and relatively peripheral (Harcourt, 1979*c*). Over the years, however, study groups at Karisoke Research Center have differed in the number of fully adult males of comparable competitive abilities, as well as in the number of related adult females (Stewart & Harcourt, 1987; Watts, 1992). Such differences could relate to variation in male–immature interactions. For example, the presence of maternal kin might dilute the value of the adult male as a social ally, while mating competition between mature males might favor close ties with infants in order for the males to appear more attractive to their mothers (van Schaik & Paul, 1996). Sicotte has shown that the number of males in a group and extent of mating competition between them can influence not only male–female interactions, but the behavior of immatures as well (Sicotte, 1994, 1995, and this volume).

This chapter presents data on the social interactions between fully adult males and immatures of various ages in three study groups. The study spans almost a decade, and, in combination with results from more recent work (Watts & Pusey, 1993), provides longitudinal data on some individuals from infancy to near adulthood. The purpose is to present a more detailed picture of silverbacks' relationships with immatures over time, to consider the various benefits to young animals in associating with adult males, and to describe the variation in male–immature relationships between groups. I discuss the results in the context of the gorillas' social system, and in light of more recent data on a much larger sample.

Methods
Study site and subjects
Data were collected during fieldwork based at the Karisoke Research Center in the Parc National des Volcans in Rwanda. For detailed descriptions of the habitat see Fossey & Harcourt (1977), Watts (1984), and McNeilage (this volume).

The main study subjects were immature gorillas between the ages of 6 months and 8 years, which I observed during two main study periods: 8 months and 6 months in 1974 and 1977, respectively, and 22 months in 1981–83. The ages used in this chapter are those for subjects at the mid-point of data collection during 1974 and 1977, and 1981–83.

The data on social interactions among adults in the study groups were collected by A.H. Harcourt during the 6-month subperiod in 1974 and the period from 1981 to 1983.

Age–sex class definitions
Previous studies have classified gorillas of 0 ≤3 years as infants, and those >3 ≤6 years as juveniles. It is not until after the age of 3, however, that major breaks occur in the maternal bond. Gorillas are usually weaned between 3 and 4 years of age, when their mothers normally give birth again (Stewart, 1988). At this time, the older offspring stops sharing its mother's nest at night, and maternal grooming decreases dramatically (Fossey, 1979; Fletcher, this volume). In this chapter, therefore, I refer to immatures of 0 ≤4 years as infants, and those > 4 ≤6 as juveniles. Individuals >6 ≤8 are adolescents (when females usually experience their first labial swelling and copulation), and those >8 <10 are young adults (when males first copulate). Females first give birth at 10, at which age they are considered fully adult. Males from 10 to about 12/13 are blackbacks (presumably sexually but not physically mature), from 13 to 15 young silverbacks, and at 15–16 fully grown silverbacks. For more details on developmental stages see Fossey (1979) and Watts & Pusey (1993).

In presentation of results, I do not distinguish sex of immatures, because there were not enough data to consider sex differences.

Data collection

While all gorillas in the study groups were habituated to human observers, the degree of habituation differed between the 1970s and the 1980s. In the earlier decade, individuals did not tolerate being followed persistently, which made focal animal sampling difficult, especially during feeding periods when the group was more spread out and moving. Therefore, a record was kept of when group members moved into and out of sight using 1-minute instantaneous sampling, and the social interactions of any immatures in sight were recorded. Behavioral rates between immatures and other individuals were calculated by taking the mean of daily values for all days when immatures and/or the partner were in sight more than 1 hour. Records of animals within 2 m and 5 m of immatures were made with instantaneous sampling every minute. Results from the 1970s are based on approximately 40 hours of direct observation on each immature.

By 1981 the study groups were more habituated and data were collected using focal animal sampling, with each sample lasting 30 minutes, no more than two focal samples per animal per day, separated by at least 2 hours. The daily order of focals was decided each day before the group was contacted. Every 5 minutes I recorded individuals within 2 m and 5 m of the focal animal, and every minute the activity of the focal animal. All affiliative and agonistic interactions (e.g. grooming, play, approaches, suckling) involving the focal animal were noted as they occurred. Behavioral rates are calculated as monthly means. Finally, *ad libitum* records were kept of close calls (Harcourt & Stewart, 1996, this volume), suckling and grooming bouts, and agonistic interactions. Results here are based on approximately 44 hours of focal data per immature, collected between October 1981 and June 1982 and between September 1982 and April 1983. Data on directed and non-directed interventions and on grooming distributions come from my focal and *ad libitum* records as well as *ad libitum* observations collected by Harcourt during his study of adult social behavior (50 hours of focal observation per adult in Groups 5-2 and NK).

Group activity

Gorilla groups exhibit synchronized activities, alternating between rest periods and travel/feeding periods (Schaller, 1963; Fossey & Harcourt, 1977). Ideally, data from different group acitivites should be analyzed separately, since the pattern of animals' interactions has been shown to differ between resting and feeding periods (Harcourt, 1978*a*; Fletcher, this volume). During the 1970s, however, there were not enough data to take into account acitivty periods. Analyses here that do separate data according to group activity come only from the 1980s data set.

Definitions and analyses

1. Analyses of *approaches and departures at 2 m and 5 m* excluded those movements associated with aggression or supplanting, play activity with another individual, or those that occurred during concerted group movement.

2. One animal was considered to *follow* another when it moved after it within 3 seconds of a departure. Follows during concerted group movement were excluded from analyses.

3. *Grooming minutes* were calculated as the number of consecutive half-minutes on which grooming occurred. *Grooming bouts* were defined as at least three consecutive half-minutes on which grooming occurred separated by at least one whole minute with no grooming.

4. *Aggressive interactions* usually consisted of a vocalization, the "cough-grunt", but occasionally escalated into lunges, chases, or physical attack such as hitting or biting, sometimes accompanied by screams. Displays such as strutting, chest-beating, or charging in a stiff-legged fashion were excluded from this analysis because this type of aggression was often directed at a group of animals and appeared to lack a specific target.

5. *Supplants* occurred when one animal approached within 2 m of another who moved away within 5 seconds. Dominance rank in the gorilla groups was assessed using supplants.

6. An act of *support* was a directed aggressive intervention on behalf of one animal, against its opponent in a fight. However, 72% of interventions were not clearly support for one opponent or another, and appeared to simply break up the conflict. These are referred to as *control interventions*.

Study groups

During the 1970s, the immature subjects were members of two groups: Group 5-1 and Group 4. During the 1980s the groups were NK and Group 5-2, which was Group 5-1 almost ten years later. Table 7.1 gives the age–sex composition of each group during the different study periods, the ages of the immature subjects, and their kin relations.

There were some important differences between the study groups in levels of consanguinity, composition, and maturity. Firstly, the degree of

Table 7.1. *Group compositions; study subjects are italicized*

Group	Adult males	Adult females[a]	Immatures	Sex	1974	1977	Maternal kin[d]
					Age class[b] and age (yrs)[c]		
4	Ub				Sb	Sb	
	Dg				Bb	Sb	
		Og				—[e]	
			Tg	m	Ado 6.5	Bb 9.1	0
		Fl					
			Ts	m	—	Inf 2.2	0
			Cl	f	Inf 2.6	Juv 5.2	0
		Pe					
			Au	f	Juv 3.6	Ado 6.2	0
		Mo			—	—	
			Kw	m	—	Inf 1.3	0
		Mi				—	
		Ps			Y Ad c. 8	—	
			Si	f	Ado 6.2	Y Ad 8.6	0
5-1	Bv				Sb	Sb	
	Ic				Bb	Sb	
		Ef					
			Po	f	—	Inf 0.6	1 in 1977
			Tu	f	Inf 2.1	Juv 4.5	1 in 1977
			Pu	f	Juv 5.4	Y Ad 8	0
		Ma					
			Zz	m	Juv 3.3	Juv 5.7	1 in 1977
		Pa			Y Ad 8.2	Ad	
		Lz					
			Pb	m	—	Inf 2.2	0
			Qu	f	Juv 3.7	Ado 6.3	0
5-2	Ic				Sb		
	Bv				Sb		
	Zz				Bb		1
		Pa					1
			Jz	f	Inf 1.3		1
			Mu	f	Juv 5.1		1
		Ef					
			Mg	f	Inf 1.8		2
			Po	f	Juv 6		2
		Tu					2
		Pu					2
			Ca	m	Juv 3.4		2
			Sh	m	Juv 5.1		2
			Pb	m	Ado 7.6		0

Table 7.1. *(cont.)*

Group	Adult males	Adult females[a]	Immatures	Sex	Age class[b] and age (yrs)[c] 1974	Age class[b] and age (yrs)[c] 1977	Maternal kin[d]
NK	Nk				Sb		
		[Ps					
			Sz	f	Inf 1.6	0	
		[Pe					
			Dy	m	Inf 3	1	
		[Au					
			Gi	f	Inf 1.6	1	
		[Pd					
			Su	m	Juv 3.6	0	
		[Fu					
			Bo	m	Juv 4	0	

[a] Horizontal black lines connect mothers to immature study subjects; vertical black lines connect mother and adult daughter.
[b] Sb = silverback; Bb = blackback; YAd = young adult; Ad = adult; Ado = adolescent; Juv = juvenile; Inf = infant.
[c] Ages given are those during the mid-point of the study periods.
[d] The number of that subject's adult maternal relatives, other than the mother, present in the group during the study period.
[e] A dash means not in the group during the study period

known maternal relatedness among adults in Group 5-2 was higher than in the other units. While nulliparous and primiparous females commonly emigrate from their natal groups to breed elsewhere (Harcourt *et al.*, 1976; Watts, 1990; Sicotte, 1993), in Group 5-2, three of the four adult females had been born in the group. Two of them were the daughters of the oldest breeding female, and another was the full sibling of the blackback. As a result, immatures had present a greater number of adult maternal kin than is usually the case for gorillas. When I speak of "kin" or "close relatives" of immatures, I mean either grandmothers, aunts/uncles, or maternal siblings, which in most cases were probably full siblings. In addition, Group 5-2 had seen no immigration of females since 1971 (Fossey, 1983). As a result of these circumstances, all of the adults had been familiar with each other for a very long time, and many of them had grown up together.

Secondly, the adult male of Group 5-2 who was the dominant breeding male since the start of observations in 1969 (Fossey, 1983) was still alive, and in fact, had only recently dropped in rank just below his presumed son, the younger, dominant male in the 1980s. Both males were observed mating in the late 1970s and early 1980s, and were central figures in the group. The paternity of the infants and younger juveniles in this group was

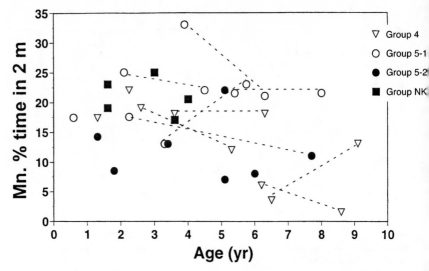

Figure 7.1. Mean percent of time that immatures spent within 2 m of the silverback in their group with whom they had the higher value. Symbols represent values for individuals. Dotted lines connect the same animals in different study periods. Data from two young adults older than 8 years are included in the figure.

not clear to observers. In contrast, in Group 4 and Group 5-1 the social differences between the adult males were much greater: the dominant silverbacks were in their prime, while the subordinate males had only just acquired their silver backs, and were the most peripheral group members. Although these young males were seen to mate with younger, natal females in the group, the dominant silverbacks performed most matings with fertile adult females, and were the likely fathers of the immatures in the sample (Harcourt, 1979*a*, *c*).

In contrast, Group NK was a "classic" one-male group, and differed from the other social units in its maturity, having been formed only in 1974 (Stewart & Harcourt, 1987). Thus, the only age classes in the group were breeding adults or infants (immatures up to 4 years of age, see above).

Results
Time in close proximity
The most common indication of affiliation in wild gorillas was close proximity. The time that immatures spent within 2 m of the dominant/oldest silverback in their group varied greatly, but tended to be highest in older infants, and lower in adolescents/young adults (Figure 7.1). Data on the two young adults are included in Figure 7.1 in order to show changes within individuals across study periods.

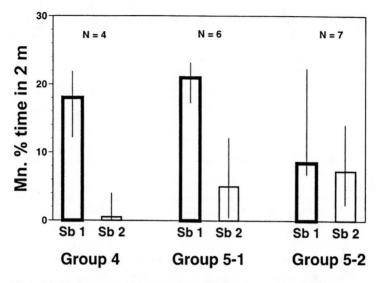

Figure 7.2. Mean percent of time that immatures aged 0–8 years in different study groups spent within 2 m of the oldest/dominant silverback in their group (heavy histograms) and the younger or subordinate silverback (light histogram). Histograms are medians from all subjects; vertical lines show ranges.

The values from Group 5-2 are those for the silverback with the higher proximity score. No immatures in the group differed significantly in the time they spent near either male (see below and Figure 7.2). For most ages, values in Group 5-2 were lower than those in other groups, possibly because immatures were splitting their time between the two silverbacks. However, even when the values for both adult males are added together, the immatures in Group 5-2 did not have higher scores than did those in the other groups (median for added scores in Group 5-2 = 17, n = 7 vs. median of 19 for Group 4, 5-1, and NK, n = 17; young adults not included).

Comparison between silverbacks
In two out of three groups with more than one silverback, there were clear differences in association between immatures and the two males (Figure 7.2). In Group 5-1 and Group 4, the subordinate silverbacks were very young (about 13 and 12 years, respectively) and relatively peripheral (see also "Methods"). All ten immatures (infants, juveniles, and adolescents) in these groups spent more time near the older silverback (their likely father) than near the younger.

By the 1980s, the younger silverback (Sb2) in Group 5-2 had become

Table 7.2. *Mean percent of time spent by immatures (0–8 years) within 2 m of the older/dominant silverback compared to time near the "favorite" non-mother adult female*

Group	n	% time near Sb	% time near favorite F
4	8	18 (4–22)	5 (1–8)
5-1	10	22 (13–33)	9 (2–16)
5-2	7	11 (2–22)	11 (7–17)
NK	5	21 (17–25)	9 (6–16)

Values given are those for the median immature in the group, those in brackets are the range across immatures.

Table 7.3. *Responsibility for 2 m proximity between dominant/oldest adult males and immatures*

Group		% approaches	% leaves	0–4 yrs n	>4–8 yrs n	+ index n
4	Median	91 (56)	75 (62)	4	2	6
	range	82–95	62–82			
5-1	Median	93 (50)	70 (54)	4	0	4
	range	83–96	68–84			
5-2						
Sb1	Median	86 (17)	53 (10)	2	3	4
	range	55–100	45–88			
Sb2	Median	86 (16)	29 (12)	2	2	4
	range	78–100	22–36			
NK	Median	88 (15)	50 (11)	5	0	5
	range	82–92	0–78			

Numbers in brackets are the median number of movements per pair.

dominant over his aging father and there were no significant differences in the overall time that immatures spent near either male, as was true also for their mothers. When the data were separated into resting and feeding periods, there were still no obvious preferences for associating with one male or the other.

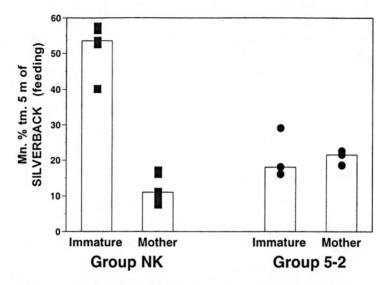

Figure 7.3. Mean percent of time that infants and their mothers spent within 5 m of the silverback during feeding periods. Values in Group 5-2 are those for the male with the higher values. Histograms are values for median immature and mother in group. Squares are values for Group NK, circles for Group 5-2.

Time close to silverbacks vs. adult females

Most immatures, except those in Group 5-2, spent more time near the older/dominant silverback of their group than they did near their "favorite" non-mother adult female, that is, the female with whom they had the highest score for mean time within 2 m (Table 7.2). The preference among infants and juveniles (15/17 spent more time near the silverback) was significant ($p < 0.002$). In contrast, in Group 5-2, four out of seven immatures spent more time near their favorite female than near their favorite silverback. In two cases, the favorite female was an older sister.

Responsibility for close proximity

In all groups, immatures for whom there were enough data (at least 15 moves per immature–silverback pair) performed the majority of approaches within and leaves beyond 2 m between themselves and the oldest/dominant adult male (Table 7.3). Using Hinde's index of the %approaches − %leaves by one partner as an indication of responsibility for proximity (Hinde & Atkinson, 1970), in 21 of 22 immature–silverback pairs (95.5%), the immature had a positive index, indicating that it performed a greater proportion of the approaches than of the departures (Table 7.3). There were no striking differences between the two silverbacks in Group 5-2 in

Table 7.4. *The ratio of time within 5 m of the adult male to time within 5 m of the mother during feeding (Ratio F) and resting (Ratio R) periods, for infants and young juveniles (ages 0–4 years)*

Group	Immature	Ratio F sb/ma	Ratio R sb/ma
5-1	Jz (Sb2)	0.6	0.5
	Mg (Sb1)	0.2	0.6
	Ca (Sb1)	0.4	0.75
NK	Sz	1.1	1.0
	Gi	1.6	1.7
	Dy	2.3	1.0
	Su	1.0	0.75
	Bo	1.6	1.3

either proportion of moves performed by immatures or responsibility indices. In Groups 4 and 5-1, there were too few data on moves between immatures and the subordinate males to include here, but the immatures performed most of the few approaches and leaves that were observed.

Group activity and comparison between mother and silverback proximity

During infancy, gorillas develop an attraction to the adult male independent of their mothers' association with him (Fossey, 1979; Fletcher, 1994, this volume). The strength of this attraction was most apparent when infants had to choose between being near their mothers or being near the male. Such was often the case during feeding periods, when distances between adults were greater than during rest periods. It was not uncommon, for example, for adult females to be more than 20 m from the silverback during feeding, while during rest periods, they were often within 5–10 m (Harcourt, 1979c). In Groups 4 and 5-1, Harcourt, using scan sampling on adults, showed that infants and young juveniles increased the time they spent within 5 m of the silverback relative to that near their mothers during feeding periods, and some immatures spent absolutely more time near him than near their mothers during feeding (Harcourt, 1978a). I examined silverback/mother proximity ratios during feeding (Ratio F) and resting (Ratio R) periods in Groups 5-2 and NK, for which I have the best data during feeding.

The two groups differed strikingly in this comparison. In Group 5-2, there was no general increase in relative time near adult males during

Table 7.5. *Immatures following silverbacks: percentage of a silverback's departures beyond 2 m that were followed by immatures*

Group	0–4 yr	% leaves followed (number of leaves)	>4–8 yr	% leaves followed (number of leaves)
4	Cl	42 (36)	Cl	33 (10)
	Au	40 (35)		
	Kw	31 (16)		
	Ts	29 (14)		
5-1	Tu	8 (24)	Tu	11 (9)
	Zz	28 (25)	Zz	13 (8)
	Qu	39 (14)	Pu	0 (20)
	Pb	47 (15)		
5-2	Mg Sb2	14 (7)	Sh Sb1	41 (17)
	Ca Sb1	20 (5)	Mu Sb1	0 (5)
	Sb2	14 (7)	Sb2	21 (14)
			Pb Sb1	20 (5)
NK	Sz	73 (11)		
	Dy	80 (10)		
	Bo	67 (12)		
Median		39		20

Pairs with fewer than five departures were excluded.

feeding, and no infant appeared to prefer the male over its mother during either group activity. In Group NK, not only did four out of five infants spend absolutely more time near the silverback than their mothers during feeding periods, but two of them did so during rest periods as well (Table 7.4).

This difference between the groups was due to the facts that infants in Group NK spent far more time associating with the adult male during feeding than did those in Group 5-2, while their mothers spent less time near the male than did the females in Group 5-2. Relatively greater distances between Group NK adults during feeding meant that the infants had to "choose" between their mothers or the silverback more often than did those in Group 5-2, and this could have resulted in a closer attachment to the male (Figure 7.3).

Following

The apparently greater attachment of infants to silverback Nk than of infants to either silverback in Group 5-2 is further suggested by their following scores: for those infants with enough data (at least five departures by the silverback beyond 2 m), the two in Group 5-2 had far lower following scores (14.3% and 20%) than did the three in Group NK (67%–73%). In fact, the two youngest infants in Group NK did not have enough data for this measure because they followed their silverback so doggedly, he was rarely observed to move beyond 2 m from them.

Considering the larger sample of infants, juveniles, and adolescents from all groups, all immatures followed the dominant/older adult male more than vice versa. Following scores were generally higher in infants than in juveniles or adolescents (Table 7.5). Scores from Group 5-2 are those for the silverback who was followed the most. Variation in following scores among infants supports the notion that the silverback might be important as an attachment figure for animals at a time when the bond with their mother is weakening (Bowlby, 1969). Six of the seven infants with following scores at or above the median of 39% had experienced some form of relatively early break in the maternal bond: three had mothers who conceived at less than the median interval of 3.1 years postpartum (Stewart *et al.*, 1988); one had a mother who emigrated during the study period; and the other two were siblings whose mother was the most peripheral female (Lz) in Group 5-1 (Harcourt, 1979*b*). Thus, they frequently had to "choose" between their mother and the adult male, as did the infants in Group NK.

Maternal orphans

The clearest evidence that the adult male serves as an attachment figure concerns maternal orphans. Immatures who lose their mothers permanently through emigration or death increase greatly their association with the dominant silverback, sleeping in or very near his nest at night, spending more time near him during the day, and more time in body contact (Fossey, 1979). For example in Group 4, the young juvenile female, Au, whose mother emigrated during 1974, increased significantly her time near her father from less than 10% in the month before her mother left, to 35% in the month after ($p < 0.05$, $U = 21$; $n = 9$ days pre-departure; 18 days post-departure); she also increased body contact and grooming with the silverback, and followed him more than did any other immature in the group. Two other infants/young juveniles, one of whom had an adult sister in the group, showed similar changes in association with the adult male in response to losing their mothers (Watts & Pusey, 1993). These orphans

Table 7.6. *Percentage of minutes grooming given to and received from silverbacks in Groups 4 and 5-1*

Group	Immature	Age (yr)	% grooming given to Sb	% grooming received from Sb	Total minutes given	Total minutes received
4	Cl	2.6	18	11	22	267
		5.3	18	0	22	5
	Au	3.6	17	36	169	53
		6.3	2	0	62	3
	Si	6.2	0	0	0	64
		8.6	0	0	0	0
	Tg	6.5	0	0	15	151
		9.1	0	100	0	95
	Kw	1.3	0	0	0	88
	Ts	2.25	0	6	0	169
5-1	Tu	2.1	0	0	19	206
		4.5	41	0	68	60
	Zz	3.3	0	2	0	50
		5.75	0	8	0	52
	Qu	3.7	75	32	137	28
		6.3	78	0	172	17
	Pu	5.5	16	36	51	11
		8	86	0	219	0
	Pb	2.2	0	2.5	51	80

Only immatures who gave or received at least 10 minutes of grooming are counted.

continued their unusually close association with the silverback thoughout their juvenescence and subadulthood, grooming him more frequently than did other group members (see below).

Grooming
Immatures primarily groomed silverbacks rather than vice versa, with the exception of Ub in Group 4 (Tables 7.6 and 7.7). He was unusual among silverbacks in the Karisoke study groups in that he performed most of the grooming between himself and all other partners, including adult females

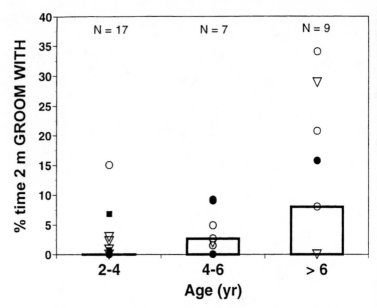

Figure 7.4. Mean percent of time within 2 m that immatures groomed with (either gave or received grooming) the older/dominant silverback. Values for Group 5-2 are those for the male with the higher value. N = number of immatures in each age class. Age class > 6 includes data for two young adults older than 8 years. Histograms are values for median immature. Symbols are values for individual animals. Open triangles are values for Group 4; open circles for Group 5-1; solid circles for Group 5-2; and solid squares for Group NK.

(Harcourt, 1979c). For Ub's most frequent grooming partners, however, including the young adult male in Figure 7.4, the groomees usually initiated the interaction with very obvious invitations (Harcourt, 1979c). Data collected by Watts 2 years after my study show that Ub was still the primary groomer in his relationships with his two juvenile sons (Watts & Pusey, 1993).

Grooming between young infants and adult males was extremely rare, but became common for some individuals after the age of 2, and, relative to time within 2 m, increased with age (Figure 7.4). I have included data for two young adults in Figure 7.4 to illustrate the increase in grooming with maturity. Of ten immatures for whom I have data in two difference study periods, seven increased their rates of grooming with the older silverback relative to time near him. The most frequent groomers of the silverback were the two males in Group 5-2, Sh and Pb, who had lost their mothers when they were young juveniles (Watts & Pusey, 1993).

In Groups 4 and 5-1, grooming between immatures and the subordinate silverbacks was extremely rare. In Group 5-2, the juvenile and

Table 7.7. *Percent grooming bouts given to and received from adult males in Groups 5-2 and NK*

Group	Immature[a]	Age (yr)	% grooming given to Sb	Sb > Ma Bouts given	Total bouts given	Total bouts received
5-2						
	Jz	1.3	0	no	0	56
W[c]	Jz[d]	6	67	yes		
	Mg	1.8	0	no	0	27
W	Mg[d]	6	50	yes		
	Ca–Sb1	3.4	33	no	6	20
W	Ca	7	46	no		
	Sh–Sb1	5.1	100	—	76	43
W	Sh	8	90	—		
	Mu–Sb1	5.1	29	yes	68	35
W	Mu–Sb1	7	67	yes		
	Po	6	0	no	32	13
W	Po[d]	8	69	yes		
	Pb–Sb1	7.6	86	—	44	6
W	Pb	11	100	—		
NK						
	Sz	1.6	0	no	0	12
W	Sz	5	79	yes		
	Gi	1.6	0	no	1	22
W	Gi	6	79	yes		
	Dy	3	25	no	12	30
W	Dy	6	88	yes		
	Su	3.5	11	no	38	22
W	Su	7	20	no		
	Bo	4	10	no	42	14
W	Bo	7.5	14	no		

[a] Only immatures who gave or received at least five bouts of grooming are included.
[b] SB > Ma bouts given indicates whether or not the immature gave a greater proportion of its grooming to the silverback than to its mother.
[c] W = values from Watts & Pusey (1993).
[d] Means that silverback was the not the same individual as in previous value, but was his son (see text).

adolescent males, Sh and Pb (respectively), concentrated their grooming on the older SB1, while the juvenile female, Mu, split her efforts between SB1 and SB2, as did her mother (Table 7.7).

While grooming relationships do not invariably develop between immatures and silverbacks, those that do are among the most extensive and persistent seen in the Karisoke study groups. For some juveniles and adolescents, the older/dominant adult male became their primary grooming partner, surpassing even their mothers and other adult kin. Data collected 3–4 years after my observations of Groups 5-2 and NK show that all five immatures in NK and all seven in 5-2 (two of whom were now blackbacks) were giving adult males a greater proportion of their grooming than when I observed them (Table 7.7). For four of the seven in Group 5-2, the silverback was the same male as in my sample. For the three younger ones, the dominant silverback was now a son of Sb1. Seven of the ten individuals with mothers ended up grooming the silverback more than their mothers.

Rates of agonistic interactions

Supplants and aggressive encounters between adults and immatures occurred most frequently over food, with aggression usually taking the form of a vocalization, the "cough-grunt" (Harcourt & Stewart, 1987, this volume). Data collected from Groups 5-2 and NK during feeding periods indicate that, on the whole, adult males were not more tolerant of immatures during feeding, relative to time spent within 5 m, than were other adults (Table 7.8). If anything, Sb1 was more frequently agonistic to juveniles and adolescents than were most other adults except for the blackback male. All seven immatures in Group 5-2 received higher rates of aggression from the older male (Sb1) than they did from the younger, but dominant male (Sb2), and six out of seven were supplanted more by Sb1.

The data from Groups 4 and 5-1 on rates of aggression per hour at 5 m during all group activities show a very similar pattern: silverbacks were as frequently or slightly more aggressive (median rate of 0.3 for Group 4 and 0.5 for Group 5-1) as were most non-maternal adult females (median rate of 0.3 in Group 4 and 0.2 in Group 5-1, and slightly more aggressive than were mothers (median of 0.1 for both groups).

Rates of agonistic encounters with all adults increased in adolescence (Watts & Pusey, 1993).

Agonistic support

Immatures in the study group received support in 4.3% of their conflicts with a dominant opponent (Harcourt & Stewart, 1987), and adult males

Table 7.8. Rates of supplanting and aggression by adults per hour within 5 m during feeding/rest-feed periods

Group	Behavior	Sb			Mother	Median non-mother adult female	Female with highest score
		Sb1	Sb2	Bb			
5-2							
0-4 yr							
median	Supplanting	0.4	0	0	0.2	0.2	0.3
range		0.2-0.5	0-0.6	0-2.5	0-0.2	0-0.3	0.1-1.2
median	Aggression	0.5	0.4	0.5	0.2	0	0.4
range		0.4-1.1	0.2-0.9	0-0.8	0.1-0.4	0-0.3	0.1-0.5
n = 3							
>4-8 yrs							
median	Supplanting	1.1	0.8	1.5	0.7	0.8	1.4
range		0.8-1.6	0.3-1.2	0.5	0.6-0.8	0.5-1.1	0.9-3.0
median	Aggression	1.0	0.5	2.7	0.5	0.7	1.2
range		0.8-1.6	0.3-1.7	1.4-4.3	0.4-0.6	0.3-1.0	0.7-2.0
n = 4							
NK							
0-4 yrs							
median	Supplanting	0.3			0.1	0.3	0.6
range		0.1-0.7			0-0.4	0-0.4	0-1.7
median	Aggression	0.5			0.3	0.3	0.4
range		0.2-1.1			0-0.4	0-0.6	0-1.2
n = 5							

Table 7.9. *Percentage of agonistic supports from adults that immatures received from silverback males (Sb), adult kin, including mothers (Kin), and unrelated adults (Non-kin)*

Group	Silverbacks			Kin	Non-kin
	Sb1	Sb2	Both Sbs		
5-2					
Median %	24	5	29	48	10
range	0–100	0–25	0–100	25–80	0–67
n=	7	7		6	
NK					
Median %	50			50	0
range	0–100			0–100	
n=	5			5	

Values are those for the median immature in the group. The median number of support interventions from adults was 10 in Group 5-2 and 2 in Group NK.

accounted for a large proportion of these directed interventions. In Group 5-2, of the 68 instances of support by adults, Sb1 and Sb2 gave, respectively, 13% and 12%. In Group NK, the adult male was responsible for 36% of 14 support interventions by adults. All three silverbacks supported the younger opponent in the large majority of cases (88%–100%). Most immatures received a greater proportion of their support from silverbacks than from other adults except maternal relatives, usually their mothers (Table 7.9).

The most frequent type of agonistic interference by silverbacks was control intervention. Because such acts served to end conflicts, they were effective in protecting younger opponents from escalated aggression and, in fact, in Groups 5-2 and NK, frequently led to a subordinate opponent retaining a contested resource (Harcourt & Stewart, 1987). The two silverbacks in Group 5-2 accounted for 46% of 170 control interventions, while the male Nk performed 50% of 16 such acts.

The data on control interventions (there were not enough data on agonistic support) in Groups 4 and 5-1 are similar: of 46 instances, the two silverbacks performed 33%. The subordinate silverbacks were not observed to intervene in the conflicts of immatures.

More recent studies at Karisoke Research Center that include data on six other adult males support the conclusion that silverbacks frequently intervene in the conflicts of immatures. Neither dominant nor subordinate silverbacks appear to favor particular individuals, instead giving support to whichever opponent is the youngest (Watts, 1997).

Discussion

The attraction to silverbacks that develops in gorillas during infancy and persists through maturation is a characteristic feature of wild gorilla groups, including groups of eastern lowland gorillas, *G. g. graueri* (Yamagiwa, 1983). While a male's primary contribution to the survival of his offspring is protection from extra-group threat, especially, at least in the Virungas, infanticide by unrelated males (Fossey, 1984; Watts, 1989), close association with mature males may provide a variety of other advantages to gorillas as they mature: silverbacks are a spatial focus for infants and juveniles; males serve as an attachment figure at a time of weakening of the maternal bond, or maternal loss; and they buffer young animals from intra-group aggression through interventions in disputes against older, more dominant animals. Early association with a silverback might provide a basis for receiving continued agonistic support from him as animals mature and start experiencing more aggression from group adults (Watts & Pusey, 1993). Analyses from a larger sample found similar overall patterns to those I did in immature animals' social interactions with adult males (but see below), and few significant sex differences before 8 years of age (Watts and Pusey, 1993; Fletcher, 1994).

The nature of interactions between gorilla males and immature animals resembles those described for some cercopithecine species in which adult males become a focus of attraction for older infants and juveniles (Kummer, 1968; Whitten, 1987; Maestripieri, 1998). While the benefits to immatures from this attraction appear to be similar to those in gorillas, various aspects of gorillas' social system might make association with the dominant adult male especially important during social development. These features include female dispersal, the power of the adult male, and long male tenure.

Although dispersal by parous females is often associated with the death of their silverback, females may also transfer while the group male is still alive and their offspring are just weaned (Watts, 1991; Watts & Pusey, 1993; Sicotte, this volume). With infanticide by unrelated males as a risk to an immature who follows its dispersing mother, it is clearly advantageous to remain behind and follow the silverback instead. While infants under 18 months are the most frequent victims of infanticide, immatures as old as 3 years have been killed or attacked by unrelated males, at least two of these because they were following their dispersing mother (Watts, 1989; Sicotte, 2000).

Female dispersal also means that infants do not often grow up with maternal kin. In Group 5-2 where adult relatives were present, the two silverbacks were a less obvious spatial focus for infants and young juveniles than were adult males in other groups. Among adult females too, the presence of close kin may be correlated with relatively fewer affinitive interactions with the silverback (Stewart & Harcourt, 1987; Watts, 1994*b*, this volume). One conclusion is that animals with close kin need the silverbacks less than do those without. In gorillas, however, the value of close relatives as social partners, for example in agonistic support, is limited by the interventions of the dominant silverback. His ability to end disputes and his tendency to support the younger/subordinate partner in contests have the effect of equalizing competitive differences among immatures and adult females alike. This reduces the value of kin-based cooperation while increasing the relative value of the dominant silverback as a partner (Stewart & Harcourt, 1987; Watts, 1994*b*, 1997). Thus, even in Group 5-2, some juveniles and adolescents developed more extensive grooming relationships with adult males than with related females, including sometimes their mother (Watts & Pusey, 1993).

It is not unlikely that a silverback will hold tenure for most of an immature's development (Robbins, 1995, this volume; Watts, 2000), increasing the potential long-term benefits of association with him. In multimale groups older silverbacks sometimes support their adolescent sons or full brothers against less closely related younger males (Watts & Pusey, 1993), and may also support their adult daughters and granddaughters against unrelated immigrant females (Watts, 1992).

The extent to which a gorilla's early relationship with its group's dominant male influences its future social relationships, dispersal decisions, and breeding success is unknown (Harcourt, 1981). Despite over 30 years of research at Karisoke, there are few data from the immaturity of animals who then dispersed from intact natal groups. For females, natal dispersal appears to be related to the availability of mates in their group who are not their putative father (Stewart & Harcourt, 1987; Sicotte, this volume). In fact, the nature of their early relationships with the younger silverbacks might be more salient than those with the dominant (their putative father). Among the males who did not disperse before adulthood, some had frequent affinitive interactions with the dominant male during their immaturity (eg. Sh and Pb) while others (e.g. Zz) did not. It is possible that the nature of a young male's relationship with the dominant silverback has little influence on his future mating tactics, which appear to be determined by the number of females in the group, the number of other adult males, and their relative ages and ranks (Robbins, 1995, this volume; Watts, 2000).

In addition, close cooperative relationships are not characteristic of mature males in gorilla groups. All males receive increasing levels of aggression from older silverbacks as they mature (Watts & Pusey, 1993). By the time they reach adulthood, regardless of the nature of their previous relationship with dominant/older silverbacks, adult males rarely affiliate with each other, and coalitionary behavior between them is rare (Robbins, 1995, this volume; Watts, 2000). Cooperative behavior between group males consists primarily of joint defense of the group against extra-group silverbacks (Sicotte, 1993; Watts, 2000).

This contrasts with chimpanzees in which sons are invariably philopatric, while daughters disperse. Sex differences in association with adult males becomes evident during social development when older juvenile males begin seeking the proximity of adult males. This is seen as a precursor to their integration into the adult male social network (Pusey, 1983, 1990; Strier, 1993). Similar arguments have been made for sex differences in the spatial relations of juveniles in the patrifocal muriqui (Strier, 1993). In chimpanzees therefore, the benefits of immatures' relationships with adult males are associated with their future adult relationships. In gorillas, they may have more to do with advantages during immaturity.

How do adult male gorillas benefit from their relationships with immature animals? At a general level, mating effort and parental effort in gorillas are inextricably linked (Clutton-Brock, 1991; Smuts & Gubernick, 1992; van Schaik & Paul, 1996). The aggression/avoidance during interunit encounters that protects a male's offspring from extra-group threat also retains his mates and might even gain him others if males' fighting ability influences females' residence decisions (Sicotte, 1993, this volume; Watts, 1994*a*). Within groups, however, the behavior of males towards younger immatures appears to be parental, albeit low-cost. Silverbacks behave as if the infants and juveniles in the group are potential offspring. Near the end of his life, the old silverback (Sb1) in Group 5 was no longer mating, but continued to protect his offspring or immature relatives from intra-group aggression, much like a "retired" male in geladas (Dunbar, 1984; Watts & Pusey, 1993).

More recent studies from large multimale gorilla groups at Karisoke have found increased levels of male–male competition, consequent changes in males' behavior towards females, and some hints of changes in their interactions with infants (Sicotte, 1994, this volume; Robbins, 1999, this volume; Watts, 2000). For example, Sicotte has described infants and females "interpositioning" themselves between adult males in situations of potential or real conflict, a behavior not observed previously at Karisoke (Sicotte, 1995). In six cases, males picked up infants (who had made the approach) and held them ventrally for up to several minutes. Although these instances are few, nevertheless they suggest a male's "use" of infants in a competitive situation (Paul *et al.*, 1996; van Schaik & Paul, 1996).

An increase in group size and number of males has also coincided with more frequent grooming of juveniles by subordinate silverbacks (Fletcher, 1994; Watts, 2000; Sicotte, personal communication). This may represent attempts by the males to increase mating access to the immatures' mothers whom they also groom. However, although females appear to exercise mate choice within multimale groups (Robbins, 1999; Sicotte, this volume) more data are needed to determine if they use a male's behavior towards their offspring as a basis for this choice. In the case of the most frequent male–juvenile grooming in Group 5, the male, Pablo, did not end up after group fission with the female whose infant he had groomed (Robbins, this volume).

In large, multimale groups, association patterns between males and females vary, as do mating patterns: females mate more often with some males than others (Robbins, 1999, this volume). To what extent is this differentiation reflected in paternity differences among males? Does male behavior towards infants ever reflect these differences? Is group fission associated with differences in males' behavior towards group infants? The answers could provide clues to processes of mate choice, kin recognition, and group formation in gorillas.

More long-term data are needed from Karisoke Research Center to answer questions raised here about the influence of early relationships with adult males on future mating tactics and breeding success, or about the effects of increased intra-group competition on males' interactions with infants. Perhaps even more crucial are data on interactions between silverbacks and immature animals in other subspecies of gorillas. We know next to nothing about these in western lowland gorillas. Yamagiwa has provided a description of a general attraction to adult males in infants and juveniles (Yamagiwa, 1983), but more data are needed on variation among silverbacks in their behavior towards immature group members. This would be especially interesting in light of the apparent lack of infanticide in eastern lowland gorillas, and the observation that females and their young may stay as a group for extended periods without being accompanied by a fully mature male (Yamagiwa, this volume). These suggestions present a major challenge to the claim that the primary contribution silverbacks make to their offsprings' fitness is protection from unrelated infanticidal males.

Summary

Adult male gorillas are crucial to the survival of their offspring because of the protection they provide from extra-group threat, especially that from unrelated and therefore infanticidal silverbacks. Males, however, are also key figures during their offspring's social development. Maturing gorillas benefit in a variety of ways from close association with dominant/older silverbacks, and such benefits may persist as gorillas approach adulthood. Immature animals are largely responsible for proximity and other affiliative interactions with dominant/older adult males, who do not actively care for infants and, in fact, are not more tolerant of them during feeding than most other group adults. While the presence of maternal relatives appears to decrease immatures' associations with the silverback, his ability to control group aggression reduces the value of maternal relatives as social partners. Males' behavior towards infants and young juveniles appears to be paternalistic, if usually low-cost. However, in multimale groups, with large numbers of sexually active females, males' relationships with infants may become more differentiated, along with their relationships with group females. More data are needed to understand the influence, if any, of an immature's relationship with the dominant male on its future mating tactics and breeding success. Research on other subspecies of gorillas is also needed to understand fully the role of paternal protection in gorillas' social evolution.

References

Anderson, C M (1992) Male investment under changing conditions among chacma baboons at Suikerbosrand. *American Journal of Physical Anthropology*, **87**, 479–96.

Bowlby, J (1969) *Attachment and Loss*. London: Hogarth Press.

Busse, C D (1985) Paternity recognition in multi-male primate groups. *American Zoologist*, **25**, 873–81.

Busse, C D & Hamilton, W J I (1981) Infant carrying by male chacma baboons. *Science*, **212**, 1282–3.

Clutton-Brock, T C B (1991) *The Evolution of Parental Care*. Princeton: Princeton University Press.

Dunbar, R I M (1984) *Reproductive Decisions: An Economic Analysis of Gelada Baboon Social Strategies*. Princeton: Princeton University Press.

Enciso, A E, Calcagno, J M & Gold, K C (1999) Social interactions between captive adult male and infant lowland gorillas: implications regarding kin selection and zoo management. *Zoo Biology*, **18**, 53–62.

Fairbanks, L A (1993) Juvenile vervet monkeys: establishing relationships and practicing skills for the future. In *Juvenile Primates*, ed. M E Pereira & L A Fairbanks, pp. 211–27. New York: Oxford University Press.

Ferrari, S F (1992) The care of infants in a wild marmoset. *American Journal of Primatology*, **26**, 109–18.

210   *Kelly J. Stewart*

Fletcher, A W (1994) The social development of immature mountain gorillas (*Gorilla gorilla beringei*), PhD thesis, University of Bristol.

Fossey, D (1979) Development of the mountain gorilla (*Gorilla gorilla beringei*): the first 36 months. In *The Great Apes*, ed. D A Hamburg & E R McCown, pp. 138–92. Menlo Park CA: Benjamin/Cummings.

Fossey, D (1983) *Gorillas in the Mist*. London: Hodder & Stoughton.

Fossey, D (1984) Infanticide in mountain gorillas (*Gorilla gorilla beringei*) with comparative notes on chimpanzees. In *Infanticide: Comparative and Evolutionary Perspectives*, ed. G Hausfater & S B Hrdy, pp. 217–36. New York: Aldine Press.

Fossey, D & Harcourt, A H (1977) Feeding ecology of free ranging mountain gorilla. In *Primate Ecology*, ed. T H Clutton-Brock, pp. 415–47. London: Academic Press.

Goldizen, A W (1987) Tamarins and marmosets: communal care of offspring. In *Primate Societies*, ed. B B Smuts, D L Cheney, R M Seyfarth, R W Wrangham & T T Struhsaker, pp. 34–43. Chicago: University of Chicago Press.

Harcourt, A H (1978a) Activity periods and patterns of social interaction: a neglected problem. *Behaviour*, **68**, 121–35.

Harcourt, A H (1978b) Strategies of emigration and transfer by primates with particular reference to gorillas. *Zeitschrift für Tierpsychologie*, **48**, 401–20.

Harcourt, A H (1979a) Contrasts between male relationships in wild gorilla groups. *Behavioral Ecology and Sociobiology*, **5**, 39–49.

Harcourt, A H (1979b) Social relationships among adult female mountain gorillas. *Animal Behaviour*, **27**, 251–64.

Harcourt, A H (1979c) Social relationships between adult male and female mountain gorillas. *Animal Behaviour*, **27**, 325–42.

Harcourt, A H (1981) Gorilla male relationships: can differences during immaturity lead to contrasting reproductive tactics in adulthood? *Animal Behaviour*, **26**, 206–10.

Harcourt, A H & Stewart, K J (1987) The influence of help in contests on dominance rank in primates: hints from gorillas. *Animal Behaviour*, **35**, 182–90.

Harcourt, A H & Stewart, K J (1989) Functions of alliances in contests within wild gorilla groups. *Behaviour*, **109**, 176–90.

Harcourt, A H & Stewart, K J (1996) Function and meaning of wild gorilla 'close' calls. II. Correlations with rank and relatedness. *Behaviour*, **133**, 827–45.

Harcourt, A H, Stewart, K J & Fossey, D (1976) Male emigration and female transfer in wild mountain gorilla. *Nature*, **263**, 226–27.

Hinde, R A & Atkinson, S (1970) Assessing the roles of social partners in maintaining mutual proximity, as exemplified by mother–infant relations in rhesus monkeys. *Animal Behaviour*, **18**, 169–76.

Horrocks, J A & Hunte, W (1993) Interactions between juveniles and adult males in vervets: implications for adult male turnover. In *Juvenile Primates*, ed. M E Pereira & L A Fairbanks. New York: Oxford University Press.

Hrdy, S B (1979) Infanticide among animals: a review, classification, and examination of the implications for reproductive strategies of females. *Ethology and Sociobiology*, **1**, 13–40.

Itani, J (1962) Paternal care in the wild Japanese monkey, *Macaca fuscata*. In *Primate Social Behaviour*, ed. C Southwick, pp. 91–7. New York: Van Nostrand Reinhold.

Keddy-Hector, A C, Seyfarth, R M & Raleigh, M J (1989) Male parental care, female choice, and the effect of an audience in vervet monkeys. *Animal Behaviour*, **38**, 262–71.

Kummer, H (1967) Tripartite relations in hamadryas baboons. In *Social Communication Among Primates*, ed. S Altmann, pp. 63–71. Chicago: University of Chicago Press.

Kummer, H (1968) *Social Organization of Hamadryas Baboons*. Chicago: University of Chicago Press.

Maestripieri, D (1998) The evolution of male–infant interactions in the tribe Papionini (Primates: Cercopithecidae). *Folia Primatologica*, **69**, 247–51.

Packer, C (1980) Male care and exploitation of infants in *Papio anubis*. *Animal Behaviour*, **28**, 512–20.

Palombit, R A (1999) Infanticide and the evolution of pair bonds in nonhuman primates. *Evolutionary Anthropology*, **7**, 117–29.

Palombit, R A, Seyfarth, R & Cheney, D L (1997) The adaptive value of friendships to female baboons. *Animal Behaviour*, **54**, 599–614.

Paul, A, Kuester, J & Arnemann, J (1996) The sociobiology of male–infant interactions in Barbary macaques, *Macaca sylvana*. *Animal Behaviour*, **51**, 155–70.

Pusey, A E (1983) Mother–offspring relationships in chimpanzees after weaning. *Animal Behaviour*, **31**, 363–77.

Pusey, A E (1990) Behavioural changes at adolescence in chimpanzees. *Behaviour*, **115**, 203–46.

Robbins, M M (1995) A demographic analysis of male life history and social structure of mountain gorillas. *Behaviour*, **132**, 21–47.

Robbins, M M (1999) Male mating patterns in wild mountain gorilla groups. *Animal Behaviour*, **57**, 1013–20.

Schaller, G B (1963) *The Mountain Gorilla: Ecology and Behavior*. Chicago: University of Chicago Press.

Sicotte, P (1993) Inter-group encounters and female transfer in mountain gorillas: influence of group composition on male behavior. *American Journal of Primatology*, **30**, 21–36.

Sicotte, P (1994) Effect of male competition on male–female relationships in bi-male groups of mountain gorillas. *Ethology*, **97**, 47–64.

Sicotte, P (1995) Interpositions in conflicts between males in bi-male groups of mountain gorillas. *Folia Primatologica*, **65**, 14–24.

Sicotte, P (2000) A case study of mother–son transfer in mountain gorillas. *Primates*, **41**, 95–103.

Smuts, B B (1985) *Sex and Friendship in Baboons*. New York: Aldine de Gruyter.

Smuts, B B & Gubernick, D J (1992) Male–infant relationships in nonhuman primates: paternal investment or mating effort? In *Father–Child Relations: Cultural and Biosocial Contexts*, ed. B S Hewlett, pp. 1–30. New York: Aldine de Gruyter.

212   *Kelly J. Stewart*

Stein, D M (1984) *The Sociobiology of Infant and Adult Male Baboons.* Norwood NJ: Albex.

Stein, D M & Stacey, P B (1981) A comparison of infant–adult male relations in a one-male group with those in a multi-male group for yellow baboons (*Papio cynocephalus*). *Folia Primatologica*, **36**, 264–76.

Sterck, E H M (1997) Determinants of female dispersal in Thomas langurs. *American Journal of Primatology*, **42**, 179–98.

Sterck, E H M, Watts, D P & van Schaik, C P (1997) The evolution of female social relationships in nonhuman primates. *Behavioral Ecology and Sociobiology*, **41**, 291–309.

Stewart, K J (1988) Suckling and lactational anoestrus in wild gorillas (*Gorilla gorilla*). *Journal of Reproduction and Fertility*, **83**, 627–34.

Stewart, K J & Harcourt, A H (1987) Gorillas: variation in female relationships. In *Primate Societies*, ed. B B Smuts, D L Cheney, R M Seyfarth, R W Wrangham & T T Struhsaker, pp. 155–64. Chicago: University of Chicago Press.

Stewart, K J, Harcourt, A H & Watts, D P (1988) Determinants of fertility in wild gorillas and other primates. In *Natural Human Fertility: Social and Biological Determinants*, ed. P Diggory, M Potts & S Teper, pp. 22–38. London: Macmillan.

Strier, K B (1993) Growing up in a patrifocal society: sex differences in spatial relations of immature muriquis. In *Juvenile Primates*, ed. M E Pereira & L A Fairbanks, pp. 138–47. New York: Oxford University Press.

van Schaik, C P (1996) Social evolution in primates: the role of ecological factors and male behaviour. *Proceedings of the British Academy*, **88**, 9–31.

van Schaik, C P & Janson, C H (2000) *Infanticide by Males and its Implications.* Cambridge: Cambridge University Press.

van Schaik, C P & Paul, A (1996) Male care in primates: does it ever reflect paternity? *Evolutionary Anthropology*, **5**, 152–56.

Watts, D P (1984) Composition and variability of mountain gorilla diets in the central Virungas. *American Journal of Primatology*, **7**, 323–56.

Watts, D P (1989) Infanticide in mountain gorillas: new cases and a reconsideration of the evidence. *Ethology*, **81**, 1–18.

Watts, D P (1990) Ecology of gorillas and its relation to female transfer in mountain gorillas. *International Journal of Primatology*, **11**, 21–44.

Watts, D P (1991) Mountain gorilla reproduction and sexual behavior. *American Journal of Primatology*, **24**, 211–25.

Watts, D P (1992) Social relationships of immigrant and resident female mountain gorillas. I. Male–female relationships. *American Journal of Primatology*, **28**, 159–81.

Watts, D P (1994*a*) The influence of male mating tactics on habitat use in mountain gorillas. *Primates*, **35**, 35–47.

Watts D P (1994*b*) Social relationships of immigrant and resident female mountain gorillas. II. Relatedness, residence, and relationships between females. *American Journal of Primatology*, **32**, 13–30.

Watts, D P (1997) Agonistic intervention in wild mountain gorilla groups. *Behaviour*, **134**, 23–57.

Watts, D P (2000) Causes and consequences of variation in male mountain gorilla life histories and group membership. In *Primate Males*, ed. P M Kappeler, pp. 169–79. Cambridge: Cambridge University Press.

Watts, D P & Pusey, A E (1993) Behavior of juvenile and adolescent great apes. In *Juvenile Primates*, ed. M E Pereira & L A Fairbanks, pp. 148–67. New York: Oxford University Press.

Whitten, P L (1987) Infants and adult males. In *Primate Societies*, ed. B B Smuts, D L Cheney, R M Seyfarth, R W Wrangham & T T Struhsaker, pp. 343–57. Chicago: University of Chicago Press.

Wrangham, R W (1979) On the evolution of ape social systems. *Social Science Information*, **18**, 334–68.

Wright, P C (1984) Biparental care in *Aotus trivirgatus* and *Callicebus moloch*. In *Female Primates: Studies by Women Primatologists*, ed. M F Small, pp. 59–75. New York: Alan R. Liss.

Yamagiwa, J (1983) Diachronic changes in two eastern lowland gorilla groups (*Gorilla gorilla graueri*) in the Mt. Kahuzi Region, Zaire. *Primates*, **24**, 174–83.

8 *Social relationships of female mountain gorillas*

DAVID P. WATTS

Effie and Maggie, mother and daughter in Group 5. (Photo by Martha M. Robbins.)

Introduction

Non-human primate socioecology and female social relationships

Female mountain gorillas (*Gorilla gorilla beringei*) are large, long-lived, slowly reproducing mammals whose diet consists mostly of vegetation that is abundant, evenly distributed, and high in structural carbohydrates. These few characteristics do much to explain the mountain gorilla social system, which we can only understand by placing long-term data on known individuals in comparative context. Variation in primate social systems depends largely on variation in life history tactics, predation risk, feeding competition, and conflicts and convergence of reproductive interest between the sexes (Sterck *et al.*, 1997; Kappeler, 1999*a*). Ecological factors are especially important to females, because they have slower maximum reproductive rates and higher parental investment than males (Bradbury & Vehrencamp, 1977). Consequently, foraging efficiency is crucial for female reproductive success and feeding competition strongly influences female social relationships, whereas male reproductive success depends crucially on gaining access to fertile females (Bradbury & Vehrencamp, 1977; Wrangham, 1980; van Schaik, 1989). The need to minimize predation risk is probably the main reason why females of most diurnal species live in social groups (van Schaik, 1983, 1989; Janson, 1992; Kappeler, 1999*a*). Variation in diet and in food distribution is the main source of variation in social relationships among females because it determines the predominant mode and intensity of feeding competition within and between groups. Male reproductive competition can also influence relationships between females, and has complicated effects on group size and composition, male–female relationships, and the costs and benefits of dispersal for females (Figure 8.1) (van Schaik, 1989, 1996; Sterck *et al.*, 1997; Nunn, 1999).

Sterck *et al.* (1997; cf. van Schaik, 1996; van Schaik & Kappeler, 1997; Kappeler, 1999*a*) recently summarized many of the ways that these factors interact, and I will follow their model in describing the social relationships of female mountain gorillas. The extent to which females can benefit by contesting access to food is centrally important in the model. Non-human primates in multifemale groups form strict or "despotic" dominance hierarchies when the potential for within-group contest competition for food is high and its outcome can influence reproductive success significantly. High-ranking females can gain indirect fitness benefits by helping female relatives acquire and maintain high rank, and high- and middle-ranking females can benefit directly by cooperating with non-relatives to achieve the same social effects. High potential gains or losses associated with contest feeding competition between groups makes high-ranking females

216

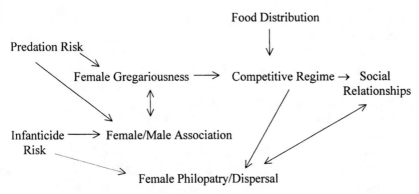

Figure 8.1. Overview of presumed influences on female association and social
relationhips, based on Sterck *et al.* (1997). The distribution of risks and of food
resources determines whether females form social groups and, if so, influences
typical group size and limits total group size. Food distribution determines the
competitive regime, which is the main influence on social relationships between
females. Male association with females may occur because males attach themselves
to female groups, or because females congregate around males who offer protection against predators and infanticide. The form and intensity of feeding competition influences the costs and benefits of female dispersal. When costs are low and
females stand to benefit from protection by males, they may transfer to join males
based on assessment of male protective ability.

more tolerant towards subordinates. High ecological and social costs
associated with dispersal, especially loss of female allies, lead to female
philopatry. When neither within-group nor between-group contest competition for food is important, agonistic relationships are more individualistic (i.e. depend on individual fighting ability, rather than coalitionary
support) and may be "egalitarian" (i.e. dyadic asymmetries are weak), and
females have either weak dominance hierarchies or none. Dispersal costs
are much lower, and females can transfer to minimize costs of scramble
feeding competition or to avoid close inbreeding.

They can also transfer to maximize their safety. Males become an issue
with regard to safety, partly because males often provide protection
against predators, but often also because they protect females against
sexual coercion, including infanticide, by other males. Also, male strategies to increase paternity certainty and to protect their offspring can
constrain them to associate continuously with certain females, and to
associate more with some females than with others in multifemale groups.

The model of Sterck *et al.* (1997) does not adequately explain
socioecological variation in lemurs (Kappeler, 1999*b*) and all platyrrhines
(e.g. Bolivian squirrel monkeys, *Saimiri boliviensis*: Boinski, 1999), and
will undoubtedly undergo future modifications. However, the basic struc-

ture enjoys broad empirical support and also finds support in several detailed between- and within-species comparative studies (e.g. Peruvian vs. Costa Rican squirrel monkeys, *Saimiri sciureus* and *S. oerstedi*: Mitchell *et al.*, 1991; Thomas langurs, *Presbytis thomasi*, vs. longtailed macaques, *Macaca fascicularis*: Sterck & Steenbeek, 1997; different populations of grey langurs, *Semnopithecus entellus*: Koenig *et al.*, 1998).

Long-term demographic and life history data from Karisoke, plus data from detailed studies of social relationships and of feeding ecology, seem to fit mountain gorillas into this socioecological model neatly (Watts, 1996; Sterck *et al.*, 1997). Females spend their entire lives in social groups, although not necessarily their natal groups. Natal and secondary transfer (*sensu* Pusey & Packer, 1987) are common in the Virungas population (Harcourt *et al.*, 1976; Harcourt, 1978; Stewart & Harcourt, 1987; Watts, 1990, 1996), and several females in the Karisoke Research Center study population have resided in at least four groups. Groups typically contain single mature males, but multimale groups are fairly common, and most research at Karisoke has been on groups with two or more sexually mature males. Mountain gorillas are highly folivorous. Food is abundant and is distributed densely and relatively evenly in most of their habitat, and its availability varies little in space and time compared to the food supplies of most other primates (Fossey & Harcourt, 1977; Vedder, 1984; Watts, 1984, 1988, 1991*a*, 1998*a*, *b*, 2000*a*; Plumptre, 1993; McNeilage, 1996). This should mean that contest competition over food is infrequent and unimportant for females, and that establishment of strong dominance relationships is similarly unimportant, as seems to be the case (Harcourt, 1979*a*; Watts, 1985, 1994*a*, 1996). It also should help to explain why females can transfer at little cost, given that resistance to immigrants is only worthwhile if their addition increases scramble feeding competition, or competition for male services, significantly. One such service is protection against males with whom the females have not mated, because infanticide by males is a major source of infant mortality (Fossey, 1984; Watts, 1989, 1990, 1996).

Karisoke data also show that female life histories and social relationships vary considerably. For example, many females have reproduced at least once in their natal groups, and most have spent at least some of their adult lives with adult female relatives (Watts, 1996). Also, many females rarely receive agonistic support from other females, but others receive support at rates comparable to those of some female cercopithecines in which females are philopatric, form strict dominance hierarchies, and "inherit" their mothers' ranks (Watts, 1997). In this chapter, I summarize some of what we know about relationships between female mountain gorillas and their connections to the relationships that females have with males. I particularly examine how genealogical relatedness structures female relationships and why females nevertheless routinely forego the benefits of residing with close female relatives. I also present some new data that address questions about long-term relationship stability and consistency. I place the Karisoke data in a comparative context with other primates and with some non-primate mammals. I conclude by noting some important questions for future research.

Social relationships between female mountain gorillas
The importance of relatedness and residence status
Like females in female-resident cercopithecines, female mountain gorillas have differentiated social relationships: affiliative interactions are more common and/or agonistic ones less so in some dyads than in others. However, they differ from cercopithecines in several ways. Notably, the effects of relatedness are weaker and less consistent, and cooperation between non-relatives is less important, for mountain gorillas, many of whom do not even live with female relatives. Data from Group 5 at several points in time and from Group 4, Group NK, and Group BM (Harcourt, 1979*a*; Watts, 1994*a*, *b*) show that when adult female relatives reside together, they are more tolerant of each other than non-relatives are and spend more time in close proximity than unrelated females do. Rates of affiliative interaction are higher between relatives than non-relatives, and aggressive interactions (at least escalated ones) are less common between relatives. In particular, fights, in which females often inflict wounds, are less common between relatives (Watts, 1994*a*, *b*). Presumed paternal half-sisters generally show levels of proximity and rates of affiliative and agonistic interaction intermediate between those of maternal relatives and non-relatives (Watts, 1994*a*).

Females with no adult female relatives in their groups groom mostly with their immature offspring and with adult males (Harcourt, 1979*a*, *b*; Watts, 1994*a*). Those with female relatives have more grooming partners, groom more with adult females, and sometimes have adult female relatives as their top grooming partners (Watts, 1994*a*). Females show reciprocity in grooming (Watts, 1994*a*), although this mostly reflects grooming between relatives, given that most unrelated females do not groom with each other.

Data on agonistic interventions in several social groups showed that females received agonistic support from other females in a median of 3.6% of all contests between females and 9.8% of all contests that involved more than mild threats (Watts, 1997). These proportions are within the range reported for cercopithecines with female philopatry and nepotistic female dominance hierarchies (Watts, 1997). However, most female coalitions occurred between relatives (especially maternal relatives, but also presumed paternal half-sisters), and most involved support that one female gave to a relative who had initiated aggression against a non-relative. More related than unrelated dyads formed coalitions, and exchanges of support occurred in more related than unrelated dyads. Females with both maternal and presumed paternal relatives available generally formed proportionately more coalitions with maternal relatives. Also, females with no female relatives faced more female coalitions in opposition than females with relatives (maternal, presumed paternal, or both) did. Maternal and some presumed paternal relatives exchanged agonistic support consistently and often enough to be considered allies. A few non-relatives (e.g. Sb and Pd in Group 5) also were allies, but most unrelated dyads did not support each other at all. Coalitions are probably uncommon, and alliances may be absent, in groups that contain only unrelated females.

Most female immigrants into Karisoke groups have met little resistance (Harcourt, 1978; Sicotte, 2000), but some who joined large groups have. Most notably, several adult and one adolescent natal females in Group 5 were highly aggressive to five immigrants who joined their group, which already had six females, within 3 months in 1984 and 1985 (Watts, 1991b). Females responsible for most of the harassment were neither pregnant nor lactating, and were thus potential emigrants by Sicotte's (1993) definition. Two eventually emigrated, whereas all immigrants stayed for at least 5 years. Because the group was so large, the potential for scramble feeding competition and for competition over access to the dominant silverback (see below) was unusually high. Large group size also allowed resident females to harass immigrants more freely than would have been possible in a smaller group, because it limited the ability of the dominant silverback to control aggression between females (Watts, 1991b; see below). This situation was also unusual because five of the six resident females had been born in Group 5, and the sixth was the mother of three of these five. All natal females were presumed paternal relatives, and two formed another mother/daughter pair. Coalitions of relatives were responsible for much of the harassment, which often involved one or more females opportunistically joining a relative who had initiated aggression at an immigrant. Harassment sometimes culminated in fights in which one or both of the initial opponents received bite wounds (Watts, 1997), so it carried some risks. However, coalitions of relatives usually outnumbered their opponents, which presumably reduced these risks and might have allowed them to score "easy victories" against intimidated opponents, as Chapais (1992) argued for macaques.

Agonistic asymmetries between females

Observers sometimes confidently identify the dominant member of a female dyad and even think a certain individual is dominant to most or all others in her group. For example, probably all Karisoke researchers who watched Group 5 in the late 1970s and the 1980s considered Ef to be dominant to the group's other females. Only in 1992, after years of observation, did I first see Ef behave submissively to another female (her daughter Pu). But do females form dominance hierarchies? The outcome of non-aggressive approach–retreat interactions is predictable in some dyads, which has led some authors to argue that we can ascribe dominance ranks to females (Harcourt, 1979a; Watts, 1985; Stewart & Harcourt, 1987; Harcourt & Stewart, 1987, 1989). However, the outcomes of aggressive interactions in many groups do not show statistical linearity, and many or most female dyads do not establish dominance relationships, where these involve one member of a dyad consistently responding submissively to aggression from the other, who consistently "wins" contests between them (Watts, 1994b). In some dyads, submission is unidirectional when it occurs (which may not be often); this justifies the terms "dominant" and "subordinate" for dyad members. However, neither female responds submissively to aggression from the other in many dyads, and in some both occasionally show submission. Those in which one female gives more submission than the other are a minority in most groups (Watts, 1994b). The most common responses to aggression in most dyads are either to ignore it or to retaliate – which rarely provokes submission – and undecided interactions constitute the majority in most dyads (Watts, 1994b). Also, aggression in both directions is common in many dyads; this can be because both females initiate aggression at high rates, or because one does so, while the other usually retaliates. Females in Group 4, Group NK, Group BM, and Group 5 during several periods showed significant bidirectionality in aggression: correlations between the rate of aggression that one female gave to another and the rate of aggression that she received from her were significantly positive (Watts, 1994b). Even in dyads in which aggression rates are high and in which one dyad member initiates most aggression (e.g. those that included Tu in Group BM during 1991 and 1992; Watts, 1994b), submission by the other may be infrequently or absent. For example, Tu and Pp had the highest dyadic rate of aggression in Group BM in 1991–92, and neither behaved submissively to the other.

Females often made agonistic interventions against other females to whom they lost some dyadic contests. The rate at which females joined coalitions against other females was positively correlated with the rate at which they received dyadic aggression from them, and the rate at which females received coalitionary opposition from other females was positively correlated with the rate at which they initiated dyadic aggressive interactions with them. Females also showed significant bidirectionality in opposition: the frequency with which they intervened against others was positively correlated with the frequency with which others intervened against them (Watts, 1997).

Bidirectionality in dyadic aggression and coalitionary opposition are not expected when females form strict dominance hierarchies (e.g. rhesus macaques: de Waal & Luttrell, 1988, 1989). Bidirectionality may be more common in species with relatively tolerant dominance relationships (e.g. stumptailed macaques, *Macaca arctoides*: de Waal & Luttrell, 1988, 1989). Significant bidirectionality in coalitionary opposition had previously been reported only among captive chimpanzees (de Waal & Luttrell, 1989), although it is not clear that this result applied specifically to females [and see Hemelrijk & Ek (1991) for a contrasting result from the same population]. Like mountain gorillas, female chimpanzees do not form clear dominance hierarchies (Pusey *et al.*, 1998). Otherwise, the agonistic relationships of female mountain gorillas show a combination of features unusual among primates: high bidirectionality in dyadic and coalitionary aggression, no dominance hierarchies, and few decided dominance relationships.

Relationship quality and stability
Female "friendships" and long-term relationship consistency
Relatedness is the most important source of variation in social relationships between females, but factors such as length of co-residence may still lead to variation in relationships between non-relatives, and relationships can change over time. Here, I examine the consistency of relationships between Group 5 females who resided together during at least two of three periods (1984–85, 1986–87, 1991–92). The 1984–85 period saw major changes in group composition (five immigrations, emigration of two natal females: Watts, 1991*b*). Two more immigrants joined by the start of observations in 1986 (Watts, 1994*a*). Two females had emigrated, one had died, one natal female had reached adulthood, and one adult and one adolescent had immigrated by 1991, and another adolescent had immigrated by 1992 (Watts, 1991*b*, 1994*a*).

For this analysis, I first used MATSQUAR (Hemelrijk, 1990) to calculate correlations between study periods for dyadic rates of affiliative interactions and of agonistic interactions, and for the percentage of time that individuals spent within 5 m of each other. For agonistic interactions, I used both absolute rates and rates corrected for time that individuals were within 5 m (cf. Watts, 1994*a*). MATSQUAR allows the calculation of several distribution-free measures of correlation between social interaction matrices. I performed K_r tests with MATSQUAR. These first generate tau$_{K_r}$ values, which are multivariate rank-order statistics that take individual variation into account by calculating Kendall rank correlations within rows (individuals) and combining these into a single measure. MAT-SQUAR then randomly permutes values within rows and columns and generates sampling distributions of tau$_{K_r}$ values against which to assess the significance of observed values. I used symmetrical matrices in which entries were the number of affiliative or agonistic interactions between the members of a given dyad per 100 hours (cf. Watts, 1994*a*, *b*) and performed 2000 matrix permutations for each test. The number of individuals in a matrix was the number present during both of the two relevant study periods. Nine females were present for both the 1984–85 and 1986–87 sample periods. Seven females present in 1984–85 were still present in 1991–92. Nine present in 1986–87 were still present in 1991–92. A significant positive correlation between two matrices indicates that relative rates of dyadic interaction, either affiliative or agonistic, were consistent between study periods, or that dyad members spent similar proportions of time in proximity, relative to their other partners. Because I made two comparisons with each data set, I used Bonferroni-adjusted alpha levels of $p = 0.025$ as the criterion for significance (Sokal & Rohlf, 1995).

I also classified dyadic relationships as good, bad, or neutral (Watts, 1995*a*). Females with good relationships had affiliative interactions at rates above the medians for both partners and aggressive interactions at rates below the medians, and they formed coalitions at least occasionally. Those with bad relationships interacted affiliatively at rates below the medians for both partners and aggressively at rates above the medians. Other relationships were neutral.

Results of K_r tests (Table 8.1) showed that time spent in close proximity and rates of affiliative interaction were consistent across all study periods (Figure 8.2). However, absolute rates of dyadic aggression were not consistent between any pair of periods. Aggression rates corrected for time spent in close proximity were significantly correlated for 1984–85 versus 1986–87 and for 1984–85 versus 1991–92, but not for 1984–85 versus 1986–87. A major reason for the inconsistency in agonistic relationships was that dyads of natal females and immigrants that interacted aggressively at extremely high rates in the first study period no longer did so by the second (Watts, 1994*a*). Nevertheless, many of these females were still intolerant of each other; they spent relatively little time in close proximity and engaged in aggression relatively often when close. Most unrelated

Table 8.1. *Results of* K_r *tests examining consistency in relative time spent in close proximity, relative rates of affiliative interaction, and relative rates of agonistic interaction for female dyads in Group 5*

	K_r	n females[a]	Tau_{K_r}	p
Proximity				
1984–85 vs. 1986–87	141	9	0.56	0.0005
1984–85 vs. 1991–92	39	7	0.37	0.0130
1986–87 vs. 1991–92	75	9	0.33	0.0135
Affiliative				
1984–85 vs. 1986–87	71	9	0.31	0.0095
1984–85 vs. 1991–92	86	7	0.37	0.0020
1986–87 vs. 1991–92	48	9	0.33	0.0075
Agonistic				
1984–85 vs. 1996–87	40	9	0.17	0.1119
1984–85 vs. 1991–92	26	7	0.25	0.0615
1986–87 vs. 1991–92	48	9	0.20	0.0575
Agonistic, 5 m[b]				
1984–85 vs. 1986–87	96	9	0.38	0.0070
1984–85 vs. 1991–92	21	7	0.20	0.1194
1986–87 vs. 1991–92	69	9	0.27	0.0155

[a] Data come from dyads that were together for at least two of three research periods.
[b] Rates of agonistic behavior, corrected for time spent within 5 m.

dyads were also relatively intolerant of each other in 1991–92. Notably, however, the rate of aggression between Sb and Pu was much lower than in earlier periods. Sb had been a main target of harassment when she immigrated in 1984 and retaliated against much of that harassment, and Pu was one of the main harassers (Watts, 1991*b*), but in 1991–92 these females actually had a good relationship. Another notable change was that aggression between Pu and Ef, her mother, was much more common in the last period than in the first two.

The quality of some relationships changed between periods. Only a small proportion of dyads, most of them related, had good relationships in any given period (Table 8.2). Four of five pairs of maternal relatives present for at least two periods had consistently good relationships. The exception was Ef and Pu, who had a neutral relationship in one of three periods (1991–92), although they still gave each other agonistic support against non-relatives. Presumed paternal relatives Pa and Mg had a consistently good relationship in two of two periods, but presumed paternal relatives Pu and Pa had a good relationship in only one of three periods,

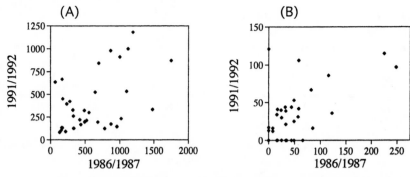

Figure 8.2. Two examples of results of analyses of how consistent female social relationships in Group 5 were across study periods. (A) Association between rates of agonistic interaction while females were in close proximity (acts per 100 hours spent within 5 m) in 1986–87 and in 1991–92; (B) association between rates of rates of affiliative interaction (acts per 100 hours) in 1986–87 and in 1991–92. Each point represents one female dyad. K_r tests gave highly significant results (see text), although the scatter shows that the quality of some dyadic relationships differed between periods.

and Tu and Pa in only one of two. However, like Ef and Pu, these females supported each other against non-relatives. Only two relationships between natal and immigrant females were good; both involved Sb. Sb and Pd, whose relationship was good in all three periods, were the only immigrants with a good relationship in more than one period. They were unrelated, but had been together in Group NK for 6 years before they joined Group 5 (cf. Watts, 1997).

Bad relationships were always more common than good ones, although most relationships were neutral in all but the first period, when harassment of immigrants was common (Table 8.2). No relatives had bad relationships; even those who interacted aggressively at relatively high rates also had relatively many affiliative interactions. All but two dyads with bad relationships included a natal female or Ef, whose natal group was unknown but who was Group 5's longest resident. However, comparable data on Group BM in 1991–92 show that unrelated immigrants can also have bad relationships, and sometimes interact aggressively at rates as high as or higher than those in most Group 5 dyads that had bad relationships (Watts, 1996). Eleven of 26 Group 5 dyads together during two or more periods and that had bad relationships during at least one also had bad relationships during at least a second. Four of 11 possible dyads had bad relationships during all three periods.

That most relationships were neutral or bad is not surprising, although the existence of good relationships between some non-relatives shows that

226   *David P. Watts*

Table 8.2. *Numbers of female dyads in Group 5 that had good or bad relationships in three research periods*

Period	Good		Bad	
	Relatives	Non-relatives	Relatives	Non-relatives
1984–85	12 (13)	4 (42)	0 (13)	23 (42)
1986–87	8 (9)	8 (57)	0 (9)	20 (57)
1991–92	7 (8)	6 (70)	0 (8)	15 (70)

Values in parentheses give total number of related or unrelated dyads, respectively.

relatedness is not the only source of differentiation. More strikingly, temporal variation in the quality of relationships between relatives reinforces the argument that these have limited long-term value.

Reconciliation

One indication that relationships between females have limited value is that females – both relatives and non-relatives – generally do not reconcile after aggressive conflicts. That is, opponents sometimes engage in friendly contact shortly after conflicts, but post-conflict rates of affiliative interaction are not consistently elevated compared to rates during "control" samples or compared to baseline rates (Watts, 1995a). In fact, females, especially non-relatives, often continue to interact aggressively at high rates after initial conflicts until they move far apart (Watts, 1995a). In contrast, females typically reconcile with males after they receive aggression from males. Minor conflicts between relatives probably do not threaten their long-term tolerance for each other and willingness to give each other agonistic support, and the same may apply to non-relatives with good relationships. In these cases, females may not need to reconcile. In most other cases, any benefits of restoring the *status quo ante* probably do not outweigh the risks of approaching opponents and possibly inciting further aggression (Watts, 1995a).

Egalitarianism and male control of female aggression
What explains egalitarianism?
In gorillas, egalitarianism means that agonistic relationships are unresolved in most female dyads and that aggression commonly goes in both directions, but not necessarily that females are tolerant of each other or have peaceful relationships. What allows high bidirectionality? The

explanation proposed for relatively high tolerance in some cercopithecines does not apply. Between-group feeding contests are rare or absent; most interactions between groups involve aggression, but this reflects male mate and infant defense and male mate acquisition tactics, not food defense (Sicotte, 1993; Watts, 1994c). Females do not form dominance hierarchies, and they suffer no obvious penalties if female group-mates emigrate. Instead, the best explanation is that females with high competitive ability – for example, those who are large or who receive agonistic support from female relatives – gain little by aggressively asserting priority of access to food (Watts, 1985, 1994b). Aggression between females was most common during feeding in most groups included in Watts's (1994b) sample, but rarely led to loss of access to food, and recipients often responded aggressively with little cost (Watts, 1994b, 1996). Poor competitors can ignore much aggression or retaliate against it, and can be assertive, when the potential gains for their opponents are low (Dunbar, 1988).

Male interventions in contests between females also facilitate egalitarianism (Watts, 1994b, 1997). Males intervene in proportionately more female–female contests than other females do, and most females receive more interventions by males than by females. Males mostly make control interventions and stop contests without either, or any, opponent clearly losing. Such neutrality, which can sometimes effectively be support for likely losers (Watts, 1997), is expected when the intervener has equally valuable relationships with both opponents; this applies to male gorillas and their mates (Watts et al., 2001). Because males stopped many polyadic contests, especially those that escalated to fights, females did not necessarily get more submission from opponents by participating in coalitions against them, despite usually outnumbering them, than they did in dyadic interactions (Watts, 1997). Targets of coalitionary aggression were somewhat more likely to retaliate against coalition members when males intervened, and sometimes used males as shields as they did so (Watts, 1997). Males' interventions in dyadic contests reduced the likelihood that aggressors directed multiple acts of aggression at their targets and could prevent prolonged sequences of aggression in which opponents often engaged otherwise (Watts, 1995a). Male control of female aggression protects females, reduces their risks if they retaliate against opponents or initiate aggression against them, and limits damaging aggression (Watts, 1994b, 1997). Given that females can transfer with little cost, this could represent male mating effort: reducing competitive disparities among females may help males to retain poorer competitors (Watts, 1994b, 1997).

Nepotism and female dispersal
What limits nepotism?
Agonistic relationships in mountain gorillas are individualistic. Even when relatives reside together, they do not influence each others' agonistic relationships like cercopithecines with nepotistic female dominance hierarchies do. Females cannot necessarily invoke submission from females who sometimes behave submissively to their mothers or sisters. Agonistic asymmetries among juveniles and adolescents seem to depend mostly on body size differences, which can outweigh any nepotistic effects. Females support their infants and juveniles against other adult females and immatures, but mostly to defend them against threats; support for offspring who initiate aggression is less common (Harcourt & Stewart, 1987, 1989; Watts, 1997). Maternal and sororal support is often ineffective against adults (Watts, 1997). Females do not continue to support offspring or sisters consistently against opponents from whom those relatives can induce submission, nor do they support unrelated immatures against opponents from whom the females themselves can induce submission (Harcourt & Stewart, 1987, 1989; Watts, 1997).

Male control of female aggression also hinders nepotism. Consistent support for immature daughters against opponents from whom mothers sometimes get submission would have limited effect, because males commonly defend offspring against larger opponents and stop conflicts between similar-sized peers, and they rarely support immature offspring who initiate aggression. This mitigates agonistic asymmetries, which would be contrary to the interests of males (Harcourt & Stewart, 1987, 1989; Watts, 1997), rather than reinforcing them. Also, male interventions in conflicts between adult females mitigate the ability of females to promote stable agonistic asymmetries between their mature daughters and unrelated females, or between unrelated females. The one clear exception involved Bv in Group 5, who supported his presumed daughters and granddaughter against unrelated immigrant females in 1984–85 (Watts, 1991*b*, 1997). Bv was subordinate to his son, Zz, who intervened much more often and typically made control interventions or supported immigrants against resident females who had targeted them (Watts, 1991*b*). Bv did not mate with the immigrants. Zz did, and his behavior may have helped him to retain them in the group (Watts, 1991*b*; see above).

Why don't female relatives stay together?
Many females have reproduced at least once in their natal groups. Relatives can sometimes associate even when they emigrate from natal groups: some have transferred with relatives or to groups to which relatives had

already transferred. Still, incentive to stay with relatives permanently are small because alliances with them have limited effectiveness and, especially, because neither within- nor between-group contest competition for food is important. The fact that females usually do not replace allies after they transfer reinforces this conclusion (Watts, 1997; cf. Lima, 1989). Other influences on female reproductive success can outweigh any nutritional or social gains offered by alliances between relatives. Avoidance of close inbreeding can explain natal transfer when a female's only potential mate is her father (Harcourt, 1978). Otherwise, the need for male protection is probably the most important influence on female dispersal (Watts, 1996, 2000b; see below).

Why do females engage in damaging aggression?

Fights between females are uncommon in small groups (Harcourt, 1979a). However, females in large groups fight fairly often, and often intervene in fights between other females (Watts, 1994b, 1997). One or more opponents, including some interveners, received bite wounds in 48% of 46 fights documented in Watts (1997). Why females contest access to herbaceous food, even at low rates, is unclear. Contests may sometimes limit group size by inducing females to emigrate or by preventing immigration; females who could achieve these effects would face less scramble competition for food (Watts, 1991b, 1994b). However, immigrants to Group 5 stayed despite frequent harassment in 1984–85 (see above). Also, aggression rates are not consistently related to female group size. Data on six groups that had from five to 13 females showed that the median rate at which individual females gave and received aggression in all contexts increased with the number of females per group, but not significantly ($r = 0.70$, d.f. = 4, n.s.). Aggression rates during feeding also were not significantly related to female group size ($r = 0.67$, d.f. = 4, n.s.).

Much competition among females may be over access to male services that may depreciate with distance, especially protection (Watts, 1994b). Fights and other aggressive interactions were disproportionately common when females clustered around males during group progressions, and often seemed to concern proximity to males during other activities (Watts, 1994b). Females who initiate such contests could impose costs on others if they can force them to stay farther from males and thereby become more exposed to risks (e.g. infanticide: Watts, 1994b). However, males constrain the ability of females to do this by intervening. Targets of aggression often stayed near males after male interventions (Watts, 1995a), and were more likely to do so than were the aggressors (Watts, 1995b, 1997).

Comparative perspectives
Mountain gorillas and other primates

The dearth of quantitative data on the social relationships of other gorilla subspecies in the wild constrains our ability to assess how generally Karisoke data represent gorillas. Greater frugivory in western lowland gorillas and in eastern lowland gorillas at low altitudes raises the possibility that they face more contest competition for food. Because fruit sources occur at lower density than herbaceous foods, the potential for scramble competiton is also higher, which could help to explain why mean group sizes are often smaller than those found in the Virungas (reviewed in Doran & McNeilage, 1998; but see Bermejo, 1997). However, female transfer also occurs in other subspecies, which indicates that the potential benefits of permanent association between female relatives are limited. Occasional inter-group contests over access to fruit patches occur in some populations (e.g. Lopé: Tutin, 1996), but whether females really cooperatively defend patches is unclear, and contests may be too infrequent to have a fundamental influence on their social relationships (for example, female transfer occurs at Lopé: Tutin, 1996). Females include abundant and evenly distributed herbaceous foods in their diets at all times, and reduce both contest and scramble competition during fruit-poor times by using these heavily as fallbacks.

Circumstantial evidence suggests that they also avoid competition by forming temporary foraging subgroups that use scattered fruit patches (reviewed in Doran & McNeilage, 1998). Even in subgroups, however, females apparently always associate with males: subgroups seem to form around adult males in multimale groups. This would be consistent with permanent male–female association based on female need for male protection in mountain gorillas. Protection against predation may be more important for other subspecies, while protection against infanticide may be less important (see chapters by Doran & McNeilage and by Yamagiwa, this volume). However, most known infanticides that have occurred in the Virungas have occurred in groups that have lost their single mature males (Watts, 1989). These circumstances have been uncommon and irregular in Karisoke history, and long periods have passed with no known infanticides. Data from Kahuzi-Biega show that female eastern lowland gorillas can sometimes escape the infanticide of their offspring in similar circumstance, but we should be wary of discounting infanticide as a force in social evolution without continuous, long-term observations of habituated groups like those available for the Karisoke population.

Female agonistic relationships could indeed vary across gorilla subspecies, given cross-population variation in other primate species associated with variation in food distribution. For example, female grey lan-gurs (*Semnopithecus entellus*) establish dominance hierarchies in habitats where much of their food occurs in monopolizable clumps (although rank acquisition and maintenance does not involve macaque-like nepotism and mutualism), but not in those where food is more evenly dispersed and contest competition is low (Koenig *et al.*, 1998; Sterck, 1998). Female chacma baboons (*Papio ursinus*) are philopatric and establish despotic, nepotistic dominance hierarchies throughout most of their range. How-ever, those in high-altitude habitats ("mountain baboons") rely on sparse-ly, but evenly, distributed foods and live in small groups centered on adult males (Byrne *et al.*, 1989). Some females have decided dominance relation-ships, but contests over food are rare and coalitions between females are rare or absent. Females groom more with males than with other females, and female transfer is common. But the wide variation in mountain gorilla female social relationships and life history tactics cautions against any assumption of fundamental subspecific differences.

A number of other primate species also use food that is evenly distrib-uted (whether in large scattered patches, or small but more densely distrib-uted ones) and have low levels of contest competition. Like mountain gorillas, they lack clear female dominance hierarchies, and many female dyads may not have established dominance relationships. Also, female coalitions are uncommon or absent. These include Costa Rican squirrel monkeys (Mitchell *et al.*, 1991), Thomas's langurs (*Presbytis thomasi*: Sterck, 1997), hamadryas baboons (*Papio hamadryas*: Abbeglen, 1984), and patas monkeys (*Erythrocebus patas*: Isbell & Pruetz, 1998). Female transfer occurs in Costa Rican squirrel monkeys and in Thomas's langurs, and in many other langurs in which contest feeding competition is prob-ably low (reviewed in Sterck *et al.*, 1997).

Research on Thomas's langurs at Ketambe (Sterck, 1997; Sterck & Steenbeek,1997; Sterck *et al.*, 1997) provides some of the clearest similari-ties to, but also some contrasts with, data on mountain gorillas. Female Thomas's langurs also have individualistic and egalitarian dominance relationships. Although females in some dyads show consistent direc-tionality in agonistic outcomes, bidirectionality is high in others and females do not form linear dominance hierarchies. Researchers have not seen female coalitions, although this could be an effect of demographic variation if the research groups contained few or no female relatives. As in mountain gorillas (and as is true more generally; de Waal, 1989), domi-nance and competitiveness are at least partly independent: despite the absence of dominance hierarchies and even of decided dominance relation-ships, rates of aggression were relatively high in the langurs, especially during feeding. No effects of contest competition on female reproductive success are evident from long-term work at Ketambe, nor are scramble effects evident; Thomas's langurs and mountain gorillas seem to be similar in these respects (Sterck et al., 1997). Female Thomas's langurs probably transfer several times during their lives, and females are not hostile to potential immigrants. However, inter-birth intervals for new immigrants are longer than those for long-term residents, which indicates that transfer may impose reproductive costs. Data on age at first reproduction show that natal transfer is not costly for female mountain gorillas (Watts, 1996), although the possibility that secondary transfer has associated costs remains open.

Sterck & Steenbeek (1997) found that rates of aggression between females feeding in the same tree were lower when their group's adult male was also in the tree than when he was not. Male proximity alone may inhibit aggression (Sterck & Steenbeek, 1997), although males in some groups intervene in most female contests, usually without choosing sides (Steenbeek, 1997). Male hamadryas baboons commonly police female aggression within their own one-male units and support their females against females from other units, and females often seek out their males for reassuring contact after aggressive interactions with other females (Kummer, 1968; Colmenares & Lazaro-Perea, 1994; Watts et al., 2000). Provision of post-conflict reassurance may be a widespread service that can influence female mate choice in species with female transfer or in which females typically have opportunities to mate with multiple males (e.g. mountain gorillas: Watts, 1995b; spectacled langurs, *Trachypithecus obscurus*: Watts et al., 2000; hamadryas baboons: Zaragoza & Colmenares, cited in Watts et al., 2000; savanna baboons: Castles et al., 1998). Like in mountain gorillas, bidirectionality in aggression between females is relatively high in hamadryas baboons; female–female grooming is limited compared to grooming between females and males; females do not reconcile with each other, but do with males; and females use males as shields behind which they can either gain protection against other females or can threaten them (Kummer, 1968; Abegglen, 1984; Castles et al., 1998). Males also intervene in conflicts between female spectacled langurs, and this species shows male–female, but not female–female, reconciliation (Watts et al., 2000). Male interventions to protect certain females and their infants, both against other females and against males, characterize "friendships" in savanna baboons (Smuts, 1985; Palombit et al., 1997). In all of these cases, male protection of females against aggression from other females could be a mating tactic. It is not associated with female transfer in savanna baboons, and so is not a mate retention tactic in the same way that male policing in mountain gorillas may be, but still apparently influences female mate choice (Smuts, 1985) and, in extreme cases, can influence female dominance ranks (Pereira, 1989). In savanna and hamadryas baboons and in Thomas's langurs (and other langurs), male protection against sexual coercion by other males, including infanticide, is clearly important to females (Smuts & Smuts, 1993; Palombit *et al.*, 1997; Sterck, 1997).

Mountain gorillas and other mammals

Female–female and female–male social relationships in mountain gorillas strongly resemble those in gregarious equids (Watts, 1994*a*, 1996). This presumably stems from similarities in feeding ecology (reliance on found sources that are small, but densely and evenly distributed and perennially available) and in female need for male protection against sexual coercion and other forms of harassment by other males who are not their mates. Male protection against predators is more important in some equids (e.g. plains zebra: Schaller, 1972). Female transfer occurs regularly in gregarious equids, and male plains zebra (Klingel, 1969; Schilder, 1990), wild horses (Berger, 1986; Rubenstein, 1986), and feral ponies (Rutberg, 1990; Rutberg & Greenberg, 1990) protect female immigrants against harassment by resident females and make control interventions in other conflicts between females. Male policing, combined with generally low incentives to contest access to food, may be responsible for the absence of clear dominance hierarchies or of rank effects on female reproduction in many equids (e.g. wild horses: Stevens, 1990; Berger, 1996; plains zebra: Schilder, 1990; wild ponies: Rutberg & Greenberg, 1990; but see Duncan, 1992, for rank effects on reproductive success in Camargue horses, and Lloyd & Rasa, 1989, for Cape Mountain zebra).

Future directions

Although we have many indications that females compete for proximity to males, and thus might be competing for access to male services, the consequences of this competition are unclear. The "male services" hypothesis predicts that females who lose contests near males relatively often are more consistently on group edges than are other females (Watts, 1994*a*). However, males may prevent consistent differences from arising directly from the outcome of contests near themselves by intervening in these (see above). In this case, females who tend to lose contests that occur out of proximity to males may be forced into less safe positions, although they may also seek male proximity immediately after contests. Quantitative data on group spread and on the positions of individual females within groups are needed to test these predictions.

The social relationships of juvenile and young adolescent females with peers and with adults of both sexes deserve considerably more attention. Such data could address questions about variation in natal transfer and about whether variation in the relationships of juvenile and adolescent females with males predicts future mating in multimale groups. More detailed analysis of data on variation in male–female relationships could address similar questions.

Natal transfer does not seem to delay age at first reproduction (Watts, 1996), although this conclusion is based on a small sample size. Whether secondary transfer leads to reproductive delays also deserves more attention.

More informative intra-specific comparisons await better data on social relationships in other gorilla subspecies. Close attention to studies of colobine monkeys and of equids, in particular, will be useful for comparative purposes.

Conclusions

1. Social relationships between female mountain gorillas are differentiated: some dyads are much more tolerant of each other, and have higher rates of affiliative interaction and lower rates of affiliative interaction, than others. Also, some females form coalitions, and some do this consistently and often enough to be considered allies.
2. Genealogical relatedness has the strongest influence on relationships, with the effects of maternal relatedness generally stronger than those of paternal relatedness. Most non-relatives have few affiliative interactions and many are also intolerant of each other, although a few have comparatively friendly relationships.
3. Agonistic relationships among females are individualistic and egalitarian. Many females have high rates of agonistic interaction. Aggression is commonly bidirectional in many female dyads, and females also often intervene in contests against those who intervene against them in turn. Many dyads do not establish dominance relationships, and relatives have no consistent effect on each other's agonistic relationships. These characteristics are expected, given that contest competition for access to food is relatively infrequent and seems not to affect female fitness. Males intervene in many contests between females and usually end these without either opponent winning. Male control of female aggression promotes egalitarianism and may help males to retain females.

4. Transfer imposes slight costs, at most, on females because alliances between relatives offer limited benefits and because resident females usually offer little resistance to potential immigrants. Immigrants to large groups are more likely to meet resistance, but male control of female aggression discourages it. The advantages of choosing males who can effectively protect females against infanticide and other risks outweigh those of remaining with female relatives.

References

Abbeglen, J J (1984) *On Socialization in Hamadryas Baboons.* Cranbury NJ: Associated University Presses.

Berger, J (1986) *Wild Horses of the Great Basin.* Chicago: University of Chicago Press.

Bermejo, M (1997) Study of western lowland gorillas in the Lossi Forest of North Congo and a pilot gorilla tourism plan. *Gorilla Conservation News*, 11, 6–7.

Boinski, S (1999) The social organization of squirrel monkeys: implications for ecological models of social organization. *Evolutionary Anthropology*, 8, 101–11.

Bradbury, J & Vehrencamp, S (1977) Social organization and foraging in emballonurid bats. *Behavioral Ecology and Sociobiology*, 2, 1–17.

Byrne, R B, Whiten, A & Henzi, S P (1989) Social relationships of mountain baboons: leadership and affiliation in a non-female bonded monkey. *American Journal of Primatology*, 18, 191–207.

Castles, D L & Whiten, A (1998) Post-conflict behavior of wild olive baboons. I. Reconciliation, redirection, and consolation. *Ethology*, 104, 126–47.

Chapais, B (1992) The role of alliances in social inheritance of rank among female primates. In *Coalitions and Alliances in Humans and Other Primates*, ed. A H Harcourt & F B M de Waal, pp. 29–59. Oxford: Oxford University Press.

Colmenares, F & Lazaro-Perea, C (1994) Greeting and grooming during social conflicts in baboons: strategic uses and social functions. In *Current Primatology*, vol. 2, *Social Development, Learning, and Behavior*, ed. J J Roeder, B Thierry, J R Anderson & N Herrenschmidt, pp. 165–74. Strasbourg: Université Louis Pasteur.

Doran, D M & McNeilage, A (1998) Gorilla ecology and behavior. *Evolutionary Anthropology*, 6, 120–31.

Dunbar, R I M (1988) *Primate Social Systems.* Ithaca NY: Cornell University Press.

Duncan, P (1992) *Horses and Grasses.* New York: Springer-Verlag.

Fossey, D (1984) Infanticide in mountain gorillas (*Gorilla gorilla beringei*) with comparative notes on chimpanzees. In *Infanticide: Comparative and Evolutionary Perspectives*, ed. G Hausfater & S B Hrdy, pp. 217–36. Chicago: Aldine Press.

236   *David P. Watts*

Fossey, D & Harcourt, A H (1977) Feeding ecology of free-ranging mountain gorillas (*Gorilla gorilla beringei*). In: *Primate Feeding Ecology*, ed. T H Clutton-Brock, pp. 415–47. New York: Academic Press.

Harcourt, A H (1978) Strategies of emigration and transfer by female primates, with special reference to mountain gorillas. *Zeitschrift für Tierpsychologie*, **48**, 401–20.

Harcourt, A H (1979a) Social relationships among adult female mountain gorillas. *Animal Behaviour*, **27**, 251–64.

Harcourt, A H (1979b) Social relationships between adult male and female mountain gorillas. *Animal Behaviour*, **27**, 325–42.

Harcourt, A H & Stewart, K J (1987) The influence of help in contests on dominance rank in primates: hints from gorillas. *Animal Behaviour*, **35**, 182–90.

Harcourt, A H & Stewart, K J (1989) Functions of alliances in contests within wild gorilla groups. *Behaviour*, **109**, 176–90.

Harcourt, A H, Stewart, K J & Fossey, D (1976) Male emigration and female transfer in wild mountain gorillas. *Nature*, **263**, 226–7.

Hemelrijk, C K (1990) Models of, and tests for, reciprocity, unidirectionality, and other social interaction patterns at group level. *Animal Behaviour*, **39**, 1013–29.

Hemelrijk, C K & Ek, A (1991) Reciprocity and interchange of grooming and 'support' in captive chimpanzees. *Animal Behaviour*, **41**, 923–35.

Isbell, L A & Pruetz, J D (1998) Differences between vervets (*Cercopithecus aethiops*) and patas monkeys (*Erythrocebus patas*) in agonistic interactions between adult females. *International Journal of Primatology*, **19**, 837–55.

Janson, C H (1992) Evolutionary ecology of primate social structure. In *Evolutionary Ecology and Human Behavior*, ed. E A Smith & B Winterhalder, pp. 95–130. Chicago: Aldine Press.

Kappeler, P K (1999a) Primate socioecology: new insights from males. *Naturwissenschaften*, **85**, 18–29.

Kappeler, P K (1999b) Lemur social structure and convergence in primate social socioecology. In *Comparative Primate Socioecology*, ed. P C Lee, pp. 273–99. Cambridge: Cambridge University Press.

Klingel, J H (1967) Sociale Organisation und Verhaltung freilebender Steppenzebras (*Equus quagga*). *Zeitschrift für Tierpsychologie*, **24**, 580–624.

Koenig, A, Beise, J, Chalise, M & Ganzhorn, J (1998) When females should contest for food: testing hypotheses about resource density, distribution, size, and quality with hanuman langurs. *Behavioral Ecology and Sociobiology*, **42**, 225–37.

Kummer, H (1968) *Social Organization of Hamadryas Baboons*. Chicago: University of Chicago Press.

Lima, S L (1989) Iterated prisoner's dilemma: an approach to evolutionarily stable cooperation. *American Naturalist*, **124**, 828–34.

Lloyd, P H & Rasa, O A E (1989) Status, reproductive success, and fitness in Cape

Mountain zebras (*Equus zebra zebra*). *Behavioral Ecology and Sociobiology*, **25**, 411–20.

McNeilage, A J (1995) Mountain gorillas in the Virunga Volcanoes: feeding ecology and carrying capacity. PhD thesis, University of Bristol.

Mitchell, C, Boinski, S & van Schaik, C P (1991) Competitive regimes and female bonding in two species of squirrel monkeys (*Saimiri oerstedi* and *S. sciureus*). *Behavioral Ecology and Sociobiology*, **28**, 55–60.

Nunn, C (1999) The number of males in primate social groups: a comparative test of the socioecological model. *Behavioral Ecology and Sociobiology*, **46**, 1–13.

Palombit, R A, Seyfarth, R M & Cheney, D L (1997) The adaptive value of friendships to female baboons: experimental and observational evidence. *Animal Behaviour*, **54**, 599–614.

Pereira, M E (1989) Agonistic interactions of juvenile savanna baboons. II. Agonistic support and rank acquisition. *Ethology*, **80**, 152–71.

Plumptre, A (1993) The effects of trampling damage by herbivores on the vegetation of the Parc National des Volcans, Rwanda. *African Journal of Ecology*, **32**, 115–29.

Pusey, A E & Packer, C (1987) Dispersal and philopatry. In *Primate Societies*, ed. B B Smuts, D L Cheney, R M Seyfarth, R W Wrangham & T T Struhsaker, pp. 250–66. Chicago: University of Chicago Press.

Pusey, A E, Williams, J & Goodall J (1998) The influence of dominance rank on the reproductive success of female chimpanzees. *Science*, **277**, 828–31.

Rubenstein, D I (1986) Ecology and sociality in zebras and horses. In *Ecological Aspects of Social Evolution*, ed. D I Rubenstein & R W Wrangham, pp. 469–87. Princeton: Princeton University Press.

Rutberg, A T (1990) Inter-group transfer in Assateague pony mares. *Animal Behaviour*, **40**, 945–52.

Rutberg, A T & Greenberg, S A (1990) Dominance, aggression frequencies, and modes of aggressive competition in feral pony mares. *Animal Behaviour*, **40**, 322–31.

Schaller, G (1972) *The Serengeti Lion*. Chicago: University of Chicago Press.

Schilder, M B H (1990) Social behaviour and social organization of a herd of plains zebra in a safari park. PhD thesis, University of Utrecht.

Sicotte, P (1993) Inter-group encounters and female transfer in mountain gorillas: influence of group composition on male behavior. *American Journal of Primatology*, **30**, 21–36.

Sicotte, P (1994) Effects of male competition on male–female relationships in bi-male groups of mountain gorillas. *Ethology*, **97**, 47–64.

Sicotte, P (2000) A case of mother–son transfer in mountain gorillas. *Primates*, **41**, 95–103.

Smuts, B B (1985) *Sex and Friendship in Baboons*. Chicago: Aldine Press.

Smuts, B B & Smuts, R W (1993) Male aggression and sexual coercion of females in nonhuman primates and other mammals: evidence and theoretical implications. *Advances in the Study of Behavior*, **22**, 1–63.

Sokal, R R & Rohlf, F J (1995) *Biometry*, 3rd edn. New York: WH Freeman.

Steenbeek, R (1997) What a maleless group can tell us about the constraints on female transfer in Thomas's langurs. *Folia Primatologica*, **67**, 169–81.

Sterck, E H M (1997) Determinants of female dispersal in Thomas langurs. *American Journal of Primatology*, **42**, 179–98.

Sterck, E H M (1998) Female dispersal, social organization, and infanticide in langurs: are they linked to human disturbance? *American Journal of Primatology*, **44**, 235–54.

Sterck, E H M & Steenbeek, R (1997) Female dominance relationships and food competition in the sympatric Thomas langur and long-tailed macaque. *Behaviour*, **134**, 749–74.

Sterck, E H M, Watts, D P & van Schaik, C P (1997) The evolution of social relationships in female primates. *Behavioral Ecology and Sociobiology*, **41**, 291–309.

Stevens, E F (1990) Instability of harems of feral horses in relation to season and presence of subordinate stallions. *Behaviour*, **112**, 149–61.

Stewart, K J & Harcourt, A H (1987) Gorillas: variation in female relationships. In *Primate Societies*, ed. B B Smuts, D L Cheney, R M Seyfarth, R W Wrangham & T T Struhsaker, pp. 155–64. Chicago: University of Chicago Press.

Tutin, C E G (1996) Ranging and social structure of lowland gorillas in the Lopé Reserve, Gabon. In *Great Ape Societies*, ed. W C McGrew, L F Marchant & T Nishida, pp. 58–70. Cambridge: Cambridge University Press.

van Schaik, C P (1983) Why are diurnal primates living in groups? *Behaviour*, **87**, 120–44.

van Schaik, C P (1989) The ecology of social relationships amongst female primates. In *Comparative Socioecology*, ed. V Standen & R Foley, pp. 195–218. Oxford: Blackwell Scientific Publications.

van Schaik, C P (1996) Social evolution in primates: the role of ecological factors and male behavior. *Proceedings of the British Academy*, **88**, 9–31.

van Schaik, C P & Kappeler, P K (1997) Infanticide risk and the evolution of male–female association in primates. *Proceedings of the Royal Society, London*, **264**, 1687–94.

Vedder, A L (1984) Movement patterns of a group of free ranging mountain gorillas (*Gorilla gorilla beringei*) and their relation to food availability. *American Journal of Primatology*, **7**, 73–88.

de Waal, F B M (1989) Dominance style and primate social organization. In *Comparative Socioecology*, ed. V. Standen & R. Foley, pp. 243–63. Oxford: Oxford University Press.

de Waal, F B M & Luttrell, L (1988) Mechanisms of social reciprocity in three primate species: symmetrical relationship characteristics or cognition? *Ethology and Sociobiology*, **9**, 101–18.

de Waal, F B M & Luttrell, L (1989) Toward a comparative socioecology of the genus *Macaca*: different dominance styles in rhesus and stumptail monkeys. *American Journal of Primatology*, **19**, 83–109.

Watts, D P (1984) Composition and variability of mountain gorilla diets in the central Virungas. *American Journal of Primatology*, **7**, 325–56.

Watts, D P (1985) Relations between group size and composition and feeding competition in mountain gorilla groups. *Animal Behaviour*, **33**, 72–85.

Watts, D P (1988) Environmental influences on mountain gorilla time budgets. *American Journal of Primatology*, **15**, 295–312.

Watts, D P (1989) Infanticide in mountain gorillas: new cases and a reconsideration of the evidence. *Ethology*, **81**, 1–18.

Watts, D P (1990) Ecology of gorillas and its relationship to female transfer in mountain gorillas. *International Journal of Primatology*, **11**, 21–45.

Watts, D P (1991a) Strategies of habitat use by mountain gorillas. *Folia Primatologica*, **56**, 1–16.

Watts, D P (1991b) Harassment of immigrant female mountain gorillas by resident females. *Ethology*, **89**, 135–53.

Watts, D P (1992) Social relationships of immigrant and resident female mountain gorillas. I. Male–female relationships. *American Journal of Primatology*, **28**, 159–81.

Watts, D P (1994a) Social relationships of immigrant and resident female mountain gorillas, II. Relatedness, residence, and relationships between females. *American Journal of Primatology*, **32**, 13–30.

Watts, D P (1994b) Agonistic relationships of female mountain gorillas. *Behavioral Ecology and Sociobiology*, **34**, 347–58.

Watts, D P (1994c) The influence of male mating tactics on habitat use in mountain gorillas (*Gorilla gorilla beringei*). *Primates*, **35**, 35–47.

Watts, D P (1995a) Post-conflict social events in wild mountain gorillas. I. Social interactions between opponents. *Ethology*, **100**, 158–74.

Watts, D P (1995b) Post-conflict social events in wild mountain gorillas. II. Redirection and consolation. *Ethology*, **100**, 175–91.

Watts, D P (1996) Comparative socioecology of gorillas. In *Great Ape Societies*, ed. W C McGrew, T Nishida & L A Marchant, pp. 16–28. Cambridge: Cambridge University Press.

Watts, D P (1997) Agonistic interventions in wild mountain gorilla groups. *Behaviour*, **134**, 23–57.

Watts, D P (1998a) Long-term habitat use by mountain gorillas (*Gorilla gorilla beringei*). I. Consistency, variation, and home range size and stability. *International Journal of Primatology*, **19**, 651–80.

Watts, D P (1998b) Long-term habitat use by mountain gorillas (*Gorilla gorilla beringei*). II. Re-use of foraging areas in relation to resource abundance, quality and depletion. *International Journal of Primatology*, **19**, 681–702.

Watts, D P (2000a) Mountain gorilla habitat use strategies and group movements. In *Group Movements: Patterns, Processes, and Cognitive Implications in Primates and Other Animals*, ed. S Boinski & P Garber, pp. 351–74. Cambridge: Cambridge University Press.

Watts, D P (2000b) Causes and consequences of variation in the number of males in mountain gorilla groups. In *Primate Males*, ed. P Kappeler. Cambridge: Cambridge University Press.

Watts, D P, Colmenares, F & Arnold, K (2000) Redirection, consolation, and male policing: how targets of aggression interact with bystanders. In *Natural*

Conflict Resolution, ed. F Aureli & F B M de Waal, pp. 281–301. Berkeley: University of California Press.

Wrangham, R W (1980) An ecological model of female-bonded primate groups. *Behaviour*, **75**, 262–300.

9 Vocal relationships of wild mountain gorillas

ALEXANDER H. HARCOURT & KELLY J. STEWART

Beetsme as a young silverback in 1981 giving hoot series and chest-beat. (Photo by Alexander H. Harcourt.)

Introduction

Schaller (1963) and Fossey (1983) pioneered field study of gorillas, including study of the gorillas' vocal communication (Schaller, 1963: ch. 5; Fossey, 1972). Their analyses were mostly of the males' calls, indeed mostly of males' loud, aggressive calls. We know now that gorillas give more or less benign within-group calls, close-calls, far more frequently than they roar, or scream, or bark (Harcourt et al., 1993). However, Fossey and Schaller were absolutely correct in identifying the long-calls and aggressive calls as frequent, and most often given by males – in the relatively unhabituated groups that they studied. Thanks to their lead, subsequent observers of gorillas could concentrate on the more subtle within-group behaviors, and do so with habituated groups.

Fossey and Schaller were both interested mainly in the social implications of the calling, the sociology of vocal communication. How did calls affect the behavior of other animals? Did they affect different animals differently? In this, they were unusual. Most analyses of non-vocal behaviors, grooming, proximity, aggression, had been sociological: how were the behaviors used to negotiate interactions, relationships, access to resources, and so on (Hinde, 1976, 1983). By contrast, the study of primates' vocal communication has been far more concerned with its semantics, with discovering connections between the nature of a call and its referent in the environment, than with its social implications (Markl, 1985; Marler & Mitani, 1988). "While broadcast communication has often been thoroughly analyzed, We know much less about that interesting realm of – private – communication which is at the very heart of complex and highly organized animal societies" (Markl, 1985: 174). While major exceptions exist (Snowdon, 1988; Cheney & Seyfarth, 1990), nevertheless, publications on social behavior, or social relationships, still rarely include consideration of vocal behavior. Indeed, within the apes, analysis of calling is largely of long-calls, or description of close-call repertoire (Harcourt et al., 1993; Mitani, 1996). Thus Goodall's (1986) qualitative description of chimpanzee close-calls has not yet been bettered, and Clark's (1993) analysis of repertoire by dominance was largely an analysis by age–sex class, although he did separate adult males into two classes of rank. This social analysis of gorilla close-calls is more feasible than for other apes' close-calls because the Virunga gorillas live in relatively stable groups that are easily observable, by contrast to the more solitary chimpanzee species and the orangutan. Furthermore, gorillas vocalize far more frequently than do chimpanzees or orangutans, presumably because the gorillas are in groups (Clark, 1993; Mitani, 1996) (we have no data on close-call rate in wild

242

bonobos, in part because political instability of the region has so curtailed field studies).

Following Schaller and Fossey, our studies of gorillas' vocal communication has been strongly sociological. Thus we here treat vocal communication like other social behaviors, as another means of negotiating with friends and enemies. All individuals' behavior is usually contingent on what others have done, are doing, and will do (Hinde, 1981; Markl, 1985; Smith, 1997). Animals, including humans, use proximity, friendly behavior, grooming, threats, submissive behaviors to negotiate their way through the day, to assess and manipulate others' intentions (potential future behavior) by providing information about their own intentions (Hinde, 1981, 1983; Markl, 1985; Hauser & Nelson, 1991; Smith, 1997; Owings & Morton, 1998). This chapter is a brief summary of our attempts to understand how gorillas use vocal communication in the same way (Hinde, 1981, 1983; Markl, 1985; Snowdon, 1988; Dunbar, 1996; Smith, 1997; Owings & Morton, 1998). We begin by summarizing the social relationships of gorilla groups, and then relate the use of close-calls with other aspects of social relationships as judged by the normal measures of proximity, grooming, supplanting aggression, and so on. If the distribution of close-calls among individuals is similar to the distribution of other behaviors, if close-calling maps onto the social system as judged by other behaviors, then presumably close-calls are being used like other behaviors, to assess and manipulate others. Rank and consanguinity are major organizing principles in understanding the behavior of animals, matching as they do contrasts in competitive ability and payoffs of competition and cooperation. We show that both rank and consanguinity correlate with the use of close-calls by gorillas, as expected if calls are used to assess and manipulate the future behavior of others. How the assessment and manipulation occur, the meaning of the call, is our final concern. Throughout, we draw comparisons where appropriate with studies of close-calling of other apes.

Data sources

All the data come from the Karisoke Research Center study groups, Group 5 and NK, in the early 1980s. Group 5 contained two silverbacks, three to four adult females, three of them maternal relatives (Ef, Pu, Tu), and several immature animals; Group NK contained one silverback, six adult females, only two of them known relatives (Pe, Au), although four of them were from a single previous group, Group 4 (Pe, Au, Ps, Sm); and several young immature animals (see chapter 7, this volume for further details).

A brief description of gorillas' close-calls

Harcourt *et al.*'s (1993) detailed description of gorilla close-calls complements Fossey's (1972) analysis of mostly long-calls of gorillas, and is complemented by Seyfarth *et al.*'s (1994) further acoustical and social analysis of one of the gorillas' close-calls, the double grunt. The nature of the close-calls will be only briefly summarized here. Harcourt *et al.* (1993) provided quantitative acoustical analysis of five close-calls, along with three overtly agonistic calls, a copulatory grunt, and the play chuckle (laugh); they provided sonograms of nine close-calls, four agonistic calls, and of the copulatory grunt, and the play chuckle. The close-call repertoire of the other apes has yet to be described in such detail. Harcourt *et al.* (1993) divided the gorilla's repertoire into syllabled calls, and non-syllabled. The former are brief, atonal grunts, sounding like a human male clearing his throat; the latter are longer grumbles and tonal calls, sounding like humans humming, for instance. Syllabled calls are divided into double grunts, single grunts, and train grunts (a train of single grunts). They last less than a second on average, the common double grunt about 850 msec. Non-syllabled calls included grumbles, hums, high hums (sometimes termed singing, and very similar to human females' humming), and whines. The grumbles sound like long second syllables of the double grunts, but are significantly acoustically distinguishable on several measures, including duration, which is of the order of 1100 msec for the non-syllabled calls, but can last more than 3 seconds in the case of singing.

The term close-call is used to distinguish the within-group calls from the loud, long-calls. The long-calls of the gorilla and of the other apes are audible at long distances, as the term implies, and obviously so adapted and used (Schaller, 1963; Fossey, 1972; Marler, 1976; Marler & Tenaza, 1977; Mitani & Nishida, 1993; Mitani, 1996). For gorillas, they include the hoot series, given with the chest-beat by males on detecting the presence of a non-group male. It appears functionally equivalent to the chimpanzees' pant-hoot, which is also used in long-distance contact, and in some ways is acoustically similar in that the calls are loud and tonal (Marler, 1976; Mitani, 1996). Gorilla long-calls (loud-calls) also include alarm calls of various sorts, for instance the "bark", which is given by both gorilla and chimpanzee on detecting potential danger, as from strange humans. And loud agonistic calls are included too, for instance the roar given by a male when charging and attacking a potential predator, such as an observer (Schaller, 1963; Fossey, 1972).

Close-calls are also distinguished from other within-group calls that have obvious contexts, if not functions. Thus, copulatory calls are not included as close-calls, even though there is debate about their function (Hauser, 1993; Manson, 1996). While copulation (and therefore presumably copulation calls) of course can be used socially (gorilla males in all-male groups will mount one another and give copulatory calls: Yamagiwa, 1987*a*), social use of mounting is rare in gorillas. Nor do we include in the analysis of close-calls immature animals' whimpers, nor any animals' play chuckles (laughs), or aggressive calls, such as the cough-grunt (Fossey, 1972), because their context and function seems obvious.

Vocal relationships
The main questions here are how the use of close-calls by individuals in gorilla groups correlates with the social interactions and social relationships of the individuals, and also with the social structure of the group or community. How do dominant individuals use close-calls by comparison to subordinate individuals? How do kin use them by comparison to non-kin? And how does the context of their use, and others' reactions to their use, correlate with interpretation of the calls as means of provision of information about the caller's intentions (situation-dependent, potential future behaviors), and assessment and manipulation of others' intentions?

Inter-group relationships and vocal communication
Gorillas live in social groups that are sometimes stable for years (Stewart & Harcourt, 1987; Yamagiwa, 1987*b*; Watts, 1994). The population is thus divided into group members and non-group members. The latter are interacted with at long distance, with loud long-calls, and the former at short distance, with quieter close-calls (Schaller, 1963; Fossey, 1972). Social groups are interpretable as existing because females cluster round a strong male for protection against predators, and because males prevent other males from accessing his group of females and their infants (Harcourt, 1979*b*; Fossey, 1984; Stewart & Harcourt, 1987; Yamagiwa, 1987*b*; Watts, 1992; Sicotte, 1993; Robbins, 1995). Both explanations for grouping imply that the dominant male(s) of the group will play a large part in non-group relationships. Indeed they do. Males, for instance, are the main defenders of females and immatures against predators (Schaller, 1963). And encounters with non-group members are mainly encounters between adult males (Schaller, 1963; Harcourt *et al.*, 1976; Harcourt, 1978*b*; Stewart & Harcourt, 1987; Watts, 1989; Sicotte, 1993). Thus males more than any other age–sex class give alarm calls, anti-predator calls, and

246   *Alexander H. Harcourt & Kelly J. Stewart*

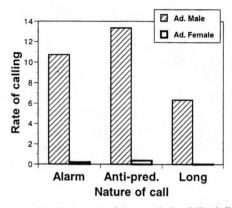

Figure 9.1. Frequency of alarm calls (loud "barks") anti-predator calls (roars, wraaghs, screams), and long-calls (hoot series) given by adult males compared to other adults. Rates expressed per 100 animal hours of observation per age–sex class. Data from Fossey (1972).

long-calls (Figure 9.1) (Schaller, 1963; Fossey, 1972). The same is true of chimpanzees (Marler, 1976; Marler & Tenaza, 1977; Mitani, 1996).

Within-group relationships
Abundant evidence, starting with Schaller, continuing with Fossey, and on up to the present day, indicates that (1) adult male gorillas are strongly dominant to all other age–sex classes; (2) among some adult females, competitive ability is sometimes not strongly differentiated or not differentiated at all, as judged by who supplants whom; and (3) adults are dominant to immature animals, among whom size determines competitive ability (Harcourt, 1979a, b; Stewart & Harcourt, 1987; Yamagiwa, 1987b; Watts, 1992, 1994). The evidence also indicates (4) little differentiation of friendly relationships among gorilla females, by comparison to the obviously differentiated relationships of, e.g. baboon *Papio*, macaque *Macaca*, and vervet *Chlorocebus* females, if not other apes (chapters in Smuts *et al.*, 1987). Nevertheless, (5) some competition for access to the male is evident in large gorilla groups; mothers with young infants, and perhaps females of long-term residence, stay closer to the male for longer than do others to the male; and (6) adult female kin are less aggressive, and more friendly with one another than they are with non-kin (e.g. groom one another more) (Smuts *et al.*, 1987).

Within-group vocal communication
Animals within a social group are, in effect, in 24-hour interaction with one another. In the case of mountain gorillas in groups of the average size of

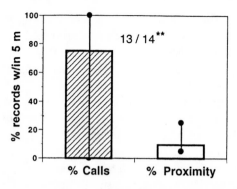

Figure 9.2. Proportion of calls given by adults when at least one other animal was within 5 m compared to proportion of time that at least one other was within 5 m. Histograms and lines show median and range of adults' values: 13 of 14 adults had higher proportion of calls within 5 m than expected by chance ($p < 0.02$, binomial test). Data are percent calls and per cent proximity records (scan samples per 5 minutes) during feeding periods when another animal was within 5 m. Results from two groups, Group 5 and NK. Details in Harcourt *et al.* (1993).

10–15 members (Schaller, 1963; Harcourt *et al.*, 1981), individuals are very rarely more than 50 m from one another, for months on end. Social interactions within groups are far more frequent than those between groups. And probably the most frequent social behavior is close-calling. Adult gorillas in two Karisoke study groups, Groups 5 and NK, called at a median rate per individual of about once every 8 minutes (Harcourt *et al.*, 1993). This rate does not include the "chuckles' of playing animals (Fossey, 1972), nor the "cough-grunts' given in mildly aggressive encounters (Fossey, 1972). These rates are far higher than for chimpanzees in parties, and certainly than for chimpanzees overall (Clark, 1993; Mitani, 1996).

Early suggestions were that the gorillas' close-calls were contact calls, given to prevent animals losing one another in the thick vegetation (Schaller, 1963). Many of the gorilla calls are individually distinctive acoustically (Seyfarth *et al.*, 1994). They could thus certainly function to keep dispersed individuals in contact in thick vegetation, especially as the adult male's calls are so distinctive. However, one of the crucial aspects of the context of close-calling by gorillas is that animals call when other animals are very close by, and often fully visible. Most grunts are given and exchanged when others are within 5 m (Figure 9.2), indeed within 2 m, as is the case for almost all other social behaviors and interactions. Furthermore, the concept of calls functioning to maintain contact between spatially separated group members has been undermined by Cheney *et al.*'s (1996) demonstration that, among other tests, baboons separated from the bulk of the group were no more likely to be answered than baboons within the group, even by their own kin. That is not to say that group cohesion is not coordinated by vocalizations, as is evident by the high rate of exchange of close-calls between a spatially tight group of gorillas just before the coordinated end of a rest period (Stewart & Harcourt, 1994). Rather, the frequent calls one hears group-living primates give are probably not primarily given to prevent separation of individuals from the group. Something else is going on. The contrast between the high rate of close-calling by individuals of the group-living gorilla and the relatively low rate by the more solitary chimpanzee along with its high rate of long-calls emphasizes the correlation between social system and the behavior of vocalizing (Mitani, 1996).

Species appear to differ in the frequency with which close-calls are exchanged. Baboons, for instance, rarely exchange calls (Cheney *et al.*, 1996). Other species so often do so that they are even described as conversing (Snowdon & Cleveland, 1984). About 60% of all calls given by gorillas were given in exchange (Harcourt & Stewart, 1996). With an average interval between calls of about half a second (spectrographically measured), the second call is clearly a response to the first, and not a coincidence arising from generally high rates of calling (Harcourt *et al.*, 1993).

In sum, if maintenance of contact is a function of the close-calls of gorillas, it is an infrequent function (cf. Cheney *et al.*, 1996). Instead, the calls are given in just the situation expected were their function to modulate social interactions, i.e. they are given and exchanged frequently when individuals are close to one another.

Various monkey calls, and not just their alarm calls, have been shown to be effectively semantic. In other words, the nature of the call is so closely tied to an event in the environment that the calls carry information about that event, independently of any other information than is carried in the call itself. Cheney & Seyfarth's (1990) work cannot be bettered for demonstrating such semantic equivalency. For instance, vervet monkeys (*Chlorocebus aethiops*) grunt in various contexts. Some of the grunts are acoustically distinguishable, and the hearers react to them as if the grunts convey information about external referents. Playback of a grunt given on noticing another group causes the hearer to look to the horizon in the direction the speaker is pointed; playback of a dominant-approaching-subordinate grunt given by a group member dominant to the hearer causes the hearer to look toward the speaker (Cheney & Seyfarth, 1982). In gorillas groups, double grunts are the close-call most often given by adults (Harcourt *et al.*, 1993), and are individually distinctive (Seyfarth *et al.*, 1994). To search for correlates of context (potential external referent) and call, and thus potential precursors to semanticity, Seyfarth *et al.* (1994) examined double-grunting in relation to maintenance activity (moving, feeding, resting), approaching dominant or subordinates, grooming, proximity to others, imminent (next 1 minute) change in activity, and in relation to position in exchange of calls. The single acoustical feature of double grunts that could be related to behavior was that double grunts given in answer to a previous grunt had a higher-pitched second formant (more than 1600 Hz) than did spontaneous double grunts, i.e. those given after a period of silence (Seyfarth *et al.*, 1994). For the most part, therefore, it seems that the meaning of gorilla close-calls is conveyed by context more than by content. In other words, call plus context, not call alone, correlates with the response to the call. The same seems true of chimpanzee calls (Clark & Wrangham, 1993; Mitani, 1996). Part of that context is the social relationships of the group members.

Within-group vocal relationships

With respect to the nature of social relationships within a group, the very dominant males often control other group members, and also reassure them. Other animals need to negotiate more often with the adult males than with any other group member, because they will suffer more if they make an error in interaction with the male than they will with one another. And the behavior of immature animals, subordinate to all, matters little to anyone except other immatures, and little to other immatures, except in play. Correspondingly, adult males give various close-calls more than do other adults, who give them more than do immature animals (Figure 9.3). The nature of the calls differs among the age–sex classes, too (Figure 9.4) (Fossey, 1972), a contrast that is considered in more detail below in relation to rank differences among individuals. Clark (1993) found no consistent differences between Kibale chimpanzee age–sex classes in frequency of all calls when all social situations were lumped, a contrast with previous findings when long-calls were the predominant call type examined. However, chimpanzee party composition correlated with frequency of calling. For instance, adult females called more often when more than one adult male was present in the party (Clark, 1993).

Rank and vocal relationships

A large proportion of the variance in social relationships of probably any species, including humans, can be explained by the two components of differences in competitive ability (dominance rank) and differences in consanguinity. In gorilla groups, the nature of close-calls varies with the dominance rank of the callers. Dominant individuals give more syllabled

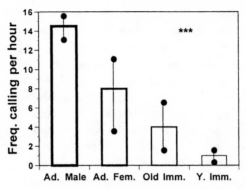

Figure 9.3. Age–sex classes differences in frequency of close-calling. Results from two groups, 5 and NK. Histograms and lines show median and range of values (Kruskal–Wallis one-way ANOVAR: d.f. = 4, H = 21.0, p <0.001). n = 3 adult males, 9 adult females, 6 older immatures (4–8 yr), and 6 younger immatures (1.25–4 yr). Details in Harcourt *et al.* (1993), including demonstration of similarity in results from the two groups.

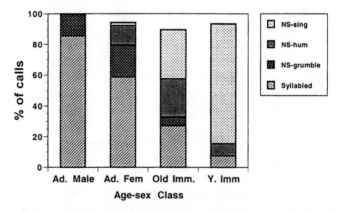

Figure 9.4. Age–sex class differences in nature of most frequent close-calls. Results from two groups, 5 and NK, showing mean of the two groups' medians. Syllabled calls are brief (< 1 s), atonal "syllabled" grunts, the most common of which is the double grunt; "non-syllabled" calls are longer (> 1 s), and some are tonal. Details in Harcourt *et al.* (1993), including demonstration of similarity in results from the two groups, and Seyfarth *et al.* (1994).

calls, especially double grunts, while subordinates give more non-syllabled calls, for instance grumbles and hums (Figure 9.5). Females grumble especially frequently and intensely when close to an adult male who has just displayed; indeed they often approach the male, grumbling intensely. In chimpanzees too, age–sex classes of different rank have different repertoires, with adult females giving more frequent submissive calls than do

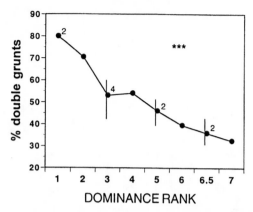

Figure 9.5. Proportion of all close-calls that are double grunts in relation to dominance rank of caller. Median and range of individual's median values shown; numbers by values are number of individuals when not one (Spearman correlation coefficient, $r_s = 0.98$, d.f. = 6, $p < 0.001$). Results from two groups, 5 and NK. Details in Harcourt & Stewart (1996), including demonstration of similarity in results from the two groups. Data for adults only.

adult males, and subordinate adult males giving submissive calls more frequently than do dominant adult males (Clark, 1993).

Despite this correlation between dominance rank and nature of calling, double grunts cannot be simply considered as dominance signals, nor non-syllabled calls as subordinance signals, because both are given by either partner. Additionally, in the case of double grunts anyway, no acoustical difference separates double grunts given by a dominant in the presence of a subordinate from those given by a subordinate in the presence of a dominant (Seyfarth *et al.*, 1994). Such lack of rank-related specificity contrasts with some of the grunts of vervet monkeys (Cheney & Seyfarth, 1982).

Adult gorillas of both sexes exchanged calls about twice as often with the fully adult males as they did with adult females (Harcourt & Stewart, 1996), because both sexes were readier to respond to males than to females (as opposed to the alternative of silverbacks being more responsive than other adults). Overall, however, rank did not correlate with rate or probability of exchanging. While rank correlated with the proportion of unexchanged calls in another animal's presence (within 5 m) that were doubles (Figure 9.6), the statistical significance of this correlation is doubtful, because we tested several possible variations in calling in relation to rank, for instance, rates of syllabled calls, non-syllabled calls, exchanged syllabled, unexchanged syllabled, and so on.

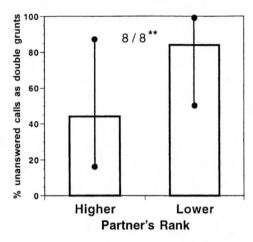

Figure 9.6. Proportion of unanswered calls given during group feeding periods
that were double grunts in relation to rank of neighbor. Median and range of
individual's median values shown. All eight individuals showed the same
direction of difference (p <0.01, binomial test). Only individuals who had both
higher and lower ranked partners are included, i.e. the dominant adult male, and
the most subordinate female are excluded. Results from two groups, 5 and NK.
Details in Harcourt & Stewart (1996), including demonstration of similarity in
results from the two groups.

Consanguinity and vocal relationships

Related individuals in gorilla groups are, like related individuals in other
primate groups, generally friendlier with one another than they are with
unrelated individuals: friendly behaviors, such as peaceful proximity and
grooming, are more frequent among kin; aggression, especially intense
aggression, is less frequent (Stewart & Harcourt, 1987). However, and in
contrast with the effect of rank of neighbor, gorillas did not vary the nature
of their close-calling (the ratio, or frequency, of syllabled and non-syllabled calls) depending on the consanguinity of their neighbor (the animal(s)
within 5 m). Nor did they vary the overall frequency of close-calling
depending on consanguinity of their neighbor. However, kin in proximity
exchanged calls at a higher rate than did neighboring non-kin (Figure 9.7).
According to the interpretation that calls are, like other social behaviors,
signals that provide and elicit information about conditional intentions
(see 'Introduction', above), related partners were not readier than were
unrelated partners to initiate exchange of information, but were readier to
negotiate once one partner had started the process of negotiation.

The outcome of vocal interactions

We have suggested that the syllabled double grunts serve primarily to draw

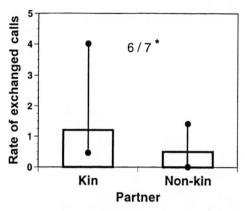

Figure 9.7. Rate of exchanged calls when within 5 m of kin compared to non-kin during group feeding periods. Median and range of individuals' median values shown. Seven of eight individuals showed the same direction of difference (Wilcoxon matched pairs, signed ranks test, $T = 27, p <0.03$). Results from two groups, 5 and NK. Details in Harcourt & Stewart (1996), including demonstration of similarity in results from the two groups.

attention to the caller, in other words to initiate the negotiation process by signaling conditional intention, and that subsequent behavior is then determined by the context (Stewart & Harcourt, 1994; Harcourt & Stewart, 1996). In the usual context of relaxed feeding, no observable reaction is the normal result, as appears to be the case with chimpanzees' close-calls (Clark, 1993). Where the context is a mother about to leave an infant, the reaction is often immediate and obvious: the infant follows the mother (Harcourt *et al.*, 1993). One context that clearly illustrates the use of close-calls, especially double grunts, as signals of conditional intention (what an individual does depends on what its partner does), is the coordinated end of day-rest-periods (Stewart & Harcourt, 1994). Gorillas have rest periods around the middle of the day, lasting an hour or more (Schaller, 1963; Fossey & Harcourt, 1977; Harcourt, 1978a). Sometimes the rest periods end as animals one by one drift off and start to feed. Sometimes the end of these rest periods is signaled by the abrupt departure of the leading adult male, with the departure emphasized by a display run through the group (Schaller, 1963; Harcourt, 1979b). Often, though, the synchronized end of the rest period, and the beginning of feeding, or travel, by the group is preceded by an increasingly frequent exchange of double grunts (Figure 9.8). Pre-departure displays, or vocal choruses, are commonplace among animals (Kummer, 1968; Smith, 1977; Black, 1988; Boinski, 1993). Group living, or flocking, or herding animals often indicate, sometimes with exaggerated behavior (i.e. signals) their readiness to

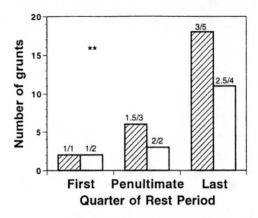

Figure 9.8. Total grunts (syllabled calls) in first, penultimate, and last quarter of rest periods that lasted at least 1 hour. Columns are median of each group's median. Results from two groups, 5 (hatched) and NK. Significantly more calls ($p < 0.01$) were given by significantly more animals ($p < 0.01$) at a significantly higher rate per animal ($p < 0.05$) in both groups in the last compared to the other periods (Wilcoxon matched pairs, signed ranks tests). Numbers above columns show grunts per vocalizer and number of vocalizers. Further details in Stewart & Harcourt (1994), including demonstration of similarity in results from the two groups.

change activity (Smith, 1965, 1977, 1997). They do so, it is argued, both to elicit from others information about their readiness to change, and to stimulate them to change (see 'Introduction', above). Group activity is thereby synchronized, and the group maintained, with the attendant advantages. This coordinated end of a rest period that results from animals in close proximity to one another exchanging calls that, in our interpretation, negotiate exact timing of the end of the rest period, obviously maintains group cohesion. However, the cohesion is equally obviously maintained in a different way than is envisaged by the use of calls as contact signals by individuals in a dispersed group.

Evidence that outcomes of interactions between members of gorilla groups vary with the calls given during the interaction is poor because, as we said, the usual observable outcome is only another call. However, when adult males approach feeding subordinate animals, and calls are exchanged, the subordinate is less likely to be feeding one minute later than if no calls are exchanged (Figure 9.9). If only one of the animals called, no change in the probability of feeding was evident. The nature of the close-calls given in this context appears to be irrelevant, but the adult male's call was almost always a double grunt because the great majority of all his calls are double grunts (see Figure 9.4). In sum, the interaction is interpretable as both animals signaling imminent change of activity. However, to some

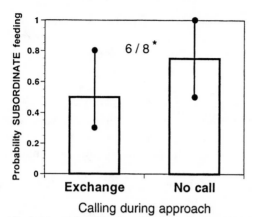

Figure 9.9. Probability of a subordinate feeding after an approach within 5 m by an adult male in relation to calls given during the approach. Six of nine animals showed the same direction of difference; one showed no difference (Wilcoxon matched pairs, signed ranks test, $T = 33$, $p < 0.04$). Data from Group 5 only. Details in Harcourt & Stewart (1996).

extent it should not matter what the subordinate signals: if the dominant signals staying, the subordinate had better stop feeding. Why then does the outcome correlate with an exchange, as opposed to only the dominant signaling? Also, in most studies of other primates, it appears that the dominant's quiet call as it approaches is associated with reassurance, and hence e.g. reduced likelihood of supplanting, or greater likelihood of friendly interaction, such as grooming (Smith *et al.*, 1982; Bauers, 1993; Hauser & Marler, 1993; Cheney *et al.*, 1995; Palombit *et al.*, 1999). However, the situation might be slightly more subtle than indicated. Thus a male baboon's (*Papio cynocephalus*) approach grunt does not influence a female's departure on his approach, but does increase the likelihood that the animals will behave in a friendly fashion (Palombit *et al.*, 1999). Our analysis of consequences of approaches and associated calling was too crude to pick up that distinction.

In the opposite situation, a (subordinate) female leaving a male, males sometimes "neighed" (Sicotte, 1994). Schaller (1963) also recorded a horse-like neigh. However, Fossey (Fossey, 1972) heard it from only one animal, and so unusual was the call, she named the animal after her term for it, "Whinny". In thousands of animal hours of observation in the 1980s in groups with either one or two males, Harcourt *et al.* (1993) never heard such a call. Clearly the use of calls varies, presumably among individuals, possibly over time, and possibly with number of males in the group. In any case, it appeared that after a departure with a neigh compared to one without, the partners were more likely to stay near one another. Thus, males were twice as likely to follow the female, watch her departure, or the female to cease departing (50% of departures compared to 25%), while the alternative of both partners ignoring one another did not vary with neighing (Sicotte, 1994). Although Harcourt & Stewart studied a group with two males in it, and never heard a "neigh", Sicotte (1994) suggests that the neigh is related to competition among males for access to females. Interpretating the neigh in terms of negotiation, it signals intention of the male to stay close, the sequel to which can be either the male or female maintaining proximity.

Discussion

The meaning and function of gorillas' loud long-calls, their alarm calls, their anti-predator calls, their obvious threat calls (Fossey, 1972), are easily interpretable. By contrast, we by no means fully understand the meaning and function of gorillas' close-calls, or indeed any ape's close-calls (Mitani, 1996), despite the correlation of the use of the calls with other measures of the nature of social relationships. As Cheney & Seyfarth (1990) so aptly describe, it is as if we are listening to conversations without knowing their content.

One of the main difficulties in understanding the function and meaning of gorillas' close-calling is that so often, the majority of the time even, no reaction other than an answering call is discernible (Fossey, 1972; Harcourt *et al.*, 1993). Yet, the lack of discernible reaction was one of the main points that Hinde (1981) made about threat displays: usually, the two birds that threatened one another at the winter feeding table remained and fed – as if there had been no exchange of threats. This lack of reaction, or rather lack of attacking or fleeing (because continuing the previous behavior is certainly a possible reaction), was a main observation used by Hinde to argue that exchange of information about conditional intention was the function of threats. The signaler continues its behavior as long as the receiver continues its uncompetitive behavior. As Hinde (1981: 535) put it, the signals (threats in the cases he was discussing) are interpretable as "Will stay [remain at the winter feeding table] unless provoked", in which case would attack or flee depending on the context and the nature of the provocation.

The hypothesis can be extended to communication in general. Almost all the time, for almost all animals, what is beneficial to do depends on what others might do, especially in response to what the actor does (Hinde, 1981; Smith, 1997; Owings & Morton, 1998). Animals thus benefit from imparting information about their intentions, especially if that information (those same signals) stimulates the onlookers to impart information in return (Hinde, 1981), or stimulates the receiver to behave in a way beneficial to the signaler. Coordinated changes of activity in group-living animals is a case in point (Stewart & Harcourt, 1994). With gorilla close-calls, the syllabled calls might function to draw attention to the caller. The functions of announcing intention and manipulating other's behavior are more evident in these calls than is the function of eliciting information about intention. By contrast, the non-syllabled calls indicate that the caller is ready to pay attention. In other words, here the eliciting function is uppermost.

Close-calls take almost no energy to produce. Some discussion has occurred of the problems of the ease of cheating if signals are not costly (Krebs & Dawkins, 1984). However, cheating (and therefore costly signals) is not really an issue when both partners benefit from information about intentions, i.e. about potential future behavior (Hinde, 1981; Hauser, 1996; Owings & Morton, 1998). Certainly, the more important the resource, and the more advantageous it is for each animal to obtain exclusive access to it, the more costly (uncheatable) will be signals about motivation and ability to get access to it. On the other hand, among animals who know one another well, who benefit from continued presence in the same group, and who would suffer from perpetual, unnecessary fighting, signals can be honest and cheap (Hinde, 1981; Cheney & Seyfarth, 1990; de Waal, 1993; Stewart & Harcourt, 1994; Cheney *et al.*, 1995).

To the extent that signaling and exchange of signals function to facilitate social interactions and social relationships in this way, they can be thought of as servicing relationships. Dunbar (1996) has suggested that vocal exchange, and especially language, is a more efficient mode of servicing relationships than is, for example, grooming, because with calls, the servicing can be done both at a distance, and while individuals are involved in other behaviors, in other words getting on with maintenance activities. The same is true of gorilla close-calls: the double grunt is often given with a mouth full of food, while reaching for more food, or moving to a new food source. It is also true of, for example, mild threats: a dog can raise its hackles while continuing to feed. However, gorilla close-calls are very far from being language-like; they seem to be of the order of complexity of threat displays, as indeed do chimpanzee calls. That simplicity raises the question of why apes, popularly considered more intelligent than monkeys, have apparently a simpler mode of communication, in the sense that they apparently do not label the environment by association of specific calls with specific contexts. Apes appear to use the same calls in a wide variety of cirucumstances (Goodall, 1986; Clark, 1993; Harcourt *et al.*,

1993; Mitani, 1996). We have no answer for the contrast, except to speculate that perhaps association of a specific call with a specific context or external referent is not as complex as it might otherwise seem (we see it as complex because that is how humans work), or that use of calls in negotiation through exchange of information is more complex than it seems at first sight. Also, of course, it might be that apes are not more intelligent than monkeys, who certainly are amazingly astute socially (Cheney & Seyfarth, 1990; de Waal & Berger, 2000).

Through the paper, we have commented on contrasts between the gorilla and chimpanzee vocal repertoires. In summary, while in both, the repertoire reflects social relationships, chimpanzees give far fewer close-calls than do gorillas. Such a contrast would be expected if close-calls were important for maintaining social relationships in a stable group (Harcourt *et al.*, 1986; Harcourt *et al.*, 1993; Stewart & Harcourt, 1994; Harcourt & Stewart, 1996; Mitani, 1996). It could be hypothesized that calls that functioned to allow assessment of intentions would be particularly important between animals that did not know one another so well, such as chimpanzees that ranged apart. However, if the animals do not know one another well, then the calls probably have to be more explicit than are the gorilla's quiet and subtle close-calls: hence the chimpanzee's apparently more exaggerated repertoire (Marler, 1976; Marler & Tenaza, 1977; de Waal, 1988; Harcourt *et al.*, 1993; Mitani, 1996).

To close, we suggest three avenues in the future study of gorilla communication. The first is more and better data on the sequelae of changes in social interaction between individuals in relation to the calls given during the interactions, and the identities of the participants. In addition, experimentation with playbacks is vital (Cheney & Seyfarth, 1990). Playbacks in the field are difficult under any circumstances, but perhaps especially difficult in the habitat of gorillas. While no mountain gorillas are in captivity, captive lowland gorillas' give, to our ears, exactly the same calls in exactly the same circumstances as do wild mountain gorillas. Thus they give grunts as they approach others; they grumble in the presence of large amounts of food; mothers grunt as they leave their infant; and so on (see Harcourt *et al.*, 1993). Hundreds of lowland gorillas are in captivity, many in groups large enough, and in facilities large enough, for manipulation of cage membership, and hence experimentation with playbacks. Finally, we know too little about the ontogeny of gorilla vocal behavior (cf. Snowdon & Hausberger, 1997).

Summary
In our work on vocalizations of gorillas, we have concentrated on under-

standing calls as a social behavior, as opposed to understanding them as precursors to language. The use of close-calls (within-group calls) by gorillas correlates closely with the use of other social behaviours by gorillas, we suggest. Thus, strong correlates of social system with calling can be detected. For instance, gorillas, living in stable groups, give far more close-calls than do the more solitary chimpanzee; the dominant adult male gorillas give the most close-calls; dominance rank correlates with the ratio of the two categories of close-call; most adults exchange calls most frequently with the adult males. Our interpretation of calls follows classic ethology in its argument that calls, like other social behaviors, are means of negotiating social interactions, i.e. are means of assessment and manipulation of others' ability and intentions (potential future behavior) by provision of information about one's own ability and intentions.

Acknowledgements

Many thanks to Chuck Snowdon for generously lengthy and detailed comments that greatly improved the paper.

References

Bauers, K A (1993) A functional analysis of staccato grunt vocalizations in the stumptailed macaque (*Macaca arctoides*). *Ethology*, **94**, 147–61.

Black, J (1988) Preflight signalling in swans: a mechanism for group cohesion and flock formation. *Ethology*, **79**, 143–57.

Boinski, S (1993) Vocal coordination of troop movement among white-faced capuchin monkeys, *Cebus capucinus*. *American Journal of Primatology*, **30**, 85–100.

Cheney, D L & Seyfarth, R M (1982) How vervet monkeys perceive their grunts: field playback experiments. *Animal Behaviour*, **30**, 739–51.

Cheney, D L & Seyfarth, R M (1990) *How Monkeys See the World*. Chicago: University of Chicago Press.

Cheney, D L, Seyfarth, R M & Silk, J B (1995) The role of grunts in reconciling opponents and facilitating interactions among female baboons. *Animal Behaviour*, **50**, 249–57.

Cheney, D L, Seyfarth, R M & Palombit, R (1996) The function and mechanisms underlying baboon 'contact' barks. *Animal Behaviour*, **52**, 507–18.

Clark, A P (1993) Rank differences in the production of vocalizations by wild chimpanzees as a function of social context. *American Journal of Primatology*, **31**, 159–79.

Clark, A P & Wrangham, R W (1993) Acoustic analysis of wild chimpanzee pant hoots: do Kibale Forest chimpanzees have an acoustically distinct food arrival pant hoot? *American Journal of Primatology*, **31**, 99–109.

Dunbar, R I M (1996) *Grooming, Gossip, and the Evolution of Language.* Cambridge MA: Harvard University Press.

Fossey, D (1972) Vocalizations of the mountain gorilla (*Gorilla gorilla beringei*). *Animal Behaviour,* **20,** 36–53.

Fossey, D (1983) *Gorillas in the Mist.* London: Hodder & Stoughton.

Fossey, D (1984) Infanticide in mountain gorillas (*Gorilla gorilla beringei*) with comparative notes on chimpanzees. In *Infanticide: Comparative and Evolutionary Perspectives,* ed. G Hausfater & S B Hrdy, pp. 217–36. New York: Aldine Press.

Fossey, D & Harcourt, A H (1977) Feeding ecology of free ranging mountain gorilla. In *Primate Ecology,* ed. T H Clutton-Brock, pp. 415–47. London: Academic Press.

Goodall, J (1986) *The Chimpanzees of Gombe.* Cambridge MA: Belknap Press of Harvard University Press.

Harcourt, A H (1978*a*) Activity periods and patterns of social interaction: a neglected problem. *Behaviour,* **66,** 121–35.

Harcourt, A H (1978*b*) Strategies of emigration and transfer by primates with particular reference to gorillas. *Zeitschrift für Tierpsychologie,* **48,** 401–20.

Harcourt, A H (1979*a*) Social relationships among adult female mountain gorillas. *Animal Behaviour,* **27,** 251–64.

Harcourt, A H (1979*b*) Social relationships between adult male and female mountain gorillas. *Animal Behaviour,* **27,** 325–42.

Harcourt, A H & Stewart, K J (1996) Function and meaning of wild gorilla 'close' calls. II. Correlations with rank and relatedness. *Behaviour,* **133,** 827–45.

Harcourt, A H, Stewart, K J & Fossey, D (1976) Male emigration and female transfer in wild mountain gorilla. *Nature,* **263,** 226–7.

Harcourt, A H, Fossey, D & Sabater Pi, J (1981) Demography of *Gorilla gorilla. Journal of Zoology, London,* **195,** 215–33.

Harcourt, A H, Stewart, K J & Harcourt, D E (1986) Vocalizations and social relationships of wild gorillas: a preliminary analysis. In *Current Perspectives in Primate Social Dynamics,* ed. D M Taub & F A King, pp. 346–56. New York: Van Nostrand Rheinhold.

Harcourt, A H, Stewart, K J & Hauser, M (1993) Functions of wild gorilla 'close' calls. I. Repertoire, context, and interspecific comparison. *Behaviour,* **124,** 89–122.

Hauser, M D (1993) Rhesus monkey copulation calls: honest signals for female choice? *Proceedings of the Royal Society, London, Series B,* **254,** 93–6.

Hauser, M D (1996) *The Evolution of Communication.* Cambridge MA: MIT Press.

Hauser, M D & Marler, P (1993) Food-associated calls in rhesus macaques (*Macaca mulatta*). II. Costs and benefits of call production and suppression. *Behavioral Ecology,* **4,** 206–12.

Hauser, M D & Nelson, D A (1991) 'Intentional' signaling in animal communication. *Trends in Evolution and Ecology,* **6,** 186–89.

Hinde, R A (1976) Interactions, relationships, and social structure. *Man,* **11,** 1–17.

Hinde, R A (1981). Animal signals: ethological and games-theory approaches are not incompatible. *Animal Behaviour,* **29,** 535–42.

Hinde, R A (1983) *Primate Social Relationships: An Integrated Approach*. Oxford: Blackwell Scientific Publications.

Krebs, J R & Dawkins, R (1984) Animal signals: mind-reading and manipulation. In *Behavioural Ecology*, ed. J R Krebs & N B Davies, pp. 380–402. Oxford: Blackwell Scientific Publications.

Kummer, H (1968) *Social Organization of Hamadryas Baboons*. Chicago: University of Chicago Press.

Manson, J H (1996) Rhesus macaque copulation calls: re-evaluating the "honest signal" hypothesis. *Primates*, 37, 145–54.

Markl, H (1985) Manipulation, modulation, information, cognition: some of the riddles of communication. In *Experimental Behavioral Ecology and Sociobiology*, ed. B Hölldobler & M Lindauer, pp. 163–94. Stuttgart: Gustav Fischer Verlag.

Marler, P (1976) Social organization, communication and graded signals: the chimpanzee and the gorilla. In *Growing Points in Ethology*, ed. P P G Bateson & R A Hinde, pp. 239–80. Cambridge: Cambridge University Press.

Marler, P & Mitani, J (1988) Vocal communication in primates and birds: parallels and contrasts. In *Primate Vocal Communication*, ed. D Todt, P Goedeking & D Symmes, pp. 3–14. Berlin: Springer-Verlag.

Marler, P & Tenaza, R (1977) Signaling behavior of apes with special reference to vocalization. In *How Animals Communicate*, ed. T Sebeok, pp. 965–1033. Bloomington IN: Indiana University Press.

Mitani, J (1996) Comparative studies of African ape vocal behavior. In *Great Ape Societies*, ed. W C McGrew, L F Marchant & T Nishida, pp. 241–54. Cambridge: Cambridge University Press.

Mitani, J C & Nishida, T (1993) Contexts and social correlates of long-distance calling by male chimpanzees. *Animal Behaviour*, 45, 735–46.

Owings, D H & Morton, E S (1998) *Animal Vocal Communication: A New Approach*. Cambridge: Cambridge University Press.

Palombit, R A, Cheney, D L & Seyfarth, R M (1999) Male grunts as mediators of social interaction with females in wild chacma baboons (*Papio cynocephalus ursinus*). *Behaviour*, 136, 221–42.

Robbins, M M (1995) A demographic analysis of male life history and social structure of mountain gorillas. *Behaviour*, 132, 21–47.

Schaller, G B (1963) *The Mountain Gorilla: Ecology and Behavior*. Chicago: University of Chicago Press.

Seyfarth, R M, Cheney, D L, Harcourt, A H & Stewart, K J (1994) The acoustic features of gorilla double-grunts and their relation to behavior. *American Journal of Primatology*, 33, 31–50.

Sicotte, P (1993) Inter-group encounters and female transfer in mountain gorillas: influence of group composition on male behavior. *American Journal of Primatology*, 30, 21–36.

Sicotte, P (1994) Effect of male competition on male–female relationships in bi-male groups of mountain gorillas. *Ethology*, 97, 47–64.

Smith, H J, Newman, J D & Symmes, D (1982) Vocal concomitants of affiliative behavior in squirrel monkeys. In *Primate Communication*, ed. C T Snowdon,

C H Brown & M R Petersen, pp. 30–49. New York: Cambridge University Press.

Smith, W J (1965). Message, meaning, and context in ethology. *American Naturalist*, **99**, 405–9.

Smith, W J (1977) *The Behavior of Communicating*. Cambridge MA: Harvard University Press.

Smith, W J (1997) The behavior of communicating, after twenty years. In *Perspectives in Ethology*, vol. 12, *Communication*, ed. D H Owings, M D Beecher & N S Thompson, pp. 7–53. New York: Plenum Press.

Smuts, B B, Cheney, D L, Seyfarth, R M, Wrangham, R W & Struhsaker, T T (1987) *Primate Societies*. Chicago: University of Chicago Press.

Snowdon, C T (1988) Communication as social interaction: its importance in ontogeny and adult behavior. In *Primate Vocal Communication*, ed. D Todt, P Goedeking & D Symmes, pp. 108–22. Berlin: Springer-Verlag.

Snowdon, C T & Cleveland, J (1984). "Conversations" among pygmy marmosets. *American Journal of Primatology*, **7**, 15–20.

Snowdon, C T & Hausberger, M (1997) *Social Influences on Vocal Development*. New York: Cambridge University Press.

Stewart, K J & Harcourt, A H (1987) Gorillas: variation in female relationships. In *Primate Societies*, ed. B B Smuts, D L Cheney, R M Seyfarth, R W Wrangham & T T Struhsaker, pp. 155–64. Chicago: University of Chicago Press.

Stewart, K J & Harcourt, A H (1994) Gorillas' vocal behaviour during rest periods: signals of impending departure. *Behaviour*, **130**, 29–40.

de Waal, F B M (1988) The communicative repertoire of captive bonobos (*Pan paniscus*) compared to that of chimpanzees. *Behaviour*, **106**, 183–251.

de Waal, F B M (1993) Reconciliation among primates: a review of empirical evidence and unresolved issues. In *Primate Social Conflict*, ed. W A Mason & S P Mendoza, pp. 111–44. Albany NY: State University of New York.

de Waal, F B M & Berger, M L (2000) Payment for labour in monkeys. *Nature*, **404**, 563.

Watts, D P (1989) Infanticide in mountain gorillas: new cases and a reconsideration of the evidence. *Ethology*, **81**, 1–18.

Watts, D P (1992) Social relationships of immigrant and resident female mountain gorillas. I. Male–female relationships. *American Journal of Primatology*, **28**, 159–81.

Watts, D P (1994) Social relationships of immigrant and resident female mountain gorillas. II. Relatedness, residence, and relationships between females. *American Journal of Primatology*, **32**, 13–30.

Yamagiwa, J (1987*a*) Intra- and inter-group interactions of an all-male group of Virunga mountain gorillas. *Primates*, **27**, 1–30.

Yamagiwa, J (1987*b*) Male life history and the social structure of wild mountain gorillas (*Gorilla gorilla beringei*). In *Evolution and Coadaptation in Biotic Communities*, ed. S Kawanao, J H Connell & T Hidaka, pp. 31–51. Tokyo: University of Tokyo Press.

Part III
Feeding behavior

10 *Diet and habitat use of two mountain gorilla groups in contrasting habitats in the Virungas*

ALASTAIR McNEILAGE

Mount Visoke with other volcanoes in the background. (Photo by Alastair J. McNeilage.)

Introduction

The diet, movement patterns, and habitat use of mountain gorillas (*Gorilla gorilla beringei*) have been extensively studied in the Karisoke study area (Fossey, 1974; Fossey & Harcourt, 1977; Vedder, 1984; Watts, 1984, 1991, 1998*a*, *b*), but little is known about the ecology of the gorillas in other parts of the Virungas. Variation in gorilla ecology could have important implications both for behavioral ecology theory and for conservation management practice. Mountain gorilla ecology has been examined in the context of foraging theory, but the Karisoke study area may only represent part of the picture. Gorillas occupying different habitats elsewhere in the Virungas, with potential differences in the abundance, distribution, and quality of resources, could further reveal how patterns of habitat and food selection tend to maximize foraging efficiency. Comparisons have been made between gorillas in the Virungas and elsewhere with implications for both the evolution of social organization (Wrangham, 1986; Tutin, 1996; Wrangham *et al.*, 1996; Doran & McNeilage, 1998) and gorilla taxonomy (Sarmiento *et al.*, 1996). However, if such comparisons are to be realistic, they must be based on more than one small part of the Virunga population, and it is important to know how much particular traits vary within the Virungas. In considering the management of the park and the conservation of the gorilla population it is important to have an understanding of what constitutes "good" gorilla habitat and therefore what habitats are available, what resources they contain, and how these are used by gorillas. For these reasons it is important to examine mountain gorilla ecology across a wider range of altitudes and habitats within the Virungas than has been covered by previous studies.

Mountain gorilla ecology in the Virungas has been a focus of several recent reviews (e.g. Watts, 1996; Doran & McNeilage, 1998, this volume) and is therefore described only briefly here. The diet is made up primarily of leaves, stems, pith, and shoots of terrestrial herbaceous vegetation (Vedder, 1984; Watts, 1984). Mountain gorillas select foods that are of high nutritional quality, with high protein content and low fiber and tannins (Watts, 1984). Most of the diet is made up from a small number of species (Vedder, 1984; Watts, 1984; McNeilage, 1995), perhaps because their favored high-quality foods are widely available. Around Karisoke, some groups have access to areas of bamboo within their ranges, and bamboo shoots are a favored food item when available (Fossey & Harcourt, 1977; Watts, 1984; McNeilage, 1995; Plumptre, 1995). Apart from this, there is little seasonality in mountain gorilla ecology (Watts, 1998*c*).

Within the Karisoke study area, the principal gorilla food plants are abundant and widely distributed, and gorilla groups move only around

266

500 m or less per day on average (Fossey & Harcourt, 1977; Watts, 1991). Group size has only a small influence on day range length (Watts, 1991). Mean annual home range size of Karisoke study groups has varied between 4 and 11 km^2, and shows no simple relationship with group size (Watts, 1998a). Home ranges vary considerably within groups and across time (Watts, 1998a). Despite the relative abundance and wide distribution of high-quality foods, food distribution is the single most important factor underlying mountain gorilla movement patterns in the Karisoke study area (Vedder, 1984; Watts, 1991, 1998a, b). Patterns of habitat use have been found to increase foraging efficiency, as gorillas select areas and habitat types where food is abundant, especially high-quality food, where day journey lengths are shortest and where nutritional requirements are met in the shortest period of daily feeding time (Vedder, 1984; Watts, 1991, 1998a, b).

Several lines of evidence suggest that there may be differences in gorilla ecology between the Karisoke study area and other parts of the Virungas. Vegetation types in tropical montane regions such as the Virungas vary considerably with altitude (Lebrun, 1960; Spinage, 1972; White, 1981). The Karisoke study area covers approximately the upper half (2700– 3600 m) of the altitudinal range occupied by mountain gorillas, which are found down to the lowest altitudes (2000 m) of the forested area on the north side in the Democratic Republic of Congo (DRC). Available vegetation maps indicate that different types exist in areas outside the Karisoke study area (Marius, 1976). Each survey made of the population has found considerable variation in gorilla density in different areas (Schaller, 1963; Harcourt & Fossey, 1981; Weber & Vedder, 1983; Aveling & Harcourt, 1984; Aveling & Aveling, 1989; Sholley, 1991). In particular, several authors have noted a difference in group size as well as population density between the eastern section of the Virungas and the remainder, which could be due to differences in habitat quality and human disturbance (Harcourt & Fossey, 1981; Harcourt et al., 1983; Weber & Vedder, 1983).

The main aim of this study was to compare the ecology of gorillas in other parts of the Virungas with those in the Karisoke study area. I compare the ecology of two groups of mountain gorillas, Group BM in the Karisoke study area, and Group 11 in the lower altitude saddle between Mount Visoke and Mount Sabinyo. The two groups occupy contrasting habitat on the Rwandan side of the Virungas. Ideally, groups would be compared over the whole array of altitudes used by gorillas within the Virungas, but this was not possible within the logistical constraints of this study. Groups appear to utilize two basic types of home range in the Virungas, larger home ranges in rich habitats at higher altitudes (above

2700 m) particularly between Mounts Visoke, Karisimbi and Mikeno (i.e. the Karisoke Research Center study area) and smaller ranges at lower altitudes in habitats with lower food availability (McNeilage, 1995). The latter may constrain group size more than the former (McNeilage, 1995). The two groups in this study represent these two types of home range, although they do not differ greatly in size. Comparisons are made between the movement patterns (home range size, day journey length, distances between feeding sites) of the two groups, and in the patterns of habitat and dietary selection shown by each.

Study site
The forested area of the Virungas covers approximately 440 km^2 and includes an altitudinal range from 2000 m in the lowest sections on the DRC side to 4500 m at the summit of Mount Karisimbi. It is protected as contiguous national parks in each of the three countries (the other two are Rwanda and Uganda), but is completely isolated by human habitation and cultivation. Annual rainfall is around 2000 mm, with a bimodal pattern (Plumptre, 1991). March through May and September through November are wet seasons, while June through August and December through February are drier. However the pattern of rainfall is variable and little evidence of seasonal variation in the biomass and growth rate of plants has been found (Plumptre, 1991).

Methods
Ranging and habitat use
Ranging and feeding data collection was concentrated on two study groups, Group BM in the Karisoke study area and Group 11 (habituated for tourist visits) in the lower altitude saddle between Mount Visoke and Mount Sabinyo. Group BM consisted of 2 silverbacks, 7 adult females, 4 juveniles, and 3 infants. Group 11 consisted of 4 silverbacks, 1 blackback, 3 adult females, 2 juveniles, and 1 infant.

Mountain gorillas generally leave clear trails as they move through the forest and feed at distinct sites that are visible along the trail. The majority of feeding and ranging data were collected by examining these trails. One hundred and twelve complete days of trail between consecutive night-nest sites of the two groups were followed between March 1991 and February 1992 (n = 59 for Group BM and n = 53 for Group 11). A total of 9986 feeding spots along 73 km of trail were examined, divided between the two groups (n = 5337 for Group BM and n = 4649 for Group 11). Trail length was measured by pacing, pace size being checked frequently against measured 50 m distances in comparable terrain. A continual note was made of the location of the trail using altitude, compass bearings, and proximity to known landmarks. The habitat type through which the trail passed was also noted. This allowed the total journey length, the mean distance between feeding sites, and the proportion of feeding sites in each habitat type to be calculated for each day's trail.

Compositional analysis (Aebischer *et al.*, 1993) was used to test whether habitat use differed from random utilization, using the proportion of feeding sites in each habitat along a trail as an index of the relative use of habitats on that day. This method circumvents the problem caused by the fact that adjacent feeding sites would not be independent samples. Logratios were calculated from the proportion of feeding sites in each habitat type for each day's trail and from the proportion of each habitat within the home range. The difference between the logratios for use and availability were calculated for each day. Multivariate ANOVA of the logratio differences provides a simultaneous test over all habitat types of the hypothesis that habitats are used at random. Wilk's lambda (λ) is used for the test. Compositional analysis was also used to rank habitat types in order of relative use and to identify where significant non-random use occurs (Aebischer *et al.*, 1993). This is equivalent to comparing all possible pairs of habitats, which can be done using logratio differences and a one-sample *t*-test. For detailed calculation procedures see Aebischer *et al.* (1993) and McNeilage (1995).

Minimum convex polygon (Southwood, 1966; Harris *et al.*, 1990) home ranges were constructed using the mid-point of each day's trail, which were located on a 100-m grid over a 1:100 000 topographic map. The polygons were entered into the Geographic Information System (GIS) described below, allowing the areas of each habitat type within each home range to be calculated. Estimates of the total biomass of food and the average food density in each home range were calculated using the density of food in each habitat, weighted by the area of each habitat type in the range.

Diet

The composition of the diets of the two groups was estimated using a combination of feeding trails, microhistological analysis of feces samples and direct observations of feeding behavior. As Group 11 was visited by tourists and not researchers, direct observations were only possible with Group BM. Indirect measurements included the relative frequency of consumption of foods. This was estimated by noting foods eaten at each individual feeding site. For certain food items, mainly stems that were peeled or split open, the biomass of that food consumed could be estimated from the length of stem peeled or split. For leaves eaten whole, microhistological analysis used the relative area of leaf cuticle in feces samples to estimate the relative biomass of different food items in the diet (Norbury & Sanson, 1992). These two methods between them could therefore provide estimates of the biomass eaten of most food types. As feces collected in one habitat could contain foods eaten elsewhere, no attempt was made to differentiate diet composition in individual habitats. The methods used are described in more detail in McNeilage (1995).

The main problem in using two separate techniques for estimating the proportion of different food items in the diet was how to combine the data from each. Results from the two different methods were combined by relating each to the frequency estimates which were made from trail signs for all foods (see McNeilage, 1995 for more details). For the few species for which biomass estimates could be made from both methods, the results from each agreed well (McNeilage, 1995). In addition, direct observations of feeding behavior in Group BM were used to evaluate the accuracy of the trail signs and fecal analysis, as well as to check the comparability of these methods. Observation techniques were based on those of Watts (1984). The relative biomass of food items in the diet estimated using the indirect methods was closely correlated with the observed diet of Group BM during the same periods (McNeilage, 1995).

Habitat mapping

A classification of habitat types was established based on those used by Watts (1984) and Plumptre (1991), simplified slightly and extended to include types found in parts of the Virungas away from the Karisoke study area (Table 10.1). The classification is subjective, but differences between habitat types are quite obvious on the ground and previous studies have found considerable differences in vegetation between habitat types (Watts, 1984; Plumptre, 1991). The classification chosen allowed each habitat to be distinguished on aerial photographs. Using personal experience of the area, the habitat types were identified on a series of 1:50 000 aerial photographs. The habitats were outlined on a 1:100 000 topographic map using a zoom transfer scope at the Center for Remote Sensing and Spatial Analysis, Rutgers University. The polygons outlining each patch of each type of habitat were digitized into the existing GIS of the Virunga area. The GIS was used to produce a map of the habitat types from which the availability of habitat types within home ranges was measured.

Food plant availability

An overall list was compiled of food plants observed to be eaten by the gorillas in the course of the study (plant names throughout this chapter

Table 10.1. *Classification of habitat types*

Type	Altitude, meters	Characteristics
1. Alpine	Above 3600	Areas above the limit of most herbaceous and woody plants, with low grass and mosses and occasional *Senecio johnstonii*. Bare rocky areas, especially on top of Mounts Mikeno and Sabinyo were also included as Alpine.
2. Subalpine	3300–3600	High altitude vegetation, up to 4–5 m high, with abundant *Senecio johnstonii*, *Lobelia stuhlmanni* and/or *L. wollostonii*, *Hypericum revolutum*, and *Rubus kirungensis*.
3. Brush Ridge	2950–3300	Dense vegetation along the ridges and ravines on the sides of the volcanoes, with abundant *Hypericum revolutum*, and shrubby growth of *Senecio mariettae*, reaching around 10 m high.
4. Herbaceous	2800–3300	Open areas with low (1–2 m), dense herbaceous vegetation, generally on the sides of volcanoes, with very few *Hagenia abyssinica* and *Hypericum revolutum* trees.
5. *Hagenia*	2750–3300	Equivalent to the "Saddle" zone of previous authors, a variable canopy woodland dominated by *Hagenia abyssinica* and *Hypericum revolutum* trees reaching up to 20 m, with a dense herbaceous or, less frequently, grassy understorey found in the saddles between certain volcanoes and on the less steep lower slopes.
6. Bamboo	2550–2950	Areas dominated by often monospecific stands of bamboo (generally 5–12 m high), mixed with a few trees and vines at lower altitudes.
7. *Mimulopsis*	2550–2800	Open herbaceous areas, differing from the Herbaceous zone in being found at lower altitudes, generally in the flat saddle between Mounts Visoke and Sabinyo and often dominated by *Mimulopsis excellens*.
8. Meadow	2200–3700	This term was used to describe open grassy areas at a variety of altitudes. These areas were often marshy and contained very little gorilla food. A large, dry, shrubby area on the east side of Muhavura, which burned extensively in 1989, was included here as Meadow.

A ninth habitat, Mixed Forest (2000–2550 m), was found in the lowest parts on the DRC side.

follow Troupin, 1977–1988). To estimate the mean biomass of gorilla food plants in each habitat type, vegetation surveys were made in a range of study sites across the Virungas. In each study site, gorilla food plants were sampled using a stratified random technique (Grieg-Smith, 1983). A baseline was measured and marked across the middle of each site, and used to establish a grid system. In the majority of habitat types a grid with cells of 500 m × 500 m was used, with 200 m × 200 m cells where the larger grid did not yield a sufficient number of sample points. Vegetation was measured at one random sampling point within each grid cell, located by pacing. A total of 405 points were sampled, between June 1992 and February 1993. The biomass of each food was estimated in concentric, circular, sample plots, at each sampling point of differing size according to plant type. All herbaceous plants were counted and the length of each stem measured in 1 m² plots. The wet weights of *Galium* spp. and of leaves of other vines were measured in the field. The length and circumference of stems of *Lobelia* spp., the lengths of stems of *Rubus* spp., and the number of bamboo stems were recorded in 10 m² plots. Finally in 100 m² plots, all species of tree were counted. For those species from which the gorillas ate the pith of stems, the total length of stem was estimated. The dry biomass of each gorilla food in each plot was estimated from these data by extracting the parts eaten by gorillas, drying, and weighing. These procedures are described in detail in McNeilage (1995).

The mean biomass of each food in habitat type, except Meadow and Alpine which contain virtually no gorilla food and are very rarely used, was calculated from the estimated biomass in each plot in that habitat. The biomass of bamboo shoots was calculated using only those plots measured in Bamboo during the shooting season (30 plots out of a total of 60). It was estimated from the Karisoke Research Center trackers' daily reports from 1988 to 1992 that bamboo shoots were available for an average of 3 months per year. The biomass estimate was therefore multiplied by 0.25, to give an estimate of effective mean annual availability.

Results
Home ranges
The areas of each habitat type in the home ranges of these two groups are given in Table 10.2. The percentage overlap in the habitat types used by the two groups was low (5.2%, calculated as described in Struhsaker, 1975). The estimated biomass of the two groups was similar, yet the home range of Group BM was found to be 2.5 times larger than that of Group 11 (Table 10.3). Both the estimated total biomass of food within the range and the density of food were greater for Group BM (Table 10.3). Given the greater availability of food resources in Group BM's range, the fact that it is so much larger than that of Group 11 is especially surprising.

Table 10.2. *Habitat types in the study group home ranges: the areas in km²*
of each habitat type in the home ranges of the two main study groups

Habitat	Group BM	Group 11
Subalpine	0.96	
Brush Ridge	1.44	
Herbaceous	1.11	0.07
Hagenia	7.75	0.17
Bamboo		1.21
Mimulopsis		3.19
Meadow	0.26	
Lake	0.06	0.02
Total area	11.58	4.66

Table 10.3. *Comparison of group size and home range parameters of the*
two study groups

	Group BM	Group 11
Group size (excluding dependent infants)	13	10
Group biomass (kg)[a]	1350	1425
Home range size (km²)	11.58	4.66
Food biomass in range (×1000 kg)	733.1	87.46
Food density (×1000 kg/km²)	63.29	18.75

[a] Estimates of the biomass of each group were made using published estimates of 200 kg for an adult male and 100 kg for an adult female (Goodall, 1977) and estimates of 75 kg and 50 kg for subadults and juveniles respectively.

Habitat selection

Feeding sites were found within four habitat types in each home range. Group BM positively selected Herbaceous and Brush Ridge habitat, using these two habitat types more than would be expected if habitat utilization was random and habitats were therefore used in proportion to their availability (Table 10.4). Group 11 showed a slight positive selection of *Mimulopsis* in the same way.

Compositional analysis (Aebischer *et al.*, 1993) showed that both Group BM and Group 11 were found to use habitats in a significantly non-random way ($\lambda = 0.541$, d.f. $= 3$, $p < 0.001$ and $\lambda = 0.190$, d.f. $= 3$, $p < 0.001$, respectively). Using compositional analysis to rank habitat types in order of relative use and to identify where significant non-random use occurs (Aebischer *et al.*, 1993) gave the rankings show below, where $>$ indicates that the first habitat was positively selected relative to the second and $> > >$ indicates a significant difference in selection (t-test, $p < 0.05$):

Table 10.4. *The use, availability, and electivity of habitat types in the home ranges of the two main study groups*

Group	Habitat	Mean daily percentage of feeding sites	Percentage of total area of home range	Ivlev's electivity index[a]
Group BM	Subalpine	7.6	8.5	−0.055
	Brush Ridge	16.7	12.7	0.136
	Herbaceous	47.8	9.8	0.659
	Hagenia	27.6	68.8	−0.427
Group 11	Herbaceous	1.3	1.5	−0.071
	Hagenia	3.7	3.7	0
	Mimulopsis	80.7	68.8	0.080
	Bamboo	14.1	26	−0.297

[a] Ivlev's electivity index is calculated as $(r_i - n_i)/(r_i + n_i)$ where r_i is proportion of feeding sites in habitat i and n_i is the proportion of habitat i in the home range. Ivlev's index returns a value between −1 and +1, with a positive electivity index indicates a positive selection of that habitat. A value of zero indicates that habitat utilization is random so that habitats are used in proportion to their availability.

Group BM: Herbaceous $>$ $>$ $>$ Brush Ridge $>$ $>$ $>$ *Hagenia* $>$ Subalpine

Group 11: *Mimulopsis* $>$ $>$ $>$ Bamboo $>$ Herbaceous $>$ *Hagenia*

The relationships between habitat rank and various parameters of habitat quality are shown in Table 10.5. Dry biomass of gorilla food available per m^2, frequency of food (the percentage of $1 \ m^2$ plots containing herbaceous food), and diversity of foods (Shanon–Wiener index of diversity of gorilla foods within each habitat) were used to reflect both the total availability and diversity of foods available in each habitat. Previous studies within the Karisoke study area have shown that habitat selection is primarily related to habitat quality. However, in this study no single habitat parameter obviously accounted for the rank position of the habitats used by either group, although some aspects of habitat selection can be related to habitat quality.

The habitat which ranked highest for Group BM, Herbaceous, had a high density and frequency of food and a high food species diversity, as would be predicted. However, Brush Ridge ranked higher and *Hagenia* ranked lower than expected on the basis of habitat quality parameters (Table 10.5). The most selected habitat for Group 11 was *Mimulopsis*. This habitat did not have as high a density of foods as Herbaceous or *Hagenia*, but did have a high frequency of food and species richness (Table 10.5). Overall, Group 11 selected *Mimulopsis* more than Bamboo, which had

Table 10.5. *Habitat preference rankings (1 is lowest, 4 highest) and habitat quality parameters for the habitat types within the home ranges of the two main study groups*

Group	Habitat type	Rank	Food density g/m^2	Frequency of food	Diversity of foods
Group BM	Herbaceous	4	74.94	100	1.85
	Brush Ridge	3	17.36	80	1.83
	Hagenia	2	75.99	92	2.03
	Subalpine	1	37.68	58	1.67
Group 11	*Mimulopsis*	4	20.08	93	2.32
	Bamboo	3	4.21	28	1.19
	(with shoots)		6.32		1.43
	Herbaceous	2	74.94	100	1.85
	Hagenia	1	75.99	92	2.03

The parameters are dry biomass of gorilla food available per m^2, frequency of food (the percentage of 1 m^2 plots containing herbaceous food), and diversity of foods (Shanon–Wiener index of diversity of gorilla foods within each habitat). Certain parameters are given separately for Bamboo during the period when shoots were present.

particularly low habitat quality parameters. Herbaceous and *Hagenia* ranked lower than expected on the basis of food density (Table 10.5).

Previous studies have found that bamboo shoots were the only major seasonal influence on mountain gorilla ecology (Vedder, 1984; Watts, 1991, 1998c). Group BM's range did not contain bamboo, and no attempt was made to analyze habitat selection by season. For Group 11, whose home range included a considerable area of bamboo, the year was split into the period when bamboo shoots were available (mid-October to mid-January) versus the rest of the year. Extending the multivariate ANOVA described above to include this distinction as a factor allowed the effect of season on the overall pattern of habitat utilization to be statistically tested. This effect was found to be significant ($\lambda = 0.614$, d.f. = 3, $p < 0.001$). The habitat ranking procedure was repeated separately for trail days during the bamboo season ($n = 11$) and the rest of the year ($n = 42$). These rankings were as follows:

Bamboo season: *Mimulopsis* > Bamboo > > > Herbaceous > *Hagenia*

Rest of year: *Mimulopsis* > > > Herbaceous > *Hagenia* > Bamboo

Bamboo habitat ranked higher during the bamboo season (not significantly different from *Mimulopsis*) than during the rest of the year. However, the total density of food in Bamboo during that season was not much

Table 10.6. *Comparison of movement parameters between the two study groups*

	Group BM	Group 11
Number of days of trail	59	52
Mean day journey length, meters (mean ± SE)	550.5 ± 37.7	756.5 ± 45.8
t-test	$t = 3.47, p < 0.001$	
Mean distance between feeding sites, meters (mean ± SE)	6.15 ± 0.39	8.82 ± 0.48
t-test	$t = 4.33, p < 0.001$	

higher than during the rest of the year, despite the presence of bamboo shoots (Table 10.5).

Movement parameters

Animals foraging efficiently should minimize the distance traveled in order to obtain sufficient resources (Altmann, 1974). Animals in poorer environments are therefore predicted to have to travel further. The difference in food density between the two ranges is reflected in the movement patterns of the groups. Group 11, in the area of lower food density, moved further each day ($t = 3.47$, $p < 0.001$) and further between feeding sites ($t = 4.33$, $p < 0.001$) than Group BM (Table 10.6).

Diet

A complete list of the food items recorded in the diet of the two study groups is given in Table 10.7, including both relative frequency from trail signs, and relative biomass estimates from trail signs and fecal analysis combined where possible. Group BM ate a total of 65 different foods items from 35 species, while Group 11 ate 72 items from 44 species. The total number of foods and food species utilized is greater for Group 11, as is the number of foods eaten per day and the number of food species per day (both $p < 0.001$, see Table 10.8).

There were considerable differences in the plants species and parts eaten by the two groups. Apart from *Galium* spp., the diets of the two groups were largely composed of different food items (Figure 10.1). The overlap in diet between the two groups (calculated as in Struhsaker, 1975) was 41.3% estimated from relative biomass and 33.2% from relative frequency. Most of this was due to the large proportion of *Galium* spp. eaten by both; excluding this food the overlap was just 6.3% by biomass and 8.8% by frequency. There were also considerable differences in the types of foods

Table 10.7. *The relative frequency and biomass of all foods recorded in the diet of Group BM and Group 11*

Species	Part[a]	Frequency, %[b]	Biomass, %[c]
Group BM			
Galium spp.	ls	24.41	34.96
Carduus nyassanus	lf	16.72	19.94
Peucedanum linderi	st	15.8	18.54
Carduus nyassanus	st	16.81	10.72
Senecio johnstonii	pi	2.69	5.07
Carduus nyassanus	rt	1.50	4.31
Laportea alatipes	lf	8.21	1.39
Lobelia stuhlmanii	pi	1.44	1.28
Urtica massaica	st	1.44	1.14
Laportea alatipes	st	0.87	0.64
Rubus spp.	lf	1.61	0.60
Helichrysum formosissimum	lf	0.39	0.41
Carex bequaertii	lb	0.73	0.23
Cyperus mannii	lf	0.58	0.20
Droguetia iners	lf	0.64	0.14
Carduus kikuyorum	lf	0.05	0.12
Cyperus mannii	lb	0.29	0.05
Laportea alatipes	rt	0.24	0.05
Carduus kikuyorum	st	0.06	0.04
Crassocephalum ducis-aprutii	st	0.63	0.03
Vernonia adolfi-fredricii	pi	0.09	0.03
Carduus leptocanthus	lf	0.03	0.02
Discopodium penninervium	pi	0.02	0.02
Crassocephalum ducis-aprutii	rt	0.11	0.01
Peucedanum linderi	rt	0.05	0.01
Arundinaria alpina	sh	0.05	0.01
Carduus leptocanthus	st	0.04	0.01
Rumex rwenzoriensis	st	0.03	0.01
Lobelia giberroa	rt	0.02	0.01
Dead stems opened [d]		1.08	
Rubus spp.	st	0.60	
Senecio johnstonii	rt	0.43	
Lobelia stuhlmanii	rt	0.28	
Lobelia stuhlmanii	lf	0.25	
Vernonia adolfi-fredricii	fl	0.19	
Plectranthus laxiflorus	lf	0.16	
Vine sp. A	lf	0.15	
Gynura scandens	lf	0.13	
Hypericum revolutum	dw	0.12	
Vernonia adolfi-fredricii	tw	0.12	
Vine sp. A	cu	0.08	
Cineraria deltoidea	lf	0.08	
Oreosyce africana	lf	0.08	
Vernonia adolfi-fredricii	lf	0.08	

Table 10.7. (*cont.*)

Species	Part[a]	Frequency, %[b]	Biomass, %[c]
Hagenia abyssinica	dw	0.07	
Clematis simensis	lf	0.06	
Arundinaria alpina	lf	0.05	
Prenanthes subpeltata	lf	0.05	
Hagenia abyssinica	ba	0.04	
Lobelia stuhlmanii	ba	0.04	
Lobelia giberroa	pi	0.04	
Cinereria deltoidea	st	0.04	
Peucedanum kerstenii	st	0.04	
Carduus nyassanus	fl	0.03	
Hypericum revolutum	ba	0.02	
Galium spp.	cu	0.02	
Urtica massaica	rt	0.02	
Plectranthus laxiflorus	st	0.02	
Pygeum africanum	ba	0.01	
Prenanthes subpeltata	cu	0.01	
Carduus kikuyorum	fl	0.01	
Echinops hoehlenii	fl	0.01	
Helichrysum sp.	lf	0.01	
Stephania abyssinica	lf	0.01	
Carduus leptocanthus	rt	0.01	
Senecio sp.	st	0.01	
Group 11			
Galium spp.	ls	31.66	53.12
Carduus leptocanthus	rt	4.02	10.50
Vernonia auriculifera	pi	4.84	6.89
Basella alba	lf	16.06	5.07
Peucedanum linderi	st	4.42	4.87
Arundinaria alpina	lf	1.34	3.32
Arundinaria alpina	sh	8.60	2.99
Droguetia iners	lf	7.66	2.92
Carduus leptocanthus	lf	3.26	2.64
Rubus spp.	lf	0.76	2.04
Carduus leptocanthus	st	2.06	1.34
Urera hypselodendron	cu	1.41	1.16
Discopodium penninervium	pi	0.78	0.61
Urera hypselodendron	lf	1.16	0.53
Fungus sp.	fu	0.03	0.44
Tacazzea apiculata	cu	0.25	0.25
Carduus kikuyorum	lf	0.15	0.23
Cynoglossum spp.	rt	0.45	0.18
Urtica massaica	st	0.44	0.18
Laportea alatipes	lf	0.72	0.16
Carduus kikuyorum	st	0.13	0.10
Carduus kikuyorum	rt	0.13	0.09
Carduus nyassanus	lf	0.05	0.09

Table 10.7. (*cont.*)

Species	Part[a]	Frequency, %[b]	Biomass, %[c]
Lobelia giberroa	rt	0.06	0.06
Carduus nyassanus	rt	0.03	0.05
Coccinea mildbraedii	fr	0.13	0.04
Vernonia adolfi-fredricii	pi	0.07	0.03
Laportea alatipes	st	0.12	0.02
Crassocephalum ducis-aprutii	rt	0.07	0.02
Carduus nyassanus	st	0.04	0.02
Vine sp. A	cu	0.12	0.01
Galium spp.	cu	0.07	0.01
Peucedanum linderi	rt	0.04	0.01
Cyanoglossum spp.	st	0.03	0.01
Mikania capensis	lf	3.05	
Oreosyce africana	lf	0.96	
Discopodium penninervium	fr	0.66	
Helichrysum maraguensis	lf	0.63	
Tacazzea apiculata	lf	0.37	
Vine sp. A	lf	0.34	
Carduus leptocanthus	fl	0.33	
Clematis simensis	lf	0.33	
Vernonia sp.	lf	0.27	
Dead stem opened [d]		0.24	
Lagenaria abyssinica	lf	0.23	
Plectranthus laxiflorus	lf	0.17	
Stephania abyssinica	lf	0.16	
Geranium aculeolatum	lf	0.11	
Mikaniopsis clematoides	lf	0.11	
Rubus spp.	st	0.07	
Tinospora caffra	lf	0.07	
Vine sp. B	lf	0.07	
Ants		0.06	
Dombeya goetzenii	ba	0.06	
Lactuca glandulifera	lf	0.06	
Cineraria deltoidea	lf	0.05	
Coccinea mildbraedii	lf	0.05	
Clerodendron johnstonii	lf	0.04	
Vernonia auriculifera	tw	0.04	
Acalypha psilostachya	lf	0.03	
Hypericum revolutum	dw	0.03	
Vine sp. C	lf	0.03	
Vernonia auriculifera	ba	0.02	
Vernonia auriculifera	lf	0.02	
Discopodium penninervium	lf	0.01	
Dombeya goetzenii	lf	0.01	
Galiniera coffeoides	lf	0.01	
Galiniera coffeoides	st	0.01	
Hagenia abyssinica	ba	0.01	
Herb sp.	cu	0.01	

Table 10.7. (*cont.*)

Species	Part[a]	Frequency, %[b]	Biomass, %[c]
Herb sp.	lf	0.01	
Laportea alatipes	rt	0.01	
Vernonia adolfi-fredricii	lf	0.01	
Vernonia adolfi-fredricii	rt	0.01	

[a]. Plant parts are: lf = leaf, ls = leaf and stem eaten together, lb = leaf base, pi = pith from branches, rt = root, fl = flowers, dw = dead wood, tw = twigs, cu = cuticle of vine stems, ba = bark, sh = shoot, fr = fruit, and fu = fungus.
[b] Calculated from the number of feeding sites at which each food was eaten,
[c] Calculated where possible either from faecal analysis or from trail signs, as described in the text.
[d] Apparently to obtain insect cocoons or egg cases.

Table 10.8. *Comparisons of the breadth and diversity of the diets of the two study groups*

	Group BM		Group 11
Number of food items	65		72
Number of food species	35		44
Foods per day	11		15
Mann–Whitney *U*		808, $p < 0.001$	
Food species per day	8		12
Mann–Whitney *U*		400, $p < 0.001$	
Diversity			
Shanon-Wiener index	1.85		1.85
Equitability	0.55		0.52

eaten by the two groups (Figure 10.2). Most notably, Group 11 ate more leaf and less stem than Group BM.

Interestingly, the dietary diversity index of the two groups is equal (Table 10.8). This index reflects both the number of species and the evenness, or equitability, of each species contributions to the diet. The number of food items in the diet of Group 11 was greater than in the diet of Group BM, but the equitability was lower for Group 11 (Table 10.8), resulting in similar diversity indices. Most of the diet of the two groups is made up of a small number of species (Figure 10.1). The top three ranking foods comprised 73.4% of the diet of Group BM and 70.5% of Group 11's diet, while the figures for the top ten foods were 98.0% and 91.4%,

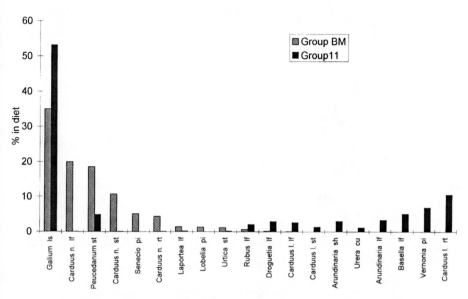

Figure 10.1. The relative biomass of the main food items in the diets of Group BM and Group 11. Foods contributing > 1% to the diet of either group are included (see Table 10.7 for full species names). Food types: lf = leaf, st =stem, ls = leaves and stem together, pi = pith, rt = root, sh = shoot.

respectively. A particularly large proportion (53.1%) of Group 11's diet was made up of the highest-ranking food (*Galium* spp.). The greater dominance of the top-ranking item is likely to be the main reason for the lower equitability of Group 11's diet.

Discussion
Previous research on mountain gorilla feeding ecology has been in the form of in-depth studies on a small number of groups in one area of the Virungas at the highest part of the altitudinal range occupied by gorillas. On the basis of ecological information from the Karisoke study area, earlier studies labeled particular areas as poor gorilla habitat (Weber & Vedder, 1983) but more recent censuses found several groups of gorillas in these areas (Aveling & Aveling, 1989; Sholley, 1991). This study has illustrated how mountain gorillas use a wider range of habitats than had previously been documented. Although mountain gorillas in different habitats are still selective feeders, concentrating on a relatively small number of food items, the specific foods exploited do vary with habitat (Watts, 1984; this study). Such dietary flexibility is probably the key factor in allowing mountain gorillas to occupy a range of different habitats within

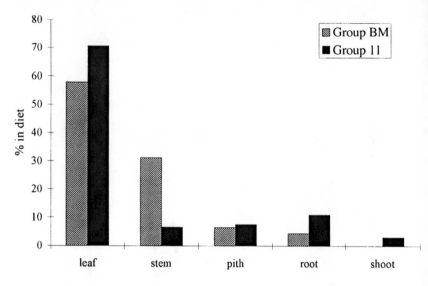

Figure 10.2. The relative biomass of the main food types in the diets of Group BM and Group 11.

the Virungas. In taking the first steps towards comparing the ecology of mountain gorillas in the Karisoke study area with those in other parts of the Virungas, this study has perhaps raised more questions than it has answered. However, the findings do have certain implications for our understanding of variation in mountain gorilla ecology in relation to foraging theory, for socioecology, and for taxonomy, as well as for the conservation management of the population.

Home range size and composition

Selection of a home range is the first level of habitat selection for any animal. The two mountain gorilla groups in this study used home ranges containing quite different habitats with different densities of food available. Several studies have found negative correlations between range size and habitat quality or productivity (e.g. rhesus macaques: Jiang *et al.*, 1991; carnivores: Gompper & Gittleman, 1991) or have found range size to be larger in areas of low habitat quality (e.g. howler monkeys: Estrada, 1984; Barbary macaques: Mehlman, 1989; wood bison: Larter & Gates, 1994). Despite the fact that Group 11 occupied an area of lower food density (and moved further each day) their home range during the year of this study was actually considerably smaller than that of Group BM, a similar-sized group in the Karisoke study area. Although range size was not found to be larger in habitats with lower food density within the range of habitats within the Virungas, much larger ranges of 31 km^2 (Casimir & Butenandt, 1973) and 34 km^2 (Goodall, 1977) have been reported for eastern lowland gorillas (*Gorilla gorilla graueri*) in Kahuzi-Biega National Park, DRC. Goodall (1977) attributes this to differences in the abundance and availability of food resources. In the Virungas, factors other than simple food abundance may influence home range size.

Firstly, the types of food may also be important. This study found that Group 11 ate more leaf and less stem than Group BM. This is a reflection of the fact that more of the foods in the home range of Group BM were herbaceous plants such as *Carduus nyassanus* and *Peucedanum linderi*, the stems of which were eaten, while Group 11 relied more on leaves from vines, including *Galium* spp. and *Basella alba*. The herbaceous plants in the Karisoke area are not only eaten by the gorillas, but flattened as the gorillas move around. Because of such trampling (or for other reasons) such plants might take longer to regenerate after a group has fed on them than the leaves of vines which make up the bulk of the food present in the lower altitude habitats where the smaller, poorer ranges are found. If food abundance was not directly correlated with productivity, this could explain how the gorillas in the latter type of habitat might need a smaller area to meet their nutritional requirements. Plumptre (1994) found that trampling by large mammalian herbivores in the Virungas, including gorillas, was unlikely to cause a significant reduction in the available biomass of vegetation. However, Watts (1987, 1998b) has found some evidence that regeneration rates affected the frequency with which gorillas revisit particular areas. More detailed studies of the regeneration rates of gorilla foods in different habitats and their role in determining ranging patterns would be required to clarify these issues.

Secondly, nutritional quality of individual foods could also be more important than simple food availability in determining ranging patterns (Watts, 1991, 1998a). Detailed nutritional analysis of the foods consumed by Group 11 in the habitats with low overall food abundance would be needed to test this hypothesis.

Thirdly, competition with other herbivores might also be expected to have an impact on home range size relative to resources available. Unfortunately, no data are available on the densities of potential herbivore competitors outside the Karisoke study area. However, there is little evidence that competition from other herbivores in the Karisoke area has a significant effect on the availability of resources for gorillas (Plumptre, 1991, 1994).

Lastly, gorilla home ranges vary considerably from year to year and may be affected by the presence of other groups in the area (Watts, 1998a). The home range sizes of the two groups measured in this study do both fall within the range previously reported for annual home ranges of gorillas in the Virungas (Watts, 1998*a*). Home ranges over one particular year may not, however, be a reliable estimate of habitat requirements on which to base comparisons between groups in different habitats. Group BM shared its entire range with another group of six individuals, and utilized some areas during the year of the study which were not previously part of their home range (personal observation). The home range size measured for this group in this study might simply have been particularly large, but within the normal range of variation in annual home range size.

Habitat selection within home ranges

If selection of a home range (and thus determination of home range size) is the first level of habitat selection, the next is the relative use of different habitats within the home range. Gorillas in the Karisoke study area have been found to be selective in the habitats they use within their ranges, preferring those with a high abundance and frequency of food (Vedder, 1984; Plumptre, 1991; Watts, 1991). This study also found that gorillas, both within and outside the Karisoke study area, used habitats selectively in a way which could be partially, but not entirely, explained in terms of measured patterns of food availability.

The habitat which ranked highest for Group BM, Herbaceous, had a high density and frequency of food and a high food species diversity, as would be predicted. However, Brush Ridge ranked higher and *Hagenia* ranked lower than expected on the basis of habitat quality parameters. This could be for several reasons. Firstly, foods vary in nutritional quality and the availability of high-quality foods in particular has been found to be most important in determining habitat selection (Watts, 1991, 1998*a*, *b*). If habitat selection were examined in terms of availability of particular high-quality foods or individual nutrients, then the relationship between habitat selection and quality might become more clear. Secondly, estimates of habitat quality in this study were made from measurements across a wider area of the Virungas, because the overall aims were to obtain a better understanding of habitat availability and quality over the whole range. However, the particular areas of certain habitats within the home range of Group BM might be more or less food rich than the average for that habitat. Finer detail of the relationship between habitat quality and selection might be more evident if the habitat measurements were made only within the range of the group concerned, but the validity of extrapolations to other parts of the Virungas would then be reduced.

The most selected habitat for Group 11 was *Mimulopsis*. This habitat did not have as high a density of foods as Herbaceous or *Hagenia*, but did have a high frequency of food and species richness (Table 10.5). Herbaceous and *Hagenia* ranked lower than expected on the basis of food density, but only a small area of each occurred at one edge of Group 11's range. The main choice of habitats available within their home range was therefore between *Mimulopsis* and Bamboo, and they selected *Mimulopsis* as would be predicted on the basis of habitat quality. Overall, Group 11 selected *Mimulopsis* more than Bamboo, which had particularly low habitat quality parameters. However, the fact that Bamboo as a habitat is selected more positively during the bamboo shooting season, despite the very small increase in the total biomass of food available relative to the rest of the year, indicates that the availability of particular high-quality foods may be important in determining habitat selection, rather than simply the total biomass of food available.

As with home range size, a number of other factors must be important in determining patterns of habitat selection, including food type and quality, and both intra- and interspecific competition. Further research will need to assess the importance of these across a wider range of habitats in the Virungas.

Diet and group movements

Foraging theory predicts that an animal will respond to reduced availability of food by expanding its diet to include lesser-quality foods, traveling further in order to obtain a given amount of food, or a combination of the two (Schoener, 1971; Altmann, 1974; Vedder, 1984). The former would be predicted if search costs are high and the value of alternative foods is not significantly lower than those already eaten, while the latter is expected if search costs are low or alternative foods are of little value. Group 11, ranging in an area of lower food availability than Group BM, showed a broader diet and moved further each day. The broadening of the diet suggests that the increased search costs associated with traveling further are sufficiently large to warrant including more food items in the diet. Another line of theory suggests that when food density is high, animals should use more energy to maximize net energy gain, while if food is scarce, they should save energetic costs (Norberg, 1977). In this study, the gorillas' response to lower food availability does not appear to reflect an energy-saving strategy. However, both western lowland gorillas (*G. g. gorilla*) and eastern lowland gorillas reduce day journey lengths when fruit is scarce (Tutin, 1996; Yamagiwa, 1999). In this case, the situation is complicated by the fact that the preferred food, fruit, has a very different, more dispersed distribution from the fallback herbaceous foods.

The difference in movement patterns in areas of lower food availability is supported by other studies. Watts (1991) found both day journey length and distance between feeding spots in each habitat type to be inversely correlated with food availability as well as certain other parameters of habitat quality for one group of mountain gorillas. Yamagiwa & Mwanza (1994) found that solitary male eastern lowland gorillas in Kahuzi-Biega National Park also moved further per day and between feeding sites than solitary mountain gorilla males in the Virungas where food is more densely and evenly distributed. Similar relationships between movement patterns and habitat quality have been found for other species, including baboons (Barton *et al.*, 1992; Henzi *et al.*, 1992), gibbons (Mukherjee, 1986), and chipmunks (Rosenberg & Anthony, 1993).

Conservation implications
This study has identified *Mimulopsis* and Bamboo as important gorilla habitats, in addition to the higher altitude habitats in the Karisoke study area. Bamboo, although a minor habitat around Karisoke, covers a large portion of the remainder of the Virungas. The fact that this habitat is selected when bamboo shoots are present, despite a low overall food availability, indicates that bamboo shoots are key resources even although they are not present year-round. *Mimulopsis* tends to occur as patches through the Bamboo zone, and contains significant gorilla foods year-round. The combination of these two habitats can support significant numbers of gorillas, providing these lower altitude habitats, which are often closer to the edge of the park, continue to be strictly protected. Mixed forest habitats in the lowest parts of the DRC side of the Virungas, although not included in this study, are also likely to fall into this category (McNeilage, 1995). Such information on habitat availability, quality, and use by gorillas has been used to estimate the number of gorillas that the Virungas could support, and indicates that there is considerable room for expansion beyond the current population level of around 300 individuals (McNeilage, 1995).

Socioecology and taxonomy
The variation found in gorilla ecology across the Virungas could influence conclusions drawn from comparisons between Virunga gorillas and other populations or subspecies. Patterns of resource availability and use are thought to be important determinants of social organization in great apes (Wrangham, 1986; Tutin, 1996; Wrangham *et al.*, 1996; Doran & McNei-

lage, 1998, this volume). The differences in diet and ranging patterns between the two gorilla groups in this study were not as great as those between mountain and western lowland gorillas (Doran & McNeilage, 1998, this volume). However, the fact that Group 11 traveled further per day and between individual feeding sites might be indicative of differences in levels of within-group feeding competition. The variation in ecology within the Virungas could be part of the same continuum of variation across subspecies which is thought to underlie differences in social organization through constraints on foraging party size (Doran & McNeilage, 1998, this volume). It is interesting to note that the largest groups in the Virungas are found in and around the Karisoke study area, where food availability is particularly high. This area might therefore represent the one extreme of the continuum, even relative to other areas within the Virungas. Other factors, such as the degree of protection from human disturbance, may also affect group size. However, the possibility that group size in other parts of the Virungas is limited by resource distribution and abundance to smaller numbers of individuals groups than those in groups around Karisoke merits further investigation.

Using gorillas in the Karisoke study area to make comparisons with other populations of mountain or eastern lowland gorillas (such as those in Bwindi and Kahuzi-Biega) may exaggerate differences in ecology. The ecology of gorillas elsewhere in the Virungas is more likely to be similar to other populations. Mountain gorilla diet is the least diverse of all subspecies (Watts, 1996), with very little fruit. However, the data on which such comparisons have been based come almost entirely from the Karisoke study area. Although fruit did not make up a large proportion of the diet of Group 11, these gorillas did show an increase in fruit consumption, reflected in over three times as many seeds per dung sample, including one species not represented in Group BM dung. Guides and trackers following the group with the lowest home range on the DRC side of the Virungas indicated 13 plant species from which they had seen gorillas eat fruit, ten of which were not found in the Karisoke site and seven of which were not found even in Group 11's range. Several of the species eaten by Group 11 but not by Group BM, notably *Basella alba* and *Urera hypselodendron*, were recorded as important foods in the diet of eastern lowland gorillas (Casimir, 1975; Goodall, 1977) and are important gorilla foods in Bwindi (personal observation). These observations do suggest that the diet of the groups in the lowest parts of the Virungas resembles that in other populations more closely than that of the Karisoke study groups, and includes a greater proportion of fruit in the diet. A detailed investigation of the diet in the lowest groups on the DRC side of the

Virungas would be needed before proper comparisons can be made between the ecology of Virunga gorillas in general and other populations or
subspecies. Without such information, conclusions on behavioral and
taxonomic differences between Virunga gorillas and other populations
based at least partly on ecological differences (Sarmiento *et al.*, 1996)
should be treated with caution.

Summary
Early research on mountain gorilla ecology in the Virungas was concentrated in the Karisoke Research Center study area, which is at the upper
end of the altitudinal range occupied by gorillas. However, mountain
gorillas elsewhere in the Virungas occupy quite different habitats in lower
altitude forest with differing levels of food availability and the gorillas
show differences in ecology which appear to reflect such habitat differences. This study investigated patterns of habitat use and diet in two groups,
in the Karisoke study area (Group BM) and in lower habitats to the east of
Mount Visoke (Group 11). The home range of Group 11 was found to
consist of different habitat types with a considerably lower density of food
than Group BM's range, but was less than half the size. There was little
overlap in the habitats or food species used by the two groups, but both
used habitats selectively, in a way which could partly be related to habitat
quality and showed similar overall patterns of dietary selection. Group 11
occupied habitats with a lower food availability and showed a broader
diet, and traveled further per day, than the group in the richer habitat, as
would be predicted by foraging theory.

This study has expanded our understanding of mountain gorilla ecology
in other parts of the Virungas. To fully understand the implications of this
variation in ecology, further research is needed including finer level studies
of habitat; nutritional analyses of foods consumed in different areas,
related to dietary selection; and variation in group spread, feeding competition, and social behavior across a wide range of habitats. Of particular
interest would be a detailed study of ecology and behavior of groups
occupying the lowest altitude areas of the Virungas in DRC. In this area,
which is of a comparable altitude to that occupied by gorillas in Bwindi,
habitats differ most from those around Karisoke and fruit eating is likely
to be more frequent.

Acknowledgements
I am grateful to the governments of Rwanda and l'Office Rwandais du
Tourisme ed des Parcs Nationaux (ORTPN) for permission to work in the
Parc National des Volcans and to the Dian Fossey Gorilla Fund Interna-

tional for access to the Karisoke Research Center. Funding for the project was from a Leverhulme Trust Study Abroad Studentship, along with the Homeland Foundation, USA, the Dian Fossey Gorilla Fund International, and Bristol University. This research benefited greatly from help and advice from my advisor, Stephen Harris, and also Andy Plumptre, Martha Robbins, Liz Rogers, Craig Sholley, Amy Vedder, David Watts, and Diane Doran. This project would have been impossible without the hard work of Uwimana Fidèle and many other Rwandan field assistants. I am grateful to Joseph Mvukiyumwami (Institut de Recherche Scientifique et Technique, Rwanda) for help with identifying plant specimens, and to Andrew Harrison and Ed Thomas (Geography Department, Bristol University) and Scott Madry and Patrick Meola (Center for Remote Sensing and Spatial Analysis, Rutgers University) for help with GIS and vegetation mapping. Finally, I am grateful to the editors of the book, Martha Robbins, Pascale Sicotte, and Kelly Stewart, for all their efforts in bringing this book together, to the editors and three anonymous reviewers for constructive comments on earlier draft of this manuscript, and to the organizers of the original meetings at the Max Planck Institute for Evolutionary Anthropology in Leipzig, Germany, on which this book is based, Martha Robbins and Christophe Boesch.

References

Aebischer, N J, Robertson, P A & Kenward, R E (1993) Compositional analysis of habitat utilisation from radiotelemetry data. *Ecology*, **74**, 1313–25.
Altmann, S A (1974) Baboons, space, time and energy. *American Zoologist*, **14**, 221–48.
Aveling, C & Aveling, R (1989) Gorilla conservation in Zaire. *Oryx*, **23**, 64–70.
Aveling, C & Harcourt, A H (1984) A census of the Virunga gorillas. *Oryx*, **18**, 8–13.
Barton, R A, Whiten, A, Strum, S C, Byrne, R W & Simpson, A J (1992) Habitat use and resource availability in baboons. *Animal Behaviour*, **43**, 831–44.
Casimir, M J (1975) Feeding ecology and nutrition of an eastern gorilla group in the Mount Kahuzi Region (République du Zaire). *Folia Primatologica*, **24**, 81–136.
Casimir, M J & Butenandt, E (1973) Migration and core area shifting in relation to some ecological factors in a mountain gorilla group (*Gorilla gorilla beringei*) in the Mount Kahuzi Region (République du Zaire). *Zeitschrift für Tierpsychologie*, **33**, 514–22.
Doran, D M & McNeilage, A (1998) Variation in behavior of gorilla subspecies: what's diet got to do with it? *Evolutionary Anthropology*, **6**, 120–31.

290   *Alastair McNeilage*

Estrada, A (1984) Resource use by howler monkeys (*Allouatta palliata*) in the rain forest of Los Tuxtlas, Veracruz, Mexico. *International Journal of Primatology*, **5**, 105–31.

Fossey, D (1974) Observations on the home range of one group of mountain gorillas (*Gorilla gorilla beringei*). *Animal Behaviour*, **22**, 568–81.

Fossey, D & Harcourt, A H (1977) Feeding ecology of free-ranging mountain gorilla (*Gorilla gorilla beringei*). In *Primate Ecology: Studies of Feeding and Ranging Behaviour in Lemurs, Monkeys and Apes*, ed. T H Clutton-Brock, pp. 415–47. London: Academic Press.

Gompper, M E & Gittleman, J L (1991) Home range scaling: intraspecific variation and comparison of trends. *Oecologia*, **87**, 343–8.

Goodall, A G (1977) Feeding and ranging behaviour of a mountain gorilla group (*Gorilla gorilla beringei*) in the Tshibinda-Kahuzi region (Zaire). In *Primate Ecology: Studies of Feeding and Ranging Behaviour in Lemurs, Monkeys and Apes*, ed. T H Clutton-Brock, pp. 459–79. London: Academic Press.

Grieg-Smith, P (1983) *Quantitative Plant Ecology*, 3rd edn. Oxford: Blackwell Scientific Publications.

Harcourt, A H & Fossey, D (1981) The Virunga gorillas: decline of an island population. *African Journal of Ecology*, **19**, 83–97.

Harcourt, A H, Kineman, J, Campbell, G, Yamagiwa, J, Redmond, I, Aveling, C & Condiotti, M (1983) Conservation of the Virunga gorilla population. *African Journal of Ecology*, **21**, 139–42.

Harris, S, Cresswell, W J, Forde, P G, Trewhella, W J, Woollard, T H & Wray, S (1990) Home-range analysis using radio-tracking data: a review of problems and techniques as applied to the study of mammals. *Mammal Review*, **20**, 97–123.

Henzi, S P, Byrne, R W & Whiten, A (1992) Patterns of movement by baboons in the Drakensberg mountains: primary responses to the environment. *International Journal of Primatology*, **13**, 601–29.

Jiang, H, Liu, Z, Zhang, Y & Southwick, C (1991) Population ecology of rhesus monkeys (*Macaca mulatta*) at Nanwan Nature Reserve, Hainan, China. *American Journal of Primatology*, **25**, 207–17.

Larter, N C & Gates, C C (1994) Home range size of wood bison: effects of age, sex and forage availability. *Journal of Mammalogy*, **75**, 142–9.

Lebrun, J (1960) Sur les horizons et étages de végétation de divers volcans du massif des Virunga (Kivu, Congo). *Bulletin du Jardin Botanique de l'Etat à Bruxelles*, **30**, 255–77.

Marius, C (1976) Cartes de la végétation 1 et 2 des Birunga. Programme de Recherche sur la Flore et la Végétation de la Chaine des Birungas et du Kahuzi-Biega. Unpublished F.K.F.O. project report.

McNeilage, A (1995) Mountain gorillas in the Virunga Volcanoes: ecology and carrying capacity. PhD thesis, University of Bristol.

Mehlman, P J (1989) Comparative density, demography and ranging behaviour of barbary macaques (*Macaca sylvana*) in marginal and prime conifer habitats. *International Journal of Primatology*, **10**, 269–92.

Mukherjee, R P (1986) The ecology of the hoolock gibbon, *Hylobates hoolock*, in

Tripura, India. In *Primate Ecology and Conservation*, ed. J G Else & P C Lee, pp. 115–23. Cambridge: Cambridge University Press.

Norberg, P A (1977) An ecological theory on foraging time and energetics and costs of optimal food searching method. *Journal of Animal Ecology*, **46**, 511–29.

Norbury, G L & Sanson, G D (1992) Problems with measuring diet selection of terrestrial mammalian herbivores. *Australian Journal of Ecology*, **17**, 1–7.

Plumptre, A J (1991) Plant–herbivore dynamics in the Birungas. PhD thesis, University of Bristol.

Plumptre, A J (1994) The effects of trampling damage by herbivores on the vegetation of the Parc National des Volcans, Rwanda. *African Journal of Ecology*, **32**, 115–29.

Plumptre, A J (1995) The chemical composition of montane plants and its influence on the diet of large mammalian herbivores in the Park National des Volcans, Rwanda. *Journal of Zoology, London*, **253**, 323–37.

Rosenberg, D K & Anthony, R G (1993) Differences in Townsend's chipmunk populations between 2nd-growth and old-growth forests in western Oregon. *Journal of Wildlife Management*, **57**, 365–73.

Sarmiento, E E, Butynski, T M & Kalina, J (1996) Gorillas of Bwindi-Impenetrable Forest and the Virunga Volcanoes: taxonomic implications of morphological and ecological differences. *American Journal of Primatology*, **40**, 1–21.

Schaller, G B (1963) *The Mountain Gorilla: Ecology and Behavior*. Chicago: University of Chicago Press.

Schoener, T W (1971) Theory of feeding strategies. *Annual Review of Ecology and Systematics*, **2**, 369–404.

Sholley, C R (1991) Conserving gorillas in the midst of guerillas. *American Association of Zoological Parks and Aquariums, Annual Conference Proceedings*, 30–7.

Southwood, T R E (1966) *Ecological Methods: With Particular Reference to the Study of Insect Populations*. London: Chapman & Hall.

Spinage, C A (1972) Ecology and problems of the Volcano National Park, Rwanda. *Biological Conservation*, **4**, 194–204.

Struhsaker, T T (1975) *The Red Colobus Monkey*. Chicago: University of Chicago Press.

Troupin, G (1977–1988) *Flore du Rwanda: Spermatophytes I–IV*. Tervuren, Belgium: Musée Royal de l'Afrique Centrale.

Tutin, C E G (1996) Ranging and social structure of western lowland gorillas in the Lopé Reserve, Gabon. In *Great Ape Societies*, ed. W C McGrew, L F Marchant & T Nishida, pp. 58–70. Cambridge: Cambridge University Press.

Vedder, A L (1984) Movement patterns of a group of free-ranging mountain gorillas (*Gorilla gorilla beringei*) and their relation to food availability. *American Journal of Primatology*, **7**, 73–88.

Watts, D P (1984) Composition and variability of mountain gorilla diets in the central Virungas. *American Journal of Primatology*, **7**, 323–56.

292   Alastair McNeilage

Watts, D P (1987) Effects of mountain gorilla foraging activities on the productivity of their food plant species. *African Journal of Ecology*, **25**, 155–63.

Watts, D P (1991) Strategies of habitat use by mountain gorillas. *Folia Primatologica*, **56**, 1–16.

Watts, D P (1996) Comparative socio-ecology of gorillas. In *Great Ape Societies*, ed. W C McGrew, L F Marchant & T Nishida, pp. 16–28. Cambridge: Cambridge University Press.

Watts, D P (1998a) Long-term habitat use by mountain gorillas (*Gorilla gorilla beringei*). I. Consistency, variation, and home range size and stability. *International Journal of Primatology*, **19**, 651–80.

Watts, D P (1998b) Long-term habitat use by mountain gorillas (*Gorilla gorilla beringei*). II. Re-use of foraging areas in relation to resource abundance, quality and depletion. *International Journal of Primatology*, **19**, 681–702.

Watts, D P (1998c) Seasonality in the ecology and life histories of mountain gorillas (*Gorilla gorilla beringei*). *International Journal of Primatology*, **19**, 929–48.

Weber, A W & Vedder, A (1983) Population dynamics of the Virunga gorillas, 1959–1978. *Biological Conservation*, **26**, 341–66.

White, F (1981) The history of the Afromontane archipelago and the scientific need for its conservation. *African Journal of Ecology*, **19**, 33–54.

Wrangham, R W (1986) Ecology and social relationships of two species of chimpanzees. In *Ecological Aspects of Social Evolution: Birds and Mammals*, ed. D I Rubenstein & R W Wrangham, pp. 352–78. Princeton: Princeton University Press.

Wrangham, R W, Chapman, C A, Clark-Arcadi, A P & Isabirye-Basuta, G (1996) Social ecology of Kanyawara chimpanzees: implications for understanding the cost of great ape groups. In *Great Ape Societies*, ed. W C McGrew, L F Marchant & T Nishida, pp. 45–57. Cambridge: Cambridge University Press.

Yamagiwa, J (1999) Socioecological factors influencing population structure of gorillas and chimpanzees. *Primates*, **40**, 87–104.

Yamagiwa, J & Mwanza, N (1994) Day-journey length and daily diet of solitary male gorillas in lowland and highland habitats. *International Journal of Primatology*, **15**, 207–24.

11 Clever hands: the food-processing skills of mountain gorillas

RICHARD W. BYRNE

Titus eating thistle. (Photo by Richard W. Byrne.)

The mentality of apes is often discussed as if it had no direct connection with their natural lives. Whereas behavioral ecology is studied by zoologists and set in relation to the foraging strategies of other species, mentality is seen as the province of psychologists and all too often studied only in the laboratory. An artificial and unhelpful divide for any species of animal, this partition is especially damaging for a proper understanding of the great apes. I will argue that the cognition of the great apes can only be understood in relation to their means of acquiring an adequate diet: that, in a very real way, great apes inhabit a "cognitive niche". To set out and evaluate this proposal, it is essential to begin by considering the foraging choices available to the apes.

The ecological niche of the great apes

Large primates with simple stomachs have a stark choice, between plant foods which are relatively nutritious but costly to acquire, and those which are more readily available but of inferior quality (Waterman, 1984). Lacking morphological adaptations for carnivory, only the very smallest primate species – such as galagos, the smaller lemurs, marmosets and tamarins – are able to specialize on animal food (insects), in some cases supplemented by tree gum (Hladik, 1978). Only the colobine monkeys possess complex stomachs, enabling high-quality nutrition to be obtained from mature leaves via bacterial action (Chivers & Hladik, 1984). For the remainder of the primate order, there is limited room to maneuver. In general, plant items which provide abundant and readily available energy, especially simple sugars and lipids, tend to be sparsely distributed and short-lasting and so require considerable locomotor effort to obtain. Plant foods which are easily obtained year-round tend to be low in energy, consequently setting tight limits on the energy that is available for further foraging. Thus primate species must balance foraging effort against energy yield (Clutton-Brock & Harvey, 1977), leading to a spectrum of strategies between the energetic and wide-ranging search for ripe fruit on the one hand, and the more sluggish foraging for relatively coarse plant items on the other. A classic illustration of these extreme options is the contrast in Neotropical forests between the foraging strategies of spider and howler monkeys, similar-sized and closely related species, both with simple stomachs (Milton, 1981). Nevertheless, in environments which offer unusual nutritional opportunities, primate species flexibly exploit alternative niches; familiar examples are the gelada specialization on gramnivory in high-altitude grass meadows (Dunbar, 1977; Iwamoto, 1979), and the reliance of some colobines on large seeds (McKey et al., 1981; Harrison & Hladik, 1986).

294

At first sight, the apes appear to fit neatly into this pattern as typical plant foragers, trading off foraging effort against energy yield; the importance of high-quality fruit items in the diet apparently dictates the extent of feeding competition, with consequent effects on foraging group sizes and day journey lengths (Wrangham, 1979*a*). Thus, the small-bodied gibbons pursue a high-energy/high-effort strategy, foraging primarily for ripe fruit and accordingly showing large day ranges and small group size (Chivers, 1974; Ellefson, 1974; Chivers *et al.*, 1975; Tenaza, 1975). The orangutan and chimpanzee are not dissimilar, also ripe fruit specialists though exploiting a wider range of plant and animal material (Rodman, 1973, 1977; Wrangham, 1977, 1979*b*; Galdikas, 1979); both frequently forage alone or in very small parties, but these fuse at times into larger groupings, more especially in the chimpanzee when day journeys are often large. At the opposite extreme, the gorilla is traditionally characterized as a folivore (Fossey & Harcourt, 1977; Goodall, 1977), and gorillas' short day ranges and maintenance of cohesive groups are seen as consequences of their abundant but low-quality forage.

Recent work on the behavioral ecology of great apes has revealed a more complex picture, however. Although energy-rich, ripe fruit is typically lacking in protein, and certain minerals and vitamins (Waterman, 1984). Fruit specialists must remedy these deficiencies by supplementing their diets in various ways, and this broadening of diet rather than simply the search for ripe fruit has been argued to be responsible for the large day ranges of frugivorous primates, including the chimpanzee (Hladik, 1975). Chimpanzee diet supplementation is evidently dependent on hunting of mammals and insect-gathering with tools, whereas orangutans are apparently able to gain sufficient protein from insects collected without tools and non-fruit plant material. But now one population of Sumatran orangutans has been found to show frequent tool-use for obtaining insects and gaining access to fruit defended by stinging hairs and hard husks (van Schaik *et al.*, 1996; Fox *et al.*, 1999) – raising the question of why other orangutan populations do not use tools. At the same time, lowland gorilla diet has proved to be more chimpanzee-like than was previously assumed, with extensive use of ripe fruit and nutritional supplementation by insect-foraging, though without the use of tools (Rogers *et al.*, 1988; Tutin *et al.*, 1997).

The use of herbaceous vegetation by African great apes is now better understood. In lowland gorillas, and also some chimpanzee populations, terrestrial herbaceous vegetation is used in periods of seasonal dearth; this vegetation is of relatively low quality and apparently forms a fallback diet (Wrangham *et al.*, 1991, 1996). In contrast, phytochemical examination of the terrestrial herbs that comprise the bulk of the mountain gorilla's diet has shown that they are by no means of low nutritional quality (Watts, 1984), as labeling gorillas "folivores" would suggest. Compared with tropical forest leaves, the staple foods of mountain gorillas contain considerable amounts of protein and relatively little indigestible fiber or poisonous secondary compounds. Studies of the behavioral ecology of bonobos have found a dietary strategy in some ways closer to that of mountain gorillas, despite their very similar body size to that of chimpanzees. Unlike chimpanzees, bonobos show little insect-gathering or hunting, but instead rely year-round on terrestrial herbaceous vegetation of relatively high nutritional quality (Badrian & Malenky, 1984; Malenky & Wrangham, 1994). Finally, this distinction between low-quality and high-quality terrestrial herbaceous vegetation has been mirrored in the case of ripe fruit by the realization that some fruit does, after all, contain an important supply of amino acids and other dietary supplementation: in the form of the bodies of fig wasps (O'Brien et al., 1998). Where chimpanzees can gain year-round access to ripe figs, and have the option of occasional fallback reliance on terrestrial herbs, they have far less need of insects or mammals in their diet, and show little tool-use or hunting (Chapman & Wrangham, 1993; Wrangham et al., 1993).

What are we to make of the now rather complex picture of great ape ecology? The simple model of linear variation, between high-effort/high-yield and low-effort/low-yield foraging, is simply insufficient to describe the differences among great apes in behavioral ecology. (The lack of any comparable revisions in our understanding of lesser ape foraging behavior may also be a function of research effort, rather than any perfect understanding of gibbon and siamang behavioral ecology.) It will be the contention of this chapter that understanding the use of *manual skills* in foraging is central to appreciating the options open to a great ape.

This is most obviously so in the case of the chimpanzee, although the attention given to the anthropological and cultural significance of tool-making and tool-use has often overshadowed the ecological significance of tool-reliant insect-gathering. Whereas in many chimpanzee populations mammalian and bird protein is eaten only sporadically and is not available to all members of the community, insect protein is obtained year-round in considerable quantities, particularly by females – the sex whose reproductive options are most limited by food availability (McGrew, 1979). It is unclear what chimpanzee life would be like if the species did not have the cognitive capacity to exploit insect food using tools, since the only populations which do not do so are buffered by the presence of abundant figs. In orangutans, the recent discovery of tool-using at one site, where standardized techniques of employing tools are widespread in the population, is evidence that the lack of tool-use in most populations is not simply a result of cognitive incapacity. Orangutan skills at tool-making in captivity have long suggested that this could not be the case (e.g. Wright, 1972; Lethmate, 1977). Instead, the key variable may be population density (van Schaik *et al.*, 1999): tool-using may occur sporadically in all orangutan populations, but fail to develop into stable traditions except at sites where meetings between individuals are frequent enough and prolonged enough to enable efficient transmission of the tool technology by social learning. Evidently, offspring can always learn from their mothers, but any such "family habit" is inherently liable to die out in the long term unless it can spread to a wider group. On this hypothesis, both species possess the innate capacity to make and use tools, but only in chimpanzees, and in one high-density orangutan population, is social contact intense enough to allow transmission of the necessary techniques through the general population, providing a basis for persistent tool-use traditions.

Given that three members of the great ape clade (Hominidae: Begun, 1999) make and use tools, human, chimpanzee, and orangutan, it would be expected that the remaining two species, gorilla and bonobo, should possess similar mental capacities. And indeed, like the orangutan, in captivity these species do show tool-making skills (McGrew, 1989; Toth *et al.*, 1993). This suggests that either the cognitive capacity that all great ape species possess evolved in response to an ecological need specifically for tool-use, in which case the common ancestor of the great apes was a tool-user and the modern distribution of tool-using in wild great apes is a secondary consequence of socioecology (Parker & Gibson, 1977); or the original evolutionary function of these skills lay elsewhere. But if so, where?

A possible answer to that question comes from study of the food-preparation skills of the mountain gorilla, which has revealed cognitive skills in many ways directly equivalent to those involved in chimpanzee and orangutan tool-use (Byrne & Byrne, 1993). This discovery has two-fold significance for our understanding of the technical abilities of great apes. Firstly, it highlights the possibility that the cognitive capacity, so famously and strikingly demonstrated in the tool cultures associated with insect feeding, has its evolutionary origin in the much wider domain of plant-gathering and plant preparation (Byrne, 1996). Secondly, it enables direct comparison to be made with the equivalent abilities of Old World monkeys, whereas the absence of tool-using cultures in all monkeys makes comparison difficult in that domain. This is important, because the monkeys form the most convenient outgroup to the great ape clade; ideally, the gibbons would be used as outgroup to the great apes, but there is little available information of their feeding skills or technological abilities in general, and their highly arboreal behavior makes it unlikely that this will be soon forthcoming. The remainder of this chapter will describe the manual skills that mountain gorillas evidence in their plant-gathering, and attempt to evaluate their cognitive and ecological implications.

Mountain gorilla plant feeding

The terrestrial herb vegetation of the Virunga Volcanoes provides a dense and even distribution of food for the mountain gorilla, making available abundant food year-round (Watts, 1996). Whereas great apes in tropical forests must search for food plants which are relatively rare, plants which are food species for mountain gorillas make up a large fraction of the total array of herbaceous plants (Watts, 1984). Furthermore, these herb foods are typically low in indigestible matter, undefended by secondary compounds, and nutritious – often particularly rich in protein (Waterman *et al.*, 1983). Infants and mothers tend to feed synchronously (Watts, 1985), and infants are showered with feeding remains from their first day of life (personal observation). Combined with the relative lack of seasonal variation in mountain gorilla diet, this makes learning *what* foods to eat a simple task for an infant gorilla – and most of the non-foods are not actually poisonous, so this learning process is not a risky one.

Knowing *how* to deal with each species of plant is more problematic. The major foods for the Karisoke study population are all plants which present physical difficulties for processing (Byrne & Byrne, 1991). Bedstraw *Galium ruwenzoriense* is adapted to clamber over other vegetation by means of tiny clinging hooks, found on the edges of the angular stem and on its small leaves, especially their edges. These hooks make the stems awkward to handle, and moreover are liable to cling to the mouth and throat, causing risk of choking. Nettle *Laportea alatipes* is defended by powerful stinging hairs, most abundant on the stems, leaf petioles, leaf edges, and uppersides. It is apparent from the behavior of immature gorillas exploring the plants that these stings hurt them, just as they do humans. Thistle *Carduus nyassanus* is a large plant of typical thistle form with numerous woody spines on the winged stems, leaf ribs, and edges. Celery *Peucedanum linderi* is a sprawling umbelliferous plant whose stems may be over 5 m long. Not only does this make it unwieldy to eat, but the pith – the edible part for a gorilla – is encased in hard outer stem casing, which is woody and evidently indigestible.

Acquisition of complex skills by gorillas

Since stinging nettles are a familiar sight in Europe and northern North America, it is instructive to consider how one might best deal with them, lacking gloves and knives. Nettles are in fact sufficiently nutritious to have been a traditional food in England – after cooking. The leaf-blades are the least sting-infested part of the plant, and the least stiffened with lignin and cellulose; the larger hindgut of the gorilla will allow more digestion of cellulose than in humans, but the extent is still limited. Logically, then, the leaves need to be picked, and the petioles removed. Since each leaf that is put in the mouth is liable to be painful, it would seem wise if a number of leaf-blades were inserted simultaneously. This approach is indeed effective, as the stinging hairs operate best in response to a gentle brush on the skin, and are deactivated by crushing; many children discover that a nettle grasped resolutely and firmly hurts less, hence the phrase "Grasping the nettle". Most adults to whom I have set this conundrum can work out most of these principles, especially if have they had experience with nettles in childhood. Their preferred solution is then to pick leaves with a firm grip at the top of their petioles, one by one, transferring each picked leaf to the other hand – since further manipulation with already-picked leaves in the same hand is liable to awkward and hence painful. Then, grasping the base of the pile of leaf-blades, the petioles can be twisted off, and with the pile held at either side it can be consumed in several large bites. This would work, and none of the naïve informants I have interrogated has suggested a better way.

Mountain gorillas use a much more efficient and effective strategy. A nettle stem is encircled near its base with a part-closed, cupped hand, and – with the roots supported if the ground is soft – the hand is pulled upwards, stripping off several whorls of leaves in one movement. The petioles of these leaves protrude from one side of the half-closed hand, and are firmly grasped with the other hand and the two hands twisted or rocked against each other (knuckle-walking adaptations apparently make twisting the forearm less easy for gorillas than humans: Aiello & Dean, 1990). The detached petioles are dropped. Finally, the bundle of leaf-blades is gently pulled out of the fist and folded over the thumb, then re-grasped to form a "sandwich" in which the great majority of the stings are enclosed in a single leaf's underside, before being popped as a whole into the mouth. This method is quick, since all the leaves of a whole stem are taken in a single movement. The expressions of young gorillas examining nettles strongly suggest that their hands are as sensitive as our own, and presumably pulling a stem through a cupped hand cannot be entirely free of pain even for juvenile and adult gorillas; however, the neat folding before ingestion minimizes the number of stings that might contact the delicate tissue of the mouth. Often several plants are dealt with to produce a single, large handful of food, the gorilla holding the part-processed leaves in the lower fingers of one hand; this iteration may involve only the first stage of the process, stripping off a whorl of leaves, or more extensive processing in which the petioles are torn off and discarded before the leaf-blades are accumulated.

How did wild mountain gorillas come by such an excellent method, one better than thoughtful humans can devise? Ideally, developmental data or field experiments would be used to answer this question; in the absence of such evidence, there are some clues in the behavior of adults. A specialized nettle-eating routine, genetically hard-wired as a complete program, is somewhat unlikely in a primate; the more so when one considers that it would be of no possible use to the vast majority of individuals of the species, since nettles are temperate plants, found in Africa only at very high altitudes. In any case, each of the major food plants is consumed with a similarly complex strategy, closely matched to the particular – and quite different – problems posed by each plant (Byrne & Byrne, 1991). The evolution of *four* complex, hard-wired routines in a small subpopulation of a widespread primate species, noted for its flexible behavior in different environments, stretches plausibility beyond the limit. Could the techniques have been learned by individual experience in every member of the population? The fine details of each technique – the precise grip with which nettle leaves are stripped from a stem, the use of left or right hand, and so on – vary idiosyncratically between individuals, even between mothers and offspring, in a way that strongly suggests that details are indeed acquired by individual exploration (Byrne & Byrne, 1993). However, the overall *organization* of the techniques is not idiosyncratic; organization here includes the linear sequence of discrete stages in the process and their grouping into larger modules, the occurrence of optional stages, the use of bimanual coordination where the two hands perform different roles in concert, and the use of "subroutines" to iterate a process to some criterion. These aspects are remarkably standardized in the population (Byrne & Byrne, 1993). The only way in which this situation could come about by individual, trial-and-error learning is if each technique were not only *optimal*, which it may well be, but also if it could be connected to the most obvious fumbling attempts of an infant by a *monotonically improving series* of intermediate techniques. Only then could a "hill-climbing" algorithm, that learns by making small changes to a technique and evaluating whether these are better or worse, reach this optimum. The argument is parallel to that in the theory of natural selection, where any structure must necessarily have evolved through a set of predecessors, each of which is close to the last yet with a selective advantage over it. Just as evolutionary saltations are statistically impossible (see Dawkins, 1986), so are major leaps in trial-and-error learning. The fumbling attempts of infant gorillas are based on picking single nettle leaves and attempting to consume them one at a time (unpublished data), a technique that could clearly develop by a series of graded intermediates into the perfectly feasible pick-and-pile method, as suggested by most humans. However, it is quite unclear how single-leaf picking, or the pick-and-pile method, could develop into the very dissimilar technique of adult gorillas without at least one major reorganization – in effect, an inspired saltation.

Since all the alternatives are highly implausible, it seems inescapable that young gorillas *do* acquire some knowledge of the adult technique by imitation; no sign of active maternal teaching has ever been recorded (Byrne, 1999*a*). Imitation in this case clearly cannot involve complete duplication of the observed actions, but rather extraction of the overall structure or program of the process; thus the term "program-level imitation" has been coined (Byrne & Byrne, 1993; Byrne & Russon, 1998). Like any complex behavior, gorilla plant preparation is hierarchically organized (Byrne, 1993). In program-level imitation, the hierarchical structure of the observed process is discovered by observation, and used to organize the future attempts of the observer, while the precise details of manual actions, for example how plants are held and moved, may be discovered more efficiently by trial-and-error.

It would be convenient for scientists if populations of gorilla were found in which exactly the same food was consumed with structurally different techniques; however, even with the opening of new study sites in the future, this is very unlikely to happen. The problem is that such a case would mean that one population persisted with an *inefficient* method. This would clearly be unstable, liable to switch to the more efficient technique as soon as one individual discovers it – and that must be presumed possible, or there could not exist another population which already used the more efficient way. Indeed, despite the many local traditions of behavior among mammals and birds maintained by social learning, no convincing example is known where an inter-population difference reflects a difference in learning history alone. These local traditions are considered to be maintained by simpler forms of social learning than imitation, in particular by stimulus enhancement, in which the behavioral method is acquired by individual learning once the animal's attention is narrowly focused on the task, as a result of seeing another individual similarly engaged (Roper, 1983; Galef, 1992). Differences in behavior between populations thus normally correlate with, and are most simply viewed as maintained by, ecological variation. Fortunately for ape researchers there is one exception to this, in the chimpanzee. The difference between east and west African chimpanzee populations in the style of "dipping" for *Dorylus* ants (McGrew, 1974; Sugiyama *et al.*, 1988; Boesch & Boesch, 1990) is a clear indicant of traditional transmission by imitation, since one method is much less efficient than the other, while the prey species and raw materials do not differ in any significant way (Boesch & Tomasello, 1998). Considering the numerous local traditions of behavior that have been described from the many long-term chimpanzee study sites (Whiten *et al.*, 1999), almost all vulnerable to explanation as the result of subtle ecological differences (Tomasello, 1990), to expect any such exception in the gorilla is optimistic. At present, feeding techniques have only been described from one site; in most studies, habituation is as yet insufficient for close observation. However, one case recorded within the Karisoke population is suggestive (Byrne, 1999*a*). The young female Pc immigrated into Group 5 in 1984 from a group whose range was at considerably lower altitude, below the zone in which nettles are common. In 1989, she still had not mastered the crucial step of folding nettle leaf-blades before consumption. Moreover, the only other individual in the study population of 38 adults and juveniles who lacked this skill was her juvenile offspring, Iz (Byrne & Byrne, 1993).

Cognitive underpinnings of gorilla manual skills

What mental apparatus is required, in order to build up the sort of complex skills that are shown in gorilla food preparation? And what sort of mind would it take to use the behavior of more expert conspecifics as a source of data about the organization of a technique, supplementing the results of individual exploration during normal development?

Part of the answer follows directly from the elaborate, hierarchical organization of the techniques. At the most basic level, the sequence of steps (i.e. subgoals) in these tasks is quite long, so a reliable *memory* is necessary. The memorized sequences cannot map directly on to motor movements, because the programs are structured hierarchically: an element in the sequence may correspond to a subroutine, itself consisting of a series of elements. For instance, in nettle processing the sequence of steps < strip up stem, grip petioles with other hand, twist/rock to detach and drop, retain leaf-blades > may optionally be repeated several times, holding the already-processed leaves with the lower fingers of one hand, until a sufficiently large bundle is accumulated. This string is treated as a subroutine, under the control of the main program, which evaluates when it has been repeated enough times. An individual's cognitive system must permit flexible handling of these hierarchically organized programs. This implies a mechanism for "keeping the place" when the sequence of action loops around an embedded subroutine, or when an optional routine occurs next in linear sequence but is omitted (Byrne, 1998): some form of *working memory* is needed for this executive place-keeping. A chain-like, linear structure of the form that would be produced by associative learning is wholly inadequate to represent conditional elements and embedded sub-routines; the memory must permit *encoding of hierarchical information.* In gorillas, the evidence points to a hierarchy with only shallow embedding, whereas in human planning much deeper structures are normal; in consequence, working-memory capacity may be less extensive than in humans.

Program-level imitation puts further demands on the cognitive system. In order to copy the higher-level organization, the fluid sequence of observed behavior must first be seen as a sequence of discrete operations: it must be *segmented* (Byrne, 1999*b*). However, since the ultimate function of such segmentation is to construct novel organizations of behavior out of building-blocks which are themselves actions in the observer's repertoire, the discrete operations that are thereby recognized may be restricted to actions that the observer can perform. A neural system that matches these requirements is already known in area F5 of rhesus monkey pre-motor cortex (Rizzolatti, 1981; Rizzolatti & Gentilucci, 1988; Gallese & Goldman, 1998). This is a population of cells, known as mirror neurons, which respond to the sight of specific goal-directed actions equally if the animal itself or another individual performs them. Mirror neurons have been found that are tuned, for instance, to pulling, pushing, or twisting an object – very much the level of specificity that would be needed as a basis for program-level imitation. Direct copying of familiar actions in the existing repertoire does not, of course, constitute imitation, and is better described as response facilitation (Byrne, 1994). For real imitation, a further process is needed to extract the underlying organizational structure from the linear sequence of these action elements: in effect, to *parse* the sequences of action elements (Byrne, 1999*b*). This process must be able to detect and use the "signature" that the underlying structure imparts to the surface form of behavior: for instance, repeated identical sub-strings in a longer sequence may be a consequence of the iterations of a subroutine, recurring pauses at the same point in identical sequences imply a disjunction between separate modules of an overall program, and so on. Note that, since these are *statistical regularities,* only multiple observations could possibly serve as the database for their extraction. The need for repeated observations in great ape imitation has been noted before

(Russon, 1996), but unfortunately most laboratory "tests" of imitation in primates have not taken this into account, with predictably disappointing results (Tomasello & Call, 1997). A population of cells has also been described recently, whose properties are consistent with their being part of this parsing mechanism. These cells are found in a part of rhesus monkey cortex richly interconnected with F5, the supplementary motor area in the medial frontal cortex, and they respond selectively to sequences of observed actions (Halsband *et al.*, 1994; Tanji & Shima, 1994; Tanji *et al.*, 1996). If this speculative identification proves correct, then the absence of any good evidence for imitation by monkeys is puzzling (Visalberghi & Fragaszy, 1990). The problem for monkeys may lie in what they can do with any statistical regularities they can detect. For effective program-level imitation, the *structure* thus deduced must be encoded in a control hierarchy, and the details of how each step can be achieved are then derived – either by (even closer) observation of a skilled model, or by individual trial-and-error learning. It may be that this creation of a novel control hierarchy is heavy on memory, and great apes crucially differ from monkeys in working-memory capacity. The reality of a monkey/ape difference in developing complex, novel manual organizations is suggested by the distribution of gestural communication among primates. Most species use a few species-typical manual gestures in communication, but chimpanzees and gorillas, at least in captivity, have been shown to build up quite elaborate repertoires of manual gestures (Tomasello *et al.*, 1985, 1989, 1994; Tanner & Patterson, 1992; Tanner & Byrne, 1996, 1999). These gestures communicate meanings to others in the social group, often as iconic representations of desired actions or indications of particular locations. In order to develop gestures not typical of the species, and to learn the intended meanings of others' gestures, some ability to analyze and synthesize gestural configurations is required. Nothing similar has been reported in monkeys, as yet.

Which other primates have comparable cognitive mechanisms?
These abilities are significant for an understanding of the evolution of human cognitive capacities, but are the gorilla skills convergent with human ones, or do they reflect common descent? If the former, what was the common environmental stimulus that led to parallel evolution of such skills in both species? And if the latter, when did they enter the human lineage? Comparative data from other species of primate should enable these questions to be answered, but at present these data are patchy.

Manual techniques of food preparation in monkeys have seldom been described in any detail, perhaps because monkeys often eat foods requiring little processing, and the procedures they then use are of a rudimentary sort: "pick up and put in mouth", or "pull off and put in mouth". African baboons (*Papio anubis*) and vervet monkeys (*Cercopithecus aethiops*) have both been observed processing foods which need more significant processing, and in each case a considerable repertoire of actions is employed, a consequence of the dextrous five-fingered primate hand and the sensitive grip allowed by fingernails and pads (Napier, 1961). Baboons typically feed while traveling, and most of their manual processing consists of single actions: twisting off a leaf, yanking out a rooted tuft, stripping leaves off a branch in a single movement (Whiten, 1988). Vervets, when feeding on larger items, sometimes employ many actions one after the other (Harrison, 1996). However, the sequence of actions is ever-changing, so that it is not possible to see any organizing principle behind the choice of what to do next. Some actions do tend to follow others, but the sequence varies on different occasions. What seems to be happening is that the food objects themselves are continually changing, as a result of the vervet's attentions, and the changing problem elicits a continually different response. Vervets have a large repertoire of manual actions available for food processing and different actions are elicited by the sight of particular configurations of the visible food stimulus: one plant-problem, one action. This "one problem, one action" style does not typically produce any sequential organization in behavior. Where a big food item changes during processing, a series of actions may be elicited one after the other, but in an unordered, unsystematic way. At present, the evidence from monkey plant-food handling suggests no comparable cognitive skills to those implied by gorilla feeding.

The strepsirhine primates have less dextrous hands than haplorhines, and in general their manual skills have been found rudimentary in laboratory testing (Tomasello & Call, 1997), and are little remarked in the field. This makes the bamboo processing of the grey bamboo lemur *Hapalemur griseus* all the more remarkable (Stafford *et al.*, 1993). These small lemurs live exclusively on one species of bamboo, which they process with a standardized, multi-stage sequence of bimanual operations. Unlike the case of gorilla food processing, bamboo lemurs possess and require only one program for their entire nutritional needs, so that a genetically encoded routine is more likely. Indeed, all the three other species of *Hapalemur* subsist on single species of giant grass: *H. simus* and *H. aureus* on two species of giant bamboo, *H. aloatrensis* on giant reeds. A single, hard-wired program could underwrite feeding skills in the whole genus, and is perhaps the most likely explanation given the lack of any comparable manual skills in the entire strepsirhine suborder (Byrne, 1999*a*).

Only among the great apes do any primate species show a whole range of standardized, multi-stage, bimanually coordinated manual skills, although comparable research on lesser apes is lacking. This is most obvious in the tool-making and tool-using of chimpanzees. In particular, the traditions of insect-fishing, ant-dipping of two different styles, and honey-probing all involve an ordered series of often delicate operations for a successful outcome, and the tool-making stages require bimanual coordination with the two hands taking different roles. Similarly, although based on found tools rather than tool-making, the traditions of stone-tool use in some west African populations often involve the prior transport of the hammer-stone, assembly of a collection of nuts to process, and use of the other hand to steady each nut for effective cracking (Boesch & Boesch, 1983; Sugiyama *et al.*, 1993). Thus when we focus on the manual actions, rather than the use of a tool *per se*, chimpanzee and gorilla feeding skills have much in common. No data are available on bonobo food-processing skills, but the first studies of orangutans dealing with natural plant foods suggest planned, hierarchically organized programs of action closely comparable to those of gorillas (Russon, 1998). In any case, the existence of two tool-using techniques in one population, and the considerable evidence of imitation in rehabilitant orangutans, suggest that their manual skills are at least as sophisticated as those of other great apes.

Much the most parsimonious interpretation of these data is that the necessary cognitive abilities, to allow the development of flexible, hier-archically organized skills dependent on social learning for their trans-mission, were present in the common ancestor of the great ape clade at least 12 million years ago, and their occurrence in modern great apes is by homology (Byrne, 2000). Precisely which of the component abilities date from earlier epochs, and are consequently shared between humans and a wider range of living primates, is a target for future research. Similarly it is unclear at present whether the ability to incorporate tools into complex routines of manual behavior requires any additional cognitive capacity, perhaps one unique among non-humans to the chimpanzee and oran-gutan, or whether the patchy distribution of tool-using traditions more reflects ecological needs (Parker & Mitchell, 1999).

The cognitive niche?
To return to the issue with which this chapter began, the behavioral ecology of the great apes, we can now view these species as differing from monkeys not simply in locomotion and body size, but in the possession of special manual skills. This helps to explain a seldom-recognized puzzle: why did great apes not become extinct long ago? Simply viewed as "big, clambering monkeys", great apes appear to be in direct niche competition

with species better adapted to survive, the Old World monkeys. Like great apes, their simple stomachs constrain them to trading off nutritional benefits and locomotor costs in their search for plant foods in the same habitats. However, these locomotor costs are considerably lower for monkeys, not only because of their smaller size, but also because of the greater efficiency of quadrupedal locomotion for long-range travel. In contrast, the advantage of large body size in allowing coarser plant material to be digested is more than offset by gut adaptations of Old World monkeys that allow them to eat fruit less ripe, and coarser leaves, than can be tackled by their great ape competitors. When it comes to feeding competition, monkeys appear to have all the aces (Byrne, 1997). Fossil evidence shows that many ape species did indeed become extinct, and exactly why has not been established. The early Miocene (around 20 million years ago) was characterized by many ape species, but very few monkeys; but by the late Miocene (around 10 million years ago, the time of origin of the clade containing all the extant great apes) to judge by the fossil record there were very few apes but an abundance of Old World monkeys (Fleagle, 1998). This is as we would expect, if monkeys in general out-compete great apes; but those few great ape species that did survive evidently had some compensatory advantage.

I suggest that this may very well have been their ability to innovate, and to transmit to others, relatively elaborate and sophisticated routines of manual actions. This enables the range of foods that may be accessed to be enlarged in crucial ways. Whereas baboons eat termites on the day of their annual emergence, chimpanzees can extract them from the mound over several months. Safari ants *Dorylus* are little used by monkeys, but the efficient dipping techniques of chimpanzees enables so many to be eaten at once that the pain of their bites becomes worthwhile. Honey and grubs of bees and wasps present a nutritious target for any primate, but only orangutans and chimpanzees can gain access to these riches. Fast-growing herbs in the temperate zone of African volcanoes offer a nutritious and chemically undefended source of food, but only mountain gorillas have developed the skills necessary to remove or avoid the physical defenses of the plants. As our knowledge improves of the feeding ecology of other populations of great apes, for example lowland gorillas and orangutans that do not employ tools, I predict that in every case it will turn out that the apes are able to exploit resources denied to monkeys, by means of the elaborate and skillful use of manual routines.

Summary

It is argued that a proper understanding of the behavioral ecology of great apes requires appreciation of the much greater importance for them of manual skills in food acquisition. Great apes may be said to inhabit a "cognitive niche". Although anthropological interest in culture has focused attention on chimpanzee tool-use, complex manual skill is even more clearly demonstrated in the plant-preparation skills of the mountain gorilla. Unlike monkeys, great apes can acquire novel manual skills with complex organization: hierarchically structured sequences of actions, with delicate bimanual coordination and flexible use of optional processes and subroutines. The cognitive underpinnings of these abilities include working memory, the ability to segment the fluid action of conspecifics into elemental units, and to parse strings of such elements in order to detect the underlying structure of the behavior. These capacities are a crucial basis to understanding how actions work, and why they are done.

Acknowledgements

For support of the field work on which this paper is based, I thank the National Geographic Society and the Carnegie Trust for the Universities of Scotland.

References

Aiello, L & Dean, C (1990) *An Introduction to Human Evolutionary Anatomy.* London: Academic Press.

Badrian, N L & Malenky, R K (1984). Feeding ecology of *Pan paniscus* in the Lomako Forest, Zaire. In *The Pygmy Chimpanzee: Evolutionary Biology and Behavior*, ed. R L Susman, pp. 275–99. New York: Plenum Press.

Begun, D R (1999) Hominid family values: morphological and molecular data on relations among the great apes and humans. In *The Mentalities of Gorillas and Orangutans: Comparative Perspectives*, ed. S T Parker, R W Mitchell & H I Miles, pp. 3–42. Cambridge: Cambridge University Press.

Boesch, C & Boesch, H (1983) Optimisation of nut-cracking with natural hammers by wild chimpanzees. *Behaviour*, **26**, 265–86.

Boesch, C & Boesch, H (1990) Tool use and tool making in wild chimpanzees. *Folia Primatologica*, **54**, 86–99.

Boesch, C & Tomasello, M (1998) Chimpanzee and human cultures. *Current Anthropology*, **39**, 591–614.

Byrne, R W (1993) Hierarchical levels of imitation. Commentary on M Tomasello, A C Kruger & H H Ratner "Cultural learning". *Behavioural and Brain Sciences*, **16**, 516–17.

Byrne, R W (1994) The evolution of intelligence. In *Behaviour and Evolution*, ed. P J

B Slater & T R Halliday, pp. 223–65. Cambridge: Cambridge University Press.

Byrne, R W (1996) The misunderstood ape: cognitive skills of the gorilla. In *Reaching into Thought: The Minds of the Great Apes*, ed. A E Russon, K A Bard & S T Parker, pp. 111–30. Cambridge: Cambridge University Press.

Byrne, R W (1997) The technical intelligence hypothesis: an additional evolutionary stimulus to intelligence? In *Machiavellian Intelligence II: Extensions and Evaluations*, ed. A Whiten & R W Byrne, pp. 289–311. Cambridge: Cambridge University Press.

Byrne, R W (1998) Imitation: the contributions of priming and program-level copying. In *Intersubjective Communication and Emotion in Early Ontogeny*, ed. S Braten, pp. 228–44. Cambridge: Cambridge University Press.

Byrne, R W (1999*a*) Cognition in great ape ecology: skill-learning ability opens up foraging opportunities. *Symposia of the Zoological Society of London*, **72**, 333–50.

Byrne, R W (1999*b*) Imitation without intentionality: using string parsing to copy the organization of behaviour. *Animal Cognition*, **2**, 63–72.

Byrne, R W (2000) The evolution of primate cognition. *Cognitive Science*, **24**, 543–70.

Byrne, R W & Byrne, J M E (1991) Hand preferences in the skilled gathering tasks of mountain gorillas (*Gorilla g. beringei*). *Cortex*, **27**, 521–46.

Byrne, R W & Byrne, J M E (1993) Complex leaf-gathering skills of mountain gorillas (*Gorilla g. beringei*): variability and standardization. *American Journal of Primatology*, **31**, 241–61.

Byrne, R W & Russon, A E (1998) Learning by imitation: a hierarchical approach. *Behavioral and Brain Sciences*, **21**, 667–721.

Chapman, C A & Wrangham, R W (1993) Range use of the forest chimpanzees of Kibale: implications for the understanding of chimpanzee social ecology. *American Journal of Primatology*, **31**, 263–73.

Chivers, D J (1974) The siamang in Malaya: a field study of a primate in a tropical rain forest. *Contributions to Primatology*, **4**, 1–331.

Chivers, D J & Hladik, C M (1984) Diet and gut morphology in primates. In *Food Acquisition and Processing in Primates*, ed. D J Chivers, B A Wood & A Bilsborough, pp. 213–30. New York: Plenum Press.

Chivers, D J, Raemakers, J J & Aldrich-Blake, F P G (1975) Long-term observations of siamang behaviour. *Folia Primatologica*, **23**, 1–49.

Clutton-Brock, T H & Harvey, P H (1977) Primate ecology and social organization. *Journal of Zoology*, **183**, 1–39.

Dawkins, R (1986) *The Blind Watchmaker*. Harlow UK: Longman.

Dunbar, R I M (1977) Feeding ecology of gelada baboons: a preliminary report. In *Primate Ecology*, ed. T H Clutton-Brock, pp. 252–73. London: Academic Press.

Ellefson, J O (1974) A natural history of white-handed gibbons in the Malayan Peninsula. *Gibbon and Siamang*, **3**, 1–136.

Fleagle, J (1998) *Primate Adaptation and Evolution*, 2nd edn. San Diego CA: Academic Press.

310   *Richard W. Byrne*

Fossey, D & Harcourt, A H (1977) Feeding ecology of free-ranging mountain gorillas (*Gorilla gorilla beringei*). In *Primate Ecology*, ed. T H Clutton-Brock, pp. 415–47. New York: Academic Press.

Fox, E, Sitompul, A & van Schaik, C P (1999) Intelligent tool use in wild Sumatran orangutans. In *The Mentality of Gorillas and Orangutans*, ed. S T Parker, H L Miles & R W Mitchell, pp. 99–116. Cambridge: Cambridge University Press.

Galdikas, B (1979) Orang-utan adaptation at Tanjung Puting Reserve: mating and ecology. In *The Great Apes*, ed. D A Hamburg & E McCown, pp. 195–233. Menlo Park CA: Addison-Wesley.

Galef, B G (1992) The question of animal culture. *Human Nature*, 3, 157–78.

Gallese, V & Goldman, A (1998) Mirror neurons and simulation theory of mind-reading. *Trends in Cognitive Sciences*, 2, 493–501.

Goodall, A G (1977) Feeding and ranging behaviour of a mountain gorilla group (*Gorilla gorilla beringei*) in the Tshibinda-Kahuzi region (Zaire). In *Primate Ecology*, ed. T H Clutton-Brock, pp. 450–79. New York: Academic Press.

Halsband, U, Matsuzaka, Y & Tanji, J (1994) Neuronal activity in the primate supplementary, pre-supplementary and pre-motor cortex during externally and internally instructed sequential movements. *Neuroscience Research*, 20, 149–55.

Harrison, K (1996) Skills used in food processing by vervet monkeys, *Cercopithecus aethiops*. PhD thesis, St Andrews University.

Harrison, M J S & Hladik, C M (1986) A seed-eating primate, the black colobus of the Gabon rain forest. *La Terre et La Vie*, 41, 281–98.

Hladik, C M (1975) Ecology, diet and social patterning in Old and New World monkeys. In *Socioecology and Psychology of Primates*, ed. R H Tuttle, pp. 3–35. Paris: Mouton.

Hladik, C M (1978) Adaptive strategies of primates in relation to leaf-eating. In *The Ecology of Arboreal Folivores*, ed. G G Montgomery, pp. 373–95. Washington DC: Smithsonian Institution Press.

Iwamoto, T (1979) Feeding ecology. In *Ecological and Sociological Studies of Gelada Baboons*, ed. M Kawai, pp. 279–330. Basel: Karger.

Lethmate, J (1977) Werkzeugherstellung eines jungen Orang-utans. *Behaviour*, 62, 174–89.

Malenky, R K & Wrangham, R W (1994) A quantitative comparison of terrestrial herbaceous food-consumption by *Pan paniscus* in the Lomake forest, Zaire, and *Pan troglodytes* in the Kibale Forest, Uganda. *American Journal of Primatology*, 32, 1–12.

McGrew, W C (1974) Tool use by wild chimpanzees feeding on driver ants. *Journal of Human Evolution*, 3, 501–8.

McGrew, W C (1979) Evolutionary implications of sex differences in chimpanzee predation and tool use. In *The Great Apes*, ed. D A Hamburg & E McCown, pp. 441–63. Menlo Park CA: Benjamin/Cummings.

McGrew, W C (1989) Why is ape tool use so confusing? In *Comparative Socioecology: The Behavioural Ecology of Humans and Other Mammals*, ed. V Standen & R A Foley, pp. 457–72. Oxford: Blackwell Scientific Publications.

McKey, D B, Gartlan, J S, Waterman, P G & Choo, G M (1981) Food selection by black colobus monkey in relation to plant chemistry. *Biological Journal of the Linnean Society*, **16**, 115–46.

Milton, K (1981) Distribution patterns of tropical plant foods as a stimulus to primate mental development. *American Anthropologist*, **83**, 534–48.

Napier, J R (1961) Prehensility and opposability in the hands of primates. *Symposia of the Zoological Society of London*, **5**, 115–32.

O'Brien, T G, Kinnaird, M F, Dierenfeld, E S, Conklin-Brittain, N L, Wrangham, R W & Silver, S C (1998) What's so special about figs? *Nature*, **392**, 668.

Parker, S T & Gibson, K R (1977) Object manipulation, tool use, and sensorimotor intelligence as feeding adaptations in cebus monkeys and great apes. *Journal of Human Evolution*, **6**, 623–41.

Parker, S T & Mitchell, R W (1999) The mentalities of gorillas and orangutans in phylogenetic perspective. In *The Mentalities of Gorillas and Orangutans in Comparative Perspective*, ed. S T Parker, R W Mitchell & H L Miles, pp. 397–411. Cambridge: Cambridge University Press.

Rizzolatti, G (1981) Afferent properties of periarcuate neurons in macaque monkey. II. Visual responses. *Behavioural Brain Research*, **2**, 147–63.

Rizzolatti, G & Gentilucci, M (1988) Motor and visual-motor functions of the premotor cortex. In *Neurobiology of Neocortex*, ed. P Rakic & W Singer, pp. 269–84. New York: John Wiley.

Rodman, P S (1973) Population composition and adaptive organisation among orangutans of the Kutai reserve. In *Comparative Ecology and Behaviour of Primates*, ed. R P Michael & J H Crook, pp. 171–209. London: Academic Press.

Rodman, P S (1977) Feeding behaviour of orang-utans of the Kutai Nature Reserve, East Kalimantan. In *Primate Ecology*, ed. T H Clutton-Brock, pp. 384–414. London: Academic Press.

Rogers, M E, Williamson, E A, Tutin, C E G & Fernandez, M (1988) Effects of the dry season on gorilla diet in Gabon. *Primate Reports*, **22**, 25–33.

Roper, T J (1983) Learning as a biological phenomenon. In *Animal Behaviour*, vol. 3, *Genes, Development and Learning*, ed. T R Halliday & P J B Slater, pp. 178–212. Oxford: Blackwell Scientific Publications.

Russon, A E (1996) Imitation in everyday use: matching and rehearsal in the spontaneous imitation of rehabilitant orangutans (*Pongo pygmaeus*). In *Reaching into Thought: The Minds of the Great Apes*, ed. A E Russon, K A Bard & S T Parker, pp. 152–76. Cambridge: Cambridge University Press.

Russon, A E (1998) The nature and evolution of intelligence in orangutans (*Pongo pygmaeus*). *Primates*, **39**, 485–503.

Stafford, D K, Milliken, G W & Ward, J P (1993) Patterns of hand and mouth lateral biases in bamboo leaf shoot feeding and simple food reaching in the gentle lemur (*Hapalemur griseus*). *American Journal of Primatology*, **29**, 195–207.

Sugiyama, Y, Koman, J & Bhoye Sow, M (1988) Ant-catching wands of wild chimpanzees at Bossou, Guinea. *Folia Primatologica*, **51**, 56–60.

Sugiyama, Y, Fushimi, T, Sakura, O & Matsuzawa, T (1993) Hand preference and tool use in wild chimpanzees. *Primates*, **34**, 151–9.

312   *Richard W. Byrne*

Tanji, J & Shima, K (1994) Role of supplementary motor cells in planning several movements ahead. *Nature*, **371**, 413–16.

Tanji, J, Shima, K & Mushiake, H (1996) Multiple cortical motor areas and temporal sequencing of movements. *Cognitive Brain Research*, **5**, 117–22.

Tanner, J E & Byrne, R W (1996) Representation of action through iconic gesture in a captive lowland gorilla. *Current Anthropology*, **37**, 162–73.

Tanner, J E & Byrne, R W (1999) The development of spontaneous gestural communication in a group of zoo-living lowland gorillas. In *The Mentalities of Gorillas and Orangutans: Comparative Perspectives*, ed. S T Parker, R W Mitchell & H L Miles, pp. 211–39. Cambridge: Cambridge University Press.

Tenaza, R R (1975) Territory and monogamy among Kloss' gibbons (*Hylobates klossii*) in Siberut Island, Indonesia. *Folia Primatologica*, **24**, 60–80.

Tomasello, M (1990) Cultural transmission in the tool use and communicatory signaling of chimpanzees? In *"Language" and Intelligence in Monkeys and Apes*, ed. S T Parker & K R Gibson, pp. 274–311. Cambridge: Cambridge University Press.

Tomasello, M & Call, J (1997) *Primate Cognition*. New York: Oxford University Press.

Tomasello, M, George, B, Kruger, A, Farrar, J & Evans, E (1985) The development of gestural communication in young chimpanzees. *Journal of Human Evolution*, **14**, 175–86.

Tomasello, M, Gust, D & Frost, T A (1989) A longitudinal investigation of gestural communication in young chimpanzees. *Primates*, **30**, 35–50.

Tomasello, M, Call, J, Nagell, C, Olguin, R & Carpenter, M (1994) The learning and use of gestural signals by young chimpanzees: a trans-generational study. *Primates*, **35**, 137–54.

Toth, N, Schick, K D, Savage-Rumbaugh, E S, Sevcik, R A & Rumbaugh, D M (1993) Pan the tool-maker: investigations into the stone-tool-making and tool-using capabilities of a bonobo (*Pan paniscus*). *Journal of Archaeological Science*, **20**, 81–91.

Tutin, C E G, Ham, R M, White, L J T & Harrison, M J S (1997) The primate community of the Lopé Reserve, Gabon: diets, responses to fruit scarcity, and effects on biomass. *American Journal of Primatology*, **42**, 1–24.

van Schaik, C P, Fox, E A & Sitompul, A F (1996) Manufacture and use of tools in wild Sumatran orangutans: implications for human evolution. *Naturwissenschaften*, **83**, 186–8.

van Schaik, C P, Deaner, R O & Merrill, M Y (1999) The conditions for tool-use in primates: implications for the evolution of material culture. *Journal of Human Evolution*, **36**, 719–41.

Visalberghi, E & Fragaszy, D M (1990) Do monkeys ape? In *"Language" and Intelligence in Monkeys and Apes*, ed. S T Parker & K R Gibson, pp. 247–73. Cambridge: Cambridge University Press.

Waterman, P G (1984) Food acquisition and processing as a function of plant chemistry. In *Food Acquisition and Processing in Primates*, ed. D J Chivers, B A Wood & A Bilsborough, pp. 177–211. New York: Plenum Press.

Waterman, P G, Choo, G M, Vedder, A L & Watts, D (1983) Digestibility, digestion-inhibitors and nutrients and herbaceous foliage and green stems from an African montane flora and comparison with other tropical flora. *Oecologia*, **60**, 244–9.

Watts, D P (1984) Composition and variability of mountain gorilla diets in the central Virungas. *American Journal of Primatology*, **7**, 323–56.

Watts, D P (1985) Observations on the ontogeny of feeding behaviour in mountain gorillas (*Gorilla gorilla beringei*). *American Journal of Primatology*, **8**, 1–10.

Watts, D P (1996) Comparative socio-ecology of gorillas. In *Great Ape Societies*, ed. W C McGrew, L F Marchant & T Nishida, pp. 16–28. Cambridge: Cambridge University Press.

Whiten, A (1988) Acquisition of foraging techniques in infant olive baboons. *International Journal of Primatology*, **8**, 469.

Whiten, A, Goodall, J, McGrew, W C, Nishida, T, Reynolds, V, Sugiyama, Y, Tutin, C E G, Wrangham, R W & Boesch, C (1999) Cultures in chimpanzees. *Nature*, **399**, 682–5.

Wrangham, R W (1977) Feeding behaviour of chimpanzees in Gombe National Park, Tanzania. In *Primate Ecology*, ed. T H Clutton-Brock, pp. 503–38. New York: Academic Press.

Wrangham, R W (1979a) On the evolution of ape social systems. *Social Sciences Information*, **18**, 335–68.

Wrangham, R W (1979b) Sex differences in chimpanzee dispersion. In *The Great Apes*, ed. D A Hamburg & E McCown, pp. 481–9. Menlo Park CA: Addison-Wesley.

Wrangham, R W, Conklin, N L, Chapman, C A & Hunt, K D (1991) The significance of fibrous foods for Kibale Forest chimpanzees. *Philosophical Transactions of the Royal Society of London, Series B*, **334**, 171–8.

Wrangham, R W, Conklin, N L, Etot, G, Obua, J, Hunt, K D, Hauser, M D & Clark, A P (1993) The value of figs to chimpanzees. *International Journal of Primatology*, **14**, 243–56.

Wrangham, R W, Conklin-Brittain, N L & Hunt, K (1996) Fallback foods of chimpanzees vs. monkeys. In *16th Congress of the International Primatological Society*, Madison, Wisconsin, August 11–16, 1996, p. 438.

Wright, R V S (1972) Imitative learning of a flaked-tool technology: the case of an orang-utan. *Mankind*, **8**, 296–306.

Part IV

Conservation and management of mountain gorillas

12 Assessment of reproduction and stress through hormone analysis in gorillas

NANCY CZEKALA & MARTHA M. ROBBINS

Titus and Tuck mating. (Photo by Martha M. Robbins.)

Introduction

Studies of behavioral endocrinology provide opportunities to examine the reproductive biology and stress responses of animals. In the case of endangered species, the study of reproductive biology has the two-fold value of contributing to both applied and pure research. In the wild, conservation efforts are enhanced by the assessment of reproductive function and population growth dynamics. It can be used as a diagnostic tool to improve the breeding management plans of captive animal populations. Exchange of information between researchers and managers of captive and wild populations can assist the efforts of both groups (Wildt & Wemmer, 1999). From the standpoint of pure research, studies of reproductive biology help us to understand the evolution of sexual behavior from both proximate and ultimate perspectives, including sex differences in behavior, sexual selection, and the diversity of reproductive mechanisms observed in the animal kingdom. Additionally, studying stress is important because it may affect the reproductive abilities and well-being of individuals which may impact lifetime reproductive success and survivorship.

The reproductive biology of gorillas has been the focus of much research with both captive and wild populations. From an applied perspective, difficulties in captive breeding led to the need for improved understanding of reproductive function. While the lowland gorilla has been kept in captivity since the late 1800s, it was not until 1956 that efforts to have captive gorillas reproduce began and initially the results were poor. Based on a survey of zoos holding gorillas, Beck & Power (1988) concluded that environmental factors such as mother rearing (as opposed to hand-rearing) and social access to conspecifics in the first year of life improved female reproductive success and that many cases of reproductive failure were due to deficits in sexual behavior. Studies of captive gorillas also allowed for more experimentally manipulative and invasive research to understand their reproductive biology than would be possible with animals in the wild. Given the endangered status of gorillas, data collection on reproductive parameters has always been a priority with the habituated Karisoke gorillas, but obtaining sufficient data to make well-founded conclusions takes many years given the long life histories of gorillas.

One particular aspect of reproductive biology, socioendocrinology, makes the link between the social environment, neuroendocrine mechanisms, sexual behavior, and reproductive success (Bercovitch & Ziegler, 1990). The goal of this chapter is to review the various ways hormones have been studied in gorillas. In particular we will discuss (1) female reproductive cycles and pregnancy, (2) male reproductive parameters, and (3) monitoring stress through corticoid measurements. Studies of and

318

comparisons between the captive lowland gorilla and the free-ranging mountain gorilla are beneficial to the management of both populations and can also be used to assess variability under differing environmental and nutritional conditions, as has been done to help understand the evolution of human reproductive ecology (Bentley, 1999). However, one problem of making comparisons between captive and wild gorillas is that all gorillas found in captivity are western lowland gorillas (*Gorilla gorilla gorilla*), whereas the majority of the behavioral studies in the wild have been performed on mountain gorillas (*G. g. beringei*). This situation has arisen from the absence of mountain gorillas in captivity along with the lack of well-habituated lowland gorillas in the wild. Therefore it is important to keep in mind that any differences reported in this chapter between captive and wild gorillas may be confounded by subspecies differences as well as by environmental variability.

Historically, the assessment of reproductive hormones required analysis of blood, which in the case of gorillas, required anesthesia for each sample (Nadler *et al.*, 1979, 1983; Nadler & Collins, 1984). While such pioneering studies have been very informative, the need for frequent blood collection precludes many facilities from using this type of analysis and the application of these techniques to wild gorillas is impossible.

Our success in understanding the reproductive parameters of any endangered species is limited by our ability to monitor gonadal and adrenal hormonal changes in a noninvasive manner on a frequent basis. Efforts to develop techniques to conduct these types of studies originally targeted urinary hormone analysis and more recently have begun to include fecal hormone analysis. Due to the vast variety of hormonal metabolites and the different excretion modes of hormones (urinary or fecal) that can vary between species, it is necessary to clearly investigate each species as a novel study. In the past decade, there has been a blossoming of the use of noninvasive techniques to monitor hormone levels of wild animals (for review see Whitten *et al.*, 1998). To give a flavor of the type of research conducted in this field, a nonexhaustive list of topics and study animals for field studies of endocrinology using either urinary or fecal analysis includes: monitoring female reproductive cycles (baboons: Wasser, 1996; vervet monkeys: Andelman *et al.*, 1985; muriquis: Ziegler *et al.*, 1997; Verreaux's sifakis: Brockman *et al.*, 1995), assessment of male reproductive hormones (elephants: Poole *et al.*, 1984; dwarf mongooses: Creel *et al.*, 1992), and using corticosteroids as indicators of stress (bighorn sheep: Miller *et al.*, 1991; wild dogs: Creel *et al.*, 1997; dwarf mongooses: Creel *et al.*, 1992; longtailed macaques: van Schaik *et al.*, 1991; muriquis: Strier *et al.*, 1999; and northern spotted owls: Wasser *et al.*, 1997).

While the results of the initial studies of captive gorillas has allowed for a better understanding of species biology and captive management, such studies only provided information on the artificial captive environment and comparable data on the truly normal or free-ranging biology was lacking. The validation of hormone analysis techniques in the captive studies has now permitted the extension of these studies to the free-ranging populations. The mountain gorillas at Karisoke are well enough habituated to allow researchers to collect urine and fecal samples frequently from known individuals in conjunction with detailed observational data.

Methods

At Karisoke, urine samples were obtained on an *ad libitum* basis by following particular individuals for up to several hours until s/he urinated after which the urine was immediately syringed from the ground or vegetation (Czekala *et al.*, 1994; Robbins & Czekala, 1997). Since only small volumes of urine are needed for the hormone analysis, this method of urine collection works well. Once the urine samples were collected, they were placed in a cold thermos until they were placed in a propane freezer upon return to the research camp. Analysis of chorionic gonadotropin for pregnancy detection or luteinizing hormone for detection of ovulation was performed at the field site. For the other hormones, the samples were tranported frozen to the San Diego laboratory for analysis. Analysis of estrogen conjugates (estrone sulfate: Czekala *et al.*, 1987), progestins (pregnanediol-3-glucuronide: Czekala *et al.*, 1991), androgens and corticoids (Robbins & Czekala, 1997) were performed as previously described by radioimmunoassay. More detailed explanations of methods used for sample collection and hormone analysis can be found in the referenced articles.

Female reproductive cycles

Female gorillas have very subtle outward signs of ovulation (only nulliparous females have obvious swellings, and even these are small) and there is no outward sign of pregnancy so hormonal analysis can add much to basic observational data. The initial goals of studying the reproductive physiology and sexual behavior of gorillas were (1) to characterize the female ovulatory cycle, (2) to determine the relationship between sexual behavior (e.g. receptivity), sexual swellings, and the ovulatory cycle, and (3) to assess the relationship between physiology, behavior, and fertility of gorillas (Nadler *et al.*, 1979, 1983). Data collected from observations alone of the captive lowland gorilla population and the Karisoke gorilla population enable us to make a comparison of between the basic reproductive par-

Table 12.1. *Comparison of female gorilla reproductive parameters for captive western lowland gorillas and wild mountain gorillas*

Reproductive parameter[a]	Captive western lowland gorilla[b]	Wild mountain gorilla[c]
Age of first observed sexual behavior/ menarche	Median = 6 yr Mean = 8.9 yr Earliest = 5 yr	Median = 6.33 yr (5.8–7.1 yr)
Period of adolescent sterility	Approx. 2 yr	Approx. 2 yr
Length of estrus cycle	30–33 d (24–44 d)	29 d (20–39 d)
Length of estrus period (receptivity)	1–2 d	1–4 d
Gestation length	258 d	255 d
Age at first parturition	10.0 yr earliest = 6 yr	10.0 yr (8–13 yr)
Time/number of estrus cycles to conception (parous females)	2 mo (2 d – 8 mo) following mother reared infants 5.5 mo (2 d – 20 mo) following infant separation	3–4 estrus cycles
Lactational anestrous with surviving infant	2.74 yr (1.5–5.2 yr)	3.2 yr (2.2–4.2 yr)
Inter-birth interval with surviving offspring	4.2 yr (2.3–6.4 yr)	3.9 yr (3.0–7.2 yr)
Inter-birth interval with death or removal of infant	1.3 yr (0.75–3.4 yr)	Median 1.0 yr (0.9–3.1 yr)

[a] Values are means unless noted as medians. Numbers in parentheses are ranges.
[b] Values for captive western lowland gorillas are derived from: Beck & Power (1988), Cousins (1976), Czekala *et al.* (1987), Dixson (1981), Keiter & Pichette (1979), Mitchell *et al.* (1982b), Nadler *et al.* (1979, 1983), Patterson (1991), and Sievert *et al.* (1991).
[c] Values for wild mountain gorillas are derived from: Harcourt *et al.* (1980, 1981), Stewart (1977), and Watts (1990, 1991).

ameters of captive lowland gorillas and wild mountain gorillas (Table 12.1). Although the earliest recorded age of menarche and conception is earlier in captivity than the wild (Keiter & Pichette, 1979; Beck & Power, 1988, Patterson *et al.*, 1991), in general there is remarkable similarity in female reproductive parameters between the two subspecies. In both captivity and the wild, female gorillas begin to exhibit estrus cycles at approximately age 6 years, but they go through a period of adolescent sterility for approximately 2 years. The average age of first parturition is 10 years. On average, parous females conceive after approximately 4–6 estrus cycles, have a gestation of 8.5 months, go through a lactational anestrus of approximately 3 years, for an inter-birth interval of 4 years (Harcourt *et al.*, 1980; Watts, 1991). The observed menstrual cycle length is similar between the two gorilla populations (captive lowland gorillas: 30–33 days and wild mountain gorillas: 29 days). Perhaps what is most notable in both captivity and the wild is the degree of variability in parameters such as the number of cycles to conception, inter-birth interval, etc. that might indicate reproductive dysfunction or the influence of some unknown factors in particular individuals for both populations.

Basic observations of a female's inability to reproduce do not tell us if it is due to the inability to ovulate, conceive, or maintain a pregnancy. Hormone measurements enable us to correlate particular behavioral patterns, such as mating, to time of ovulation and pregnancy, and can be used for detection of early miscarriages. Pioneering research at the Yerkes Primate Center used serum analysis in conjunction with behavioral observations (Nadler *et al.*, 1979, 1983; Gould & Faulkner, 1981; Nadler & Collins, 1984). These studies provided the basic correlation of reproductive behaviors, labial tumescence, and reproductive hormones. These studies provided the first information on the total cycle length and cycle phase length based on hormone analysis. They found that the cyclic variation in sex skin swellings correlated positively to serum estradiol and copulation, and that the luteinizing hormone (LH) peak occurred during maximum sex skin swelling.

To assess the relationship between the female ovulatory cycle and reproductive success, several studies were initiated with captive gorillas using urine that was collected noninvasively. Urinary estrogens are an excellent tool to determine how the ovary progresses as it advances towards ovulation. Estrogen increases during this time (the follicular phase: day of first menstruation to day of peak estrogen) as a result of the growth of the follicle. Estrogen also increases during the luteal phase (day of peak estrogen to next menstruation) to a lesser extent. In the lowland gorilla, progestins (in this case measured by pregnanediol-3-glucuronide: PDG) increase after ovulation during the luteal phase and decrease to baseline followed by menstruation. PDG patterns in the lowland gorilla indicate the functioning of the corpus luteum. Detection of the time of ovulation has been assessed by the measurement of urinary estrogen conjugates, estrone-3-sulfate (EC) and urinary LH (Mitchell *et al.*, 1982b; Czekala *et al.*, 1988a). The cycle length was determined to be 32 days by urinary estrogen analysis (Mitchell *et al.*, 1982) and 30 days (Czekala *et al.*, 1987). These results compare well with the cycle length determined by serum analysis of 33 days (Nadler *et al.*, 1983). In addition, specific infertility assessment associated with low reproductive rates has been related to subnormal concentrations of the progesterone metabolite PDG (Mitchell *et al.*, 1982a; Czekala *et al.*, 1991). These studies on captive gorillas have allowed for the technique validation for several urinary hormones that have subsequently been applied to the free-ranging mountain gorillas.

Pregnancy is an important reproductive feature to be able to detect in the free-ranging gorillas. Since there are no external physical signs of pregnancy in gorillas, it is necessary to use a hormonal test for pregnancy. Previous studies on the captive population provided assurance that a test kit for human pregnancy detection would allow for pregnancy detection from urine in the gorilla. Antibodies produced against human chorionic gonadotropin used in this test kit cross-react with gorilla chorionic gonadotropin (CG). Pregnancy can be detected as early as 10 days post-conception using these simple kits. To evaluate the validity of the pregnancy test for mountain gorillas, 12 females that were not carrying infants or had infants greater than 2 years of age were monitored for EC, PDG, and CG over several months. In humans and lowland gorillas estrogen and PDG concentrations are elevated above normal cycle levels during most of pregnancy. The results for the mountain gorilla indicate that positive pregnancy tests (CG) occur from the first month of pregnancy until the last month of pregnancy. Five females were pregnant during this sampling time. No false positives or negatives were recorded. The EC concentrations were above the nonpregnant levels from month 7 before delivery until delivery, but the PDG concentrations were not indicative of pregnancy. Thus, pregnancy can be detected in field conditions using the human test kit for CG and in the laboratory using analysis of estrogens; however PDG concentrations are not an adequate test for pregnancy in the mountain gorilla.

A preliminary investigation of female reproductive cycles at Karisoke was performed in July 1990, by focusing on three nonpregnant females from Group BM (Czekala & Sicotte, 2000). These females were Pp (multi-parous, approximately 26 years), Mw (nulliparous, 8 years 4 months), and Sg (nulliparous, 9 years 8 months). Urine samples from these females were collected on an approximately daily basis for four weeks. Urine was tested at the research station for blood in the urine indicating menstruation (Hemistix, Bayer Corp., Elkhart IN) and for LH to detect ovulation or CG to determine pregnancy (Ovuquick, Quidel Corp., San Diego CA). Urinary EC and PDG were measured by radioimmunoassay as previously described (Czekala *et al.*, 1991).

Results from the urine analysis indicate that the multiparous female, Pp, became pregnant during the study. The estrogen concentrations of Pp were higher than those of the two younger females (Figure 12.1). In addition, this conceptive cycle had much higher estrogen concentrations than the fertile lowland gorillas (Figure 12.2) (Czekala *et al.*, 1991). Mating

324   *Nancy Czekala & Martha M. Robbins*

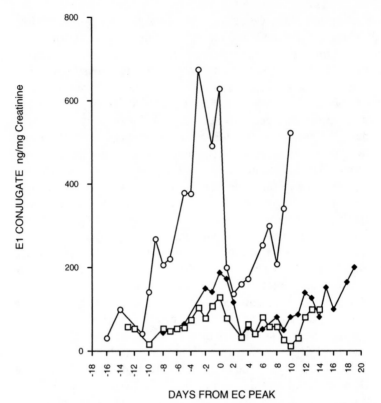

Figure 12.1. Urinary estrone conjugates (EC) values for three wild mountain gorillas, aligned to day of estrogen peak. Open circles = Pp, solid triangles = Mw, open squares = Sg.

occurred on days 0, 1, and 2 from the estrogen peak for the Pp conceptive cycle. The length of the follicular phase was 17 days. This is within the normal range of the reported values of the lowland gorilla (Mitchell *et al.*, 1982*b*). A positive LH test resulted on day 0 which also coincided with the EC peak. A positive test for LH/CG on day 10 suggests the CG rise of pregnancy. The length of gestation for this pregnancy was 254 days which is similar to those previously reported for the mountain gorilla (255 days: Harcourt *et al.*, 1980) and the captive lowland gorilla (mean 258 days: Cousins, 1976).

The female Mw had a cycle that may have been conceptive but did not reach term (Figure 12.1). The days of observed matings of Mw were days −3, 0, and 2. Sample collection began 9 days before the EC peak, thus the length of the follicular phase could not be determined. There are several factors that suggest the cycle of Mw was conceptive: no menstruation was

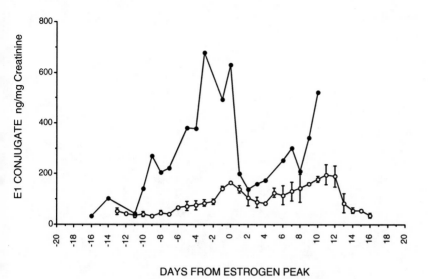

DAYS FROM ESTROGEN PEAK

Figure 12.2. Urinary estrone conjugates (EC) values comparing conceptive cycle of wild mountain gorilla (Pp = solid circles) to reproductively fertile lowland gorilla nonconceptive cycles (open circles, mean ± S E).

detected up to 18 days post EC peak; slight positive CG test was obtained on days 17 and 18; EC concentrations increased at the time of the end of the luteal phase instead of decreasing as would be expected in a nonconceptive cycle. However no miscarriages were observed and no full-term infant was delivered until August 1991 which would mean conception for this pregnancy occurred in December 1990, 5 months after our study ended.

The cycle of Sg had EC concentrations similar to the Mw cycle at peak levels (Figure 12.1). The follicular phase was 13 days in length, which is in the normal range for lowland gorillas, and was followed by a short 10-day luteal phase. Mating was observed on day 0, the day of the EC peak. No pregnancy resulted from this cycle. Sg did finally deliver in March 1994 when she was 12 years 4 months, which is one of the latest recorded first births in the Karisoke population, although the possibility of a missed miscarriage cannot be ruled out.

Although only a preliminary study, the results of both the EC profile of the conceptive mountain gorilla cycle and the PDG analysis suggest a possible difference between the captive lowland gorilla and the mountain gorilla menstrual cycle. The PDG concentrations were 4 times lower in the mountain gorilla conceptive than in the lowland gorilla mean conceptive cycle (Figure 12.3). The concentrations only reached comparable levels at

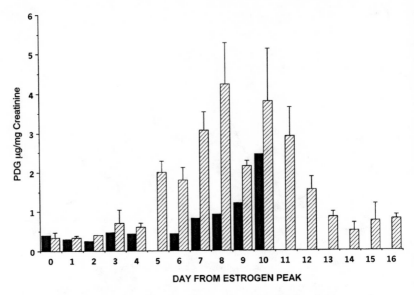

Figure 12.3. Urinary pregnanediol-3-glucuronide (PDG) values comparing conceptive cycle of wild mountain gorilla (Pp = solid bars) to reproductively fertile lowland gorilla nonconceptive cycles (shaded bars, mean ± SE).

day 10 post EC peak when urine tested positive for CG, the approximate time of implantation. This suggests that the production or metabolism of progestins during the luteal phase of the cycle may differ between the two subspecies or that the environmental differences of the two populations play a role in the type or amount of excreted progestin. Figure 12.3 indicates that there is a divergence in the PDG pattern beginning on day 4 post EC peak between the mountain gorilla conceptive cycle and the lowland gorilla mean conceptive cycle. The PDG patterns of the other two mountain gorilla females show a similar pattern of low PDG during the luteal phase with little to no distinction between the concentrations during the follicular phase and the luteal phase.

Variations of progestin excretion during the luteal phase have been reported between the different hominoid species (Czekala *et al.*, 1988*b*). While the urinary estrogen patterns and concentrations are very similar between captive orangutans, chimpanzees, gorillas, and humans, the PDG concentrations are much lower for the orangutan. Levels of PDG in the orangutan are so low as to suggest a different metabolism pattern of progestins for this species. The mountain gorilla has PDG concentrations mid-way between the lowland gorilla and the orangutan. Although sub-specific differences in the excretion of urinary steroids have not been reported, based on these findings of the decreased PDG concentrations in the mountain gorilla as compared to the lowland gorilla, such a possibility exists. If there are different metabolism products between the lowland and mountain gorilla subspecies then an alternate progesterone endpoint should be present.

Alternatively, these differences could be related to the differences in physical and social environments found between the captive and wild situations. Wasser (1996), for example, suggests that in free-ranging baboons dominance status and environmental conditions may influence fecal steroid excretion. Dominant baboons have higher estrogen excretion and lower progestin excretion than low-ranking females, and baboons under better environmental conditions have lower progesterone excretion. The lower progestins in the free-ranging mountain gorilla may be an indicator that their environment is better or more conducive to normal reproductive function than that of captive gorillas. Although based on only three individuals, it is interesting to note that multiparous Pp was dominant to Sg and Mw and she had higher estrogen concentrations; however, age and parity may also be influential.

While much can be learned about a species' reproductive parameters through routine observation of habituated animals (see Wallis, 1997 for chimpanzees), the correlation of the reproductive hormone patterns with these observations provides a new perspective to our understanding of natural reproductive biology. Although the results presented here are based on limited data from three individuals, the information obtained provides for some interesting speculation on the comparative reproductive hormones of the lowland gorilla and the mountain gorilla. Additional field studies must be conducted for longer periods of time and with more individual study subjects to confirm these speculations.

Male reproductive parameters
Maturation and fertility

Male fertility is generally more difficult to assess than female fertility using only observational techniques. Males do not exhibit well-defined physiological events to mark certain points in maturation such as the onset of menarche that are observed in females. Because in most mammals it is difficult to determine when males initially become fertile, the classification of males into age categories is typically based on somewhat arbitrary time-frames of changes in external morphological features or behavioral characteristics. Another measure of the maturation process is to determine when males go through hormonal puberty in relation to outward signs of reproductive behavior and the development of secondary sexual characteristics.

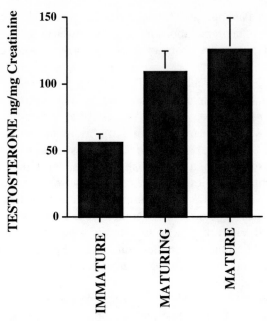

Figure 12.4. Testosterone levels of wild male mountain gorillas (immature males, $n = 8$; maturing males, $n = 5$; mature males, $n = 5$). (Mean ± SE.)

In captive lowland gorillas (Beck, 1982), chimpanzees (Marson *et al.*, 1991), and orangutans (Kingsley, 1982, 1988), males are fertile prior to the attainment of final adult size and fully developed secondary sexual characteristics. Male mountain gorillas begin the development of secondary sexual characteristics at approximately age 10. They can be considered silverbacks by age 12–13 but they may not reach final adult size until 15 years of age (Watts, 1991; Watts & Pusey, 1993). Captive gorillas are capable of siring offspring by 7 years of age (Beck, 1982; Kingsley, 1988), but it is unknown at what age wild male mountain gorillas are fertile. Male mountain gorillas exhibit sociosexual behavior as juveniles and they have been observed to first start copulating at the age of 9–10 years (Fossey, 1976; Watts, 1991). However, the mating behavior of young males may be limited or prevented by older males making it difficult to estimate when males reach sexual maturity.

Our study of wild male mountain gorillas showed that immature male gorillas (3–7 years) were found to have significantly lower levels of testosterone than both maturing males (10–13 years) and mature males (> 13 years) (Robbins & Czekala, 1997). No difference was observed between the testosterone levels of the maturing and mature males (Figure 12.4). These results suggest that the increase in testosterone associated with puberty occurs prior to any outward sign of development of secondary sexual characteristics, as has been observed in captive chimpanzees (Young *et al.*, 1993), orangutans (Kingsley, 1988; Maggioncalda *et al.*, 1999), and lowland gorillas (Kingsley, 1988). As with any study of wild animals, this study was limited by the ages of the animals in the study groups. In particular, there were no males between the ages of 7 and 10 years during the study period; it would be useful to have hormonal measurements for males of this age to determine when the onset of the testosterone rise in males occurs. The longitudinal study of the hormonal profiles of particular individuals through their lifetimes would offer a good control for individual variation as particular males mature and go through various life history changes.

Assessment of fertility in adult males usually requires more invasive diagnostics than simply a lack of newborns, which may be due to female infertility; furthermore, in multimale groups, paternity cannot be determined by behavioral observations alone. While noninvasively obtained hormone measurements may indicate abnormalities, infertile males may have normal levels of hormones but irregularities in sperm production. Even in captivity, studying fertility is difficult because the invasive nature of obtaining semen samples through electroejaculation may provide semen of poor characteristics (Brown & Loskutoff, 1998), and because sometimes males with particularly low overall sperm counts or high numbers of abnormal sperm may still sire offspring. Gould & King (1982) showed that in captive male gorillas there was no significant difference in testosterone levels between fertile and infertile males. There have been no reported cases of suspected infertility in wild male mountain gorillas, although there are reports of decreased sexual activity in older wild male gorillas which is probably linked to a decrease in dominance status (Watts, 1990).

Hormones and social behavior

In the wild, male gorillas have the option of being philopatric or emigrating, which can have a large impact on lifetime reproductive success (Robbins, 1995, this volume; Watts, 2000). The factors influencing dispersal decisions are not well understood but on the proximate level they may be triggered by particular hormonal cues associated with maturation or a certain threshold of social stress. Male gorillas also can live in a variety of social units: solitary males, one-male or multimale heterosexual groups, or all-male groups (for review see Robbins, this volume), and the behavior exhibited by males will vary depending on group membership (Robbins,

330   *Nancy Czekala & Martha M. Robbins*

1996). Dominance relationships are exhibited between adult male mountain gorillas in heterosexual and all-male mountain gorilla groups, but the level of agonistic and affiliative interactions among males varies depending on group type (Robbins, 1996). Examining steroid hormone levels may reveal if there is a physiological indicator of social rank or group type. Chronically high or low hormone levels may have implications for an individual's reproductive success and survival. Studies of hormone levels of gorillas in various social situations in the wild can also assist in determining the feasibility of various housing options for captive gorillas (e.g. all-male and multimale groups).

The relationship between male dominance rank and hormone levels, specifically testosterone and cortisol, has been examined in many species which exhibit high levels of male–male competition. A review of the literature suggests that there is no clear pattern between dominance rank and testosterone levels; some species exhibit a positive relationship and in other species there is no relationship (for reviews see Creel *et al.*, 1992; Sapolsky, 1993; Wingfield *et al.*, 1994; von Holst, 1998). Many other variables such as age, sexual activity, group stability, and individual style of dominance rank (e.g. level of aggression, response to aggression, etc.) may confound a clear relationship between testosterone and dominance rank (Sapolsky, 1991, 1993).

As part of a larger study examining male social behavior, urinary steroid measurements of wild male mountain gorillas were used to determine how levels of testosterone and cortisol corresponded with social rank (see Robbins, 1996, this volume; Robbins & Czekala, 1997). Cortisol results are reported in the section below on cortisol and stress. Comparison of three dominant and seven subordinate males in the three study groups revealed a trend toward the dominant males having higher levels of testosterone (Figure 12.5). In both heterosexual groups, the dominant male had the highest level of testosterone, but in the all male group the dominant male group had the second highest level. Subordinate males in the heterosexual groups did not have significantly different levels of testosterone from the subordinate males in the all-male group. Competition and aggression among the males may be related to testosterone levels. A continuation of hormone measurements from individuals previously used in this study would indicate how hormone levels of particular adult males have changed as a result of changes in age, dominance status, and group structure.

Increasing numbers of studies on male great apes in captivity and the wild will allow for useful comparative studies. A study examining testosterone and cortisol levels of captive male gorillas in differing social environments (heterosexual and all-male groups) is under way (T. Stoinski &

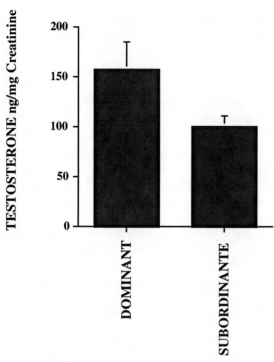

Figure 12.5. Testosterone levels of dominant males (*n* = 3) and subordinate males (*n* = 7) in wild mountain gorillas. (Mean ± SE.)

N.M. Czekala, unpublished data). A study of captive male orangutan urinary hormone profiles was conducted in the same laboratory using the same methodology as the male mountain gorilla study (Maggioncalda *et al.*, 1999). The concentrations of androgens were higher for the mountain gorilla than for the orangutan (mountain gorilla 120 ± 20 ng/mg cr; orangutan 48 ± 2 ng/mg cr;), but it is important to keep in mind that direct comparisons of mean steroid concentrations may be compromised by the potential species differences in metabolism. Immunoassays for testosterone that are not specific for this steroid also measure cross-reacting androgens to varying degrees (L. Hagey *et al.*, unpublished data). A study of dominance status and testosterone in wild chimpanzees in Kibale Forest, Uganda has revealed that dominant males do have significantly higher levels of testosterone than do subordinate animals, and their testosterone levels decrease significantly when they lose their high-ranking positions (M. Muller, unpublished data).

Stress and corticosteroids

Wild animals may face stress in the form of nutritional, energetic, or thermoregulatory constraints as well as during social or predatory situations. Captivity also has its own array of stressors for animals including the lack of ability to regulate the social and spatial environment and observations by humans. Researchers have hypothesized that attaining or maintaining high dominance status may be a stressful condition, or alternatively, that subordinate individuals who are the recipients of much aggression may experience high levels of stress, measured by cortisol (Sapolsky, 1993; Creel *et al.*, 1996; von Holst, 1998). Serum corticoids have been used as an indicator of adrenal activity to assess the effects of different stressors in primates (Clarke *et al.*, 1988; Sapolsky & Ray, 1989) and to diagnose abnormal adrenal function in humans (Eddy *et al.*, 1973). Since frequent serum collection is inappropriate for either the free-ranging gorilla or the captive gorilla, this study focused on evaluation of urinary corticoids.

To determine if a diurnal pattern of urinary corticoids could be observed, urine samples were collected at various times throughout the day. Humans, captive lowland gorillas, and the free-ranging mountain gorillas (female gorillas and male and female humans) all showed evidence of a diurnal pattern of urinary corticoids (Czekala *et al.*, 1994). Concentrations were higher during the morning hours and gradually decreased through the rest of the day. Human and lowland gorilla had similar mean concentrations during the morning hours of 0600 to 1200 (humans: 56.6 ± 7.8; lowland gorillas: 63 ± 7.9 ng/mg cr). However, the mountain gorilla had concentrations during this time that were half that of the lowland gorilla (29.7 ± 1.23 ng/mg cr). These lower levels may be due to the slightly skewed collection times (lowlands skewed towards early collection and mountain gorillas skewed towards noon), to differences between captivity and the wild environment (e.g. dietary), or to possible subspecific differences in metabolism.

Studies of corticosteroids often focus on males because interpreting corticosteroid measurements in females may be confounded by variation in their reproductive status. For example, corticoids increase during pregnancy. Although the relationship between cortisol levels and reproductive status of females in the wild has not been examined, the relationship between maternal stress and cortisol levels has been studied in captive lowland gorillas (Bahr *et al.*, 1998). As expected, cortisol levels decrease postpartum, but mothers with higher postpartum cortisol indices (ratio of pre- to postpartum cortisol levels) also had higher levels of disrupted maternal behavior (carrying and holding infants), indicating maternal

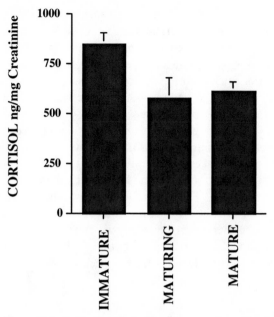

Figure 12.6. Cortisol levels of wild male mountain gorillas (immature males, *n* = 8; maturing males, *n* = 5; mature males, *n* = 5). (Mean ± SE.)

stress. Monitoring cortisol levels in captive female gorillas pre- and postpartum may be a useful tool to improve the survivorship of infants.

The study of wild male mountain gorillas (Robbins & Czekala, 1997) also examined levels of corticoids in relation to age and dominance rank. Interestingly, immature males exhibited higher corticoid values than both maturing and mature males in the morning (Figure 12.6), but no differences between the three age classes were observed in samples collected in the afternoon. No difference in corticoid levels was found between dominant and subordinate individuals suggesting that subordinate males were not socially stressed, at least as measured by corticoids. Unlike the mountain gorillas, the corticoid values for captive orangutans were not different between the juveniles and the mature group (Maggioncalda *et al.*, 1999). Mature male corticoid concentrations were similar between the two species (orangutan 0.6 ± 0.02 μg/mg cr; mountain gorilla 0.65 ± 0.02 μg/mg cr).

Stress is also a topic that cannot be ignored when considering the management of a small population of endangered species. A seemingly healthy population may actually be experiencing stress, the effects of which may not be noticed for some time (Hofer & East, 1998). Despite indications that the mountain gorilla groups that are monitored for research and tourism actually experience higher growth rates than their unhabituated peers (Weber & Vedder, 1983), there is always the concern that habituation and regular contact by humans have the potential to alter their behavior and/or cause additional stress to these animals (Butynski & Kalina, 1998). It is obviously impossible to directly test the effects of habituation on gorilla behavior because we cannot observe unhabituated gorillas. However, comparing the levels of corticoids in feces collected from the night nests of habituated and unhabituated gorillas offers one method to address this issue. In Bwindi Impenetrable National Park, Uganda, fecal samples have been collected repeatedly from the silverback nest of the research group (Kyagurilo Group) to determine how much individual fecal cortisol values vary over time. Additionally, during the 1997 gorilla census in Bwindi (McNeilage *et al.*, 1998), fecal samples were collected from the silverback nests of the entire gorilla population. Analyses of these samples is under way (N.M. Czekala & M.M. Robbins, unpublished data).

Summary
Studies on the behavioral endocrinology of gorillas in both captivity and the wild offer a unique approach to examine the reproductive biology of gorillas. Considerable progress has been made towards developing techniques to monitor reproduction, maturation, and adrenal status in the free-ranging mountain gorilla. However, despite excellent conditions in the wild, routine sample collection involves intense effort on the part of the researcher and data are limited by the *ad libitum* nature of sample collection. Sample storage and transport also present difficult challenges. Therefore we have only partial data on many of the topics that have been addressed. Our results indicate that female estrus cycles can be monitored in the wild and they are suggestive of differences in particular aspects of hormone levels pre- and post-conception. Although it is particularly difficult to assess fertility in males, monitoring hormone levels can contribute to our understanding of the maturation process and the relationship between underlying physiology and behavior. Urinary and fecal corticosteroid levels may be a useful tool to noninvasively monitor stress levels of gorillas. Applications of these techniques to evaluate reproductive status of females and males needs to be continued to determine if some of the described preliminary findings can be confirmed. The results of the studies discussed also raise many interesting questions concerning variation in hormone levels between subspecies and between animals in different environments. Additional studies at other gorilla field sites and on other great

ape species will help us understand such variation in hormone excretion. Such work will assist in the management of both the captive and wild gorilla populations as well as contribute to models of comparative reproductive biology.

Acknowledgements
Funding for the field research for N. Czekala was provided by the Scott Fund, National Geographic Society, L.S.B. Leakey Foundation, Kelco Merck, and the Lee Romney Foundation. M. Robbins received funding from a National Science Foundation predoctoral fellowship and a research assistantship from the Dian Fossey Gorilla Fund International. We thank all zoos that contributed samples for these analyses. Thanks also go to A. Fletcher, A. McNeilage, D. Watts, and S. Allen for assisting with urine collection, storage, and transport in Rwanda. We thank P. Sicotte, K. Stewart, and one anonymous reviewer for helpful comments on this manuscript.

References

Andelman, S J, Else, J G, Hearn, J P & Hodges, J K (1985) The non-invasive monitoring of reproductive events in wild Vervet monkeys (*Cercopithecus aethiops*) using urinary pregnanediol-3a-glucuronide and its correlation with behavioural observations. *Journal of Zoology, London*, **205**, 467–77.

Bahr, N L, Pryce, C R, Dobeli, M & R D Martin (1998) Evidence from urinary cortisol that maternal behavior is related to stress in gorillas. *Physiology and Behavior*, **64**, 429–37.

Beck, B B (1982) Fertility in North American lowland gorillas. *American Journal of Primatology*, Supplement 1, 7–11.

Beck, B B & Power, M (1988) Correlates of sexual and maternal competence in captive gorillas. *Zoo Biology*, **7**, 339–50.

Bentley, G R (1999) Aping our ancestors: comparative aspects of reproductive ecology. *Evolutionary Anthropology*, **7**, 175–85.

Bercovitch, F B & Ziegler, T E (1990) Introduction to socioendocrinology. *Socioendocrinology of Primate Reproduction*, ed. T E Ziegler & F B Bercovitch, pp. 1–9. New York: Wiley–Liss.

Brockman, D K, Whitten, P L, Russell, E, Richard, A F & Izard, M K (1995) Application of fecal steroid techniques to the reproductive endocrinology of female Verreaux's sifaka (*Propithecus verreauxi*). *American Journal of Primatology*, **36**, 313–25.

Brown, C S & Loskutoff, N M (1998) A training program for noninvasive semen collection in captive western lowland gorillas (*Gorilla gorilla gorilla*). *Zoo Biology*, **17**, 143–51.

336   *Nancy Czekala & Martha M. Robbins*

Butynski, T M & Kalina, J (1998) Gorilla tourism: a critical look. In *Conservation of Biological Resources*, ed. E J Milner-Gulland & R Mace, pp. 280–300. Oxford: Blackwell Scientific Publications.

Clarke, A S, Mason, W A & Moberg, G P (1988) Differential behavioral and adrenocortical responses to stress among three macaque species. *American Journal of Primatology*, **14**, 37–52.

Cousins, D (1976) Censuses of gorillas in zoological collections with notes on numerical status and conservation. *International Zoo News*, **23**, 18–20.

Creel, S, Creel, N M, Wildt, D E & Monfort, S L (1992) Behavioural and endocrine mechanisms of reproductive suppression in Serengeti dwarf mongooses. *Animal Behaviour*, **43**, 231–45.

Creel, S, Creel, N M & Monfort, S L (1996) Social stress and dominance. *Nature*, **379**, 212.

Creel, S, Creel, N M, Mills, M G L & Monfort, S L (1997) Rank and reproduction in cooperatively breeding African wild dogs: behavioral and endocrine correlates. *Behavioral Ecology*, **8**, 298–306.

Czekala, N M & Sicotte, P (2000) Reproductive monitoring of free-ranging female mountain gorillas by urinary hormone analysis. *American Journal of Primatology*, **51**, 209–15.

Czekala, N M, Mitchell, W R & Lasley, B L (1987) Direct measurements of urinary estrone conjugates during the normal menstrual cycle of the gorilla (*Gorilla gorilla*). *American Journal of Primatology*, **12**, 223–9.

Czekala, N M, Roser, J F, Mortensen, R B, Reichard, T &Lasley, B L (1988*a*) Urinary hormone analysis as a diagonostic tool to evaluate the ovarian function of female gorilla (*Gorilla gorilla*). *Journal of Reproduction and Fertility*, **82**, 255–61.

Czekala, N M, Shideler, S E & Lasley, B L (1988*b*) Comparisons of female reproductive hormone patterns in the hominoids. In *Orang-utan Biology*, ed. J H Schwartz, pp. 117–22. Oxford: Oxford University Press.

Czekala, N M, Reichard, T & Lasley, B L (1991) Assessment of luteal competency by urinary hormone evaluation in the captive female gorilla. *American Journal of Primatology*, **24**, 283–8.

Czekala, N M, Lance, V A & Sutherland-Smith, M (1994) Diurnal urinary corticoid excretion in the human and gorilla. *American Journal of Primatology*, **34**, 29–34.

Dixson, A F (1981) *The Natural History of the Gorilla*. London: Weidenfeld & Nicolson.

Eddy, R L, Jones, A L, Gilliland, P F, Ibarra, J D, Thompson, J Q & McMurry, J F (1973) Cushing's syndrome: a prospective study of diagnostic methods. *American Journal of Medicine*, **55**, 621–30.

Fossey, D (1976) The behaviour of the mountain gorilla. PhD thesis, University of Cambridge.

Gould, K G & Faulkner, J R (1981) Development, validation, and application of a rapid method for detection of ovulation in great apes and in women. *Fertility and Sterility*, **35**, 676–82.

Gould, K G & King, O R (1982) Fertility in the male gorilla (*Gorilla gorilla*):

relationships to semen parameters and serum hormones. *American Journal of Primatology*, **2**, 311–16.

Graham, C E, Collins, D C, Robinson, H & Preedy, J R K (1972) Urinary levels of estrogens and pregnanediol and plasma levels of progesterone during the menstrual cycle of the chimpanzee: relationship to the sexual swelling. *Endocrinology*, **91**, 13–24.

Harcourt, A H, Fossey, D, Stewart, K J & Watts, D P (1980) Reproduction in wild gorillas and some comparisons with chimpanzees. *Journal of Reproduction and Fertility,* Supplement, **28**, 59–70.

Harcourt, A H, Stewart, K J & Fossey, D (1981) Gorilla reproduction in the wild. In *Reproductive Biology of the Great Apes*, ed. C E Graham, pp. 265–79. New York: Academic Press.

Hofer, H & East, M L (1998) Biological conservation and stress. *Advances in the Study of Behavior*, **27**, 405–525.

Keiter, M & Pichette, L P (1979) Reproductive behavior in captive subadult lowland gorillas. *Zoological Garten*, **49**, 215–37.

Kingsley, S (1982) Causes of non-breeding and the development of the secondary sexual characteristics in the male orangutan: a hormonal study. In *The Orangutan: Its Biology and Conservation*, ed. L E M de Boer, pp. 215–29. The Hague: Junk Publishers.

Kingsley, S R (1988) Physiological development of male orang-utans and gorillas. In *Orangutan Biology*, ed. J H Schwartz, pp. 123–31. New York: Oxford University Press.

Maggioncalda, A N (1995) The socioendocrinology of orangutan growth, development, and reproduction: an analysis of endocrine profiles of juvenile, developing adolescent, developmentally arrested adolescent, adult, and aged captive male orangutans. PhD thesis, Duke University.

Maggioncalda, A N, Sapolsky, R M & Czekala, N M (1999) Reproductive hormone profiles in captive male orangutans: implications for understanding developmental arrest. *American Journal of Physical Anthropology*, **109**, 19–32.

Marson, J, Meuris, S, Cooper, R W & Jouannet, P (1991) Puberty in the male chimpanzee: progressive maturation of some characteristics. *Biology of Reproduction*, **44**, 448–55.

McNeilage, A, Plumptre, A J, Brock-Doyle, A & Vedder, A (1998) Bwindi Impenetrable National Park, Uganda Gorilla and Large Mammal Census, 1997. Wildlife Conservation Society, Working Document No. 14.

Miller, M W, Hobbs, N T & Sousa, M (1991) Detecting stress responses in Rocky Mountain bighorn sheep (*Ovis canadensis canadensis*): reliability of cortisol concentrations in urine and feces. *Canadian Journal of Zoology*, **69**, 15–24.

Mitchell, W R, Loskutoff, N M, Czekala, N M & Lasley, B L (1982*a*) Abnormal menstrual cycles in the female gorilla (*Gorilla gorilla*). *Journal of Zoo Animal Medicine*, **13**, 143–7.

Mitchell, W R, Presley, S, Czekala, N M & Lasley, B L (1982*b*) Urinary immunoreactive estrogen and pregnanediol-3–glucuronide during the normal menstrual cycle of the female lowland gorilla (*Gorilla gorilla*). *American Journal of Primatology*, **2**, 167–75.

Nadler, R D & Collins, D C (1984) Research on reproductive biology of gorillas. *Zoo Biology*, 3, 13–25.

Nadler, R D, Graham, C E, Collins, D C & Gould, K G (1979) Plasma gonadotropins, prolactin, gonadal steroids, and genital swelling during the menstrual cycle of lowland gorillas. *Endocrinology*, 105, 290–6.

Nadler, R D, Collins, D C, Miller, L C & Graham, C E (1983) Menstrual cycle patterns of hormones and sexual behavior in gorillas. *Hormones and Behavior*, 17, 1–17.

Patterson, F, Holts, C L & Saphire, L (1991) Cyclic changes in hormonal, physical, behavioral and linguistic measures in a female lowland gorilla. *American Journal of Primatology*, 24, 181–94.

Poole, J P, Kasman, L H, Ramsay, E C & Lasley, B H (1984). Musth and urinary testosterone concentrations in the African elephant (*Loxodonta africana*). *Journal of Reproduction and Fertility*, 70, 255–60.

Robbins, M M (1995) A demographic analysis of male life history and social structure of mountain gorillas. *Behaviour*, 132, 21–47.

Robbins, M M (1996) Male–male interactions in heterosexual and all-male wild mountain gorilla groups. *Ethology*, 102, 942–65.

Robbins, M M & Czekala, N M (1997) A preliminary investigation of urinary testosterone and cortisol levels in wild male mountain gorillas. *American Journal of Primatology*, 43, 51–64.

Sapolsky, R M (1991) Testicular function, social rank, and personality among wild baboons. *Psychoneuroendocrinology*, 16, 1–13.

Sapolsky, R M (1993) The physiology of dominance in stable versus unstable social hierarchies. In *Primate Social Conflict*, ed. W A Mason & S P Mendoza, pp. 171–204. Albany NY: State University of New York Press.

Sapolsky, R M & Ray, R L (1989) Styles of dominance and their endocrine correlates among wild olive baboons. *American Journal of Primatology*, 18, 1–13.

Shideler, S E, Shackleton, C H L, Moran, F M, Stauffer, P, Lohstroh, P N & Lasley, B L (1993) Enzyme immunoassays for ovarian steroid metabolites in the urine of *Macaca fascicularis*. *Journal of Medical Primatology*, 22, 301–12.

Sievert, J, Karesh, W B & Sunde, V (1991) Reproductive intervals in captive female western lowland gorillas with a comparison to wild mountain gorillas. *American Journal of Primatology*, 24, 227–34.

Stewart, K J (1977) The birth of a wild mountain gorilla (*Gorilla gorilla beringei*). *Primates*, 18, 965–76.

Strier, K B, Ziegler, T E & Wittwer, D J (1999) Seasonal and social correlates of fecal testosterone and cortisol levels in wild male muriquis (*Brachyteles arachnoides*). *Hormones and Behavior*, 35, 125–34.

van Schaik, C P, van Noordwijk, M A & Blankenstein, M A (1991) A pilot study of social correlates of levels of urinary cortisol, prolactin, and testosterone in wild long-tailed macaques (*Macaca fascicularis*). *Primates*, 32, 345–54.

von Holst, D (1998) The concept of stress and its relevance for animal behavior. *Advances in the Study of Behavior*, 27, 1–131.

Wallis, J (1997) A survey of reproductive parameters in the free-ranging chimpanzees of Gombe National Park. *Journal of Reproduction and Fertility*, **109**, 297–307.

Wasser, S (1996) Reproductive control in wild baboons measured by fecal steroids. *Biology of Reproduction*, **55**, 393–9.

Wasser, S K, Bevis, K, King, G & Hanson, E (1997) Noninvasive physiological measures of disturbance in the northern spotted owl. *Conservation Biology*, **11**, 1019–22.

Watts, D P (1990) Mountain gorilla life histories, reproductive competition, and sociosexual behavior and some implications for captive husbandry. *Zoo Biology*, **9**, 195–200.

Watts, D P (1991) Mountain gorilla reproduction and sexual behavior. *American Journal of Primatology*, **24**, 211–25.

Watts, D P (2000) Causes and consequences of variation in male mountain gorilla life histories and group membership. In *Primate Males*, ed. P M Kappeler, pp. 169–79. Cambridge: Cambridge University Press.

Watts, D P & Pusey, A E (1993) Behavior of juvenile and adolescent great apes. In *Juvenile Primates*, ed. M E Pereira & L A Fairbanks, pp. 148–72. Oxford: Oxford University Press.

Weber, A W & Vedder, A (1983) Population dynamics of the Virunga Gorillas: 1959–1978. *Biological Conservation* **46**, 341–66.

Whitten, P L, Brockman, D K & Stavisky, R C (1998) Recent advances in noninvasive techniques to monitor hormone–behavior interactions. *Yearbook of Physical Anthropology*, **41**, 1–23.

Wildt, D E & Wemmer, C (1999) Sex and wildlife: the role of reproductive science in conservation. *Biodiversity and Conservation*, **8**, 965–76.

Wingfield, J C (1994) Hormone–behavior interactions and mating systems in male and female birds. In *The Differences between the Sexes*, ed. R V Short & E Balaban, pp. 303–30. Cambridge: Cambridge University Press.

Young, L G, Gould, K G & Southwick, E B (1993) Selected endocrine parameters of the adult male chimpanzee. *American Journal of Primatology*, **31**, 287–97.

Ziegler, T E, Santos, C V, Pissinatti, A & Strier, K B (1997) Steroid excretion during the ovarian cycle in captive and wild muriquis, *Brachyteles arachnoides*. *American Journal of Primatology* **42**, 311–21.

13 Clinical medicine, preventive health care and research on mountain gorillas in the Virunga Volcanoes region

ANTOINE B. MUDAKIKWA, MICHAEL R. CRANFIELD,
JONATHAN M. SLEEMAN & UTE EILENBERGER

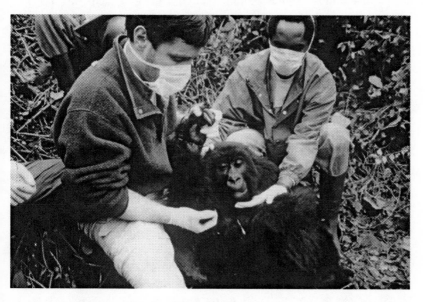

Veterinarians and park staff during a snare removal.

Introduction

Modern conservation efforts are associated with the use of multidisciplinary approaches to resolve biological and socioeconomic issues. Such conservation efforts include veterinary medicine. However, veterinary participation from inception to completion on conservation projects is frequently far below its potential (Karesh & Cook, 1995). The fact that veterinarians are trained and actively involved with population, environmental, and epidemiological approaches to medical management of domestic and captive wildlife is often overlooked in wild populations.

Since its discovery, the mountain gorilla (*Gorilla gorilla beringei*) has been under many pressures, including habitat encroachment, direct poaching, and trauma from snares set for duiker and bushbuck. Recently, direct and indirect consequences of war, as well as increasing levels of ecotourism (70% of the gorilla population is habituated to humans) with their potential for the introduction of human pathogens to the gorillas, have emerged as significant threats. The intrusion of poachers, soldiers, and domestic animals in fragmented patches facilitates the influx of new pathogens. These factors, combined with the limited area available to the animals, have reduced the ability of wild gorillas to avoid contact with humans. The situation will require increasing veterinary input to help secure the future of the mountain gorillas.

Disease management becomes even more important in light of the small size of the mountain gorilla population. With less than 620 animals in two separate populations, the survival of individual gorillas, especially females, becomes a priority, not only to minimize the risk of loss of the entire population in one catastrophic event but also for the maintenance of genetic diversity (Hill, 1999). The Mountain Gorilla Veterinary Project (MGVP) has been monitoring the health of mountain gorillas for over 13 years, and is a pioneer in health management of an endangered species.

The Mountain Gorilla Veterinary Project (formerly called the Volcano Veterinary Centre) is part of a strong multifaceted conservation program, which arose from the early efforts of George Schaller and Dian Fossey. At the request of Dian Fossey in 1985, Ruth Keesling and the Morris Animal Foundation initiated a long-term veterinary program in Rwanda for the mountain gorilla. Its primary purpose has been to provide emergency medicine to injured and severely ill gorillas. The program, however, has expanded the scope of its work to include preventive health monitoring and health research of the Virunga population.

Veterinary input is viewed as one of the major conservation activities contributing to the maintenance of the mountain gorilla population

(Butynski & Kalina, 1998). In this paper we describe clinical medicine, preventive health care, and research activities of the MGVP.

Clinical medicine
Emergency treatment or "intervention"
Mountain gorillas need to be treated in their habitat. Removal of an animal, even for a short period, causes social disruption to individuals and the group, not to mention the problems of subsequent reintroduction. Upon recovery, the gorilla must be capable of functioning in the wild since there are no captive or sanctuary situations to allow slow rehabilitation or long-term holding.

The project has a conservative policy restricting emergency treatment or interventions to cases judged to be potentially life-threatening to an individual or that threaten the survival of the population. Emphasis is placed on human-induced problems such as snare wounds and infectious diseases. The primary reason for this conservative policy is the belief that this population, although impacted by human activities, should still remain subject to the laws of natural selection. Human interference, therefore, should be kept to a minimum in order to preserve, as nearly as possible, the mountain gorillas' natural state. To date, interventions have occurred only with habituated gorillas.

Another important reason to be conservative is that emergency treatment involving anesthesia carries moderate risks, including disruption not only to the individual treated but also to the rest of the group. The decision to intervene is not an easy one. It is made jointly by many of the conservation team members, with the final decision by the local conservator and his department. It is often based on limited observation, historical data, personal experience, and the examination of limited biological specimens, such as feces or urine. The veterinarian then does his/her own risk/benefit analysis of the situation before proceeding.

All interventions require a cooperative team approach and usually involve researchers from the Karisoke Research Center and the International Gorilla Conservation Programme (IGCP). Park staff and gorilla trackers from the protected areas are paramount to the success of the intervention. Also necessary are porters hired to carry essential equipment in the field.

After deciding to intervene, medical, behavioral, environmental, and human safety aspects must be considered. Although the mean induction time of ketamine anesthesia is short at an average of 5 minutes (Sleeman *et al.*, 1998), this is enough time for the animal to get into an emergency

344   *Antoine B. Mudakikwa* et al.

situation like falling into a ravine or drowning in a river. Waiting for the gorilla groups to be on "safe" terrain during the procedure is often necessary, as is the management of other group members (i.e. it is important to keep other gorillas away from anesthetized individuals during a procedure).

Different age and sex classes pose different problems. Infants are a particular challenge because their mothers may carry them off while anesthetized, and the group is very protective of these young individuals. Since the gorillas have been habituated to lose their fear of humans, it may be difficult to keep other members of the group away from the medical team during the emergency treatment. Positioning of the animal being anesthetized is important. It should be isolated as much as possible to reduce disturbance and aggression within the group towards the medical team. There has been only one report of human injury during an intervention: a bite wound to a veterinarian.

There have been numerous and various types of veterinary interventions from the start of MGVP. Between 1986 and 1998, 26 situations required field anesthesia (Sleeman *et al.*, 1998). Drug deliveries were performed using the Telinject® system with lightweight plastic darts. Although Telazol (tiletamine and zolazepam) has been utilized safely, ketamine (average dose 7 mg/kg), with its shorter recovery time (average 42 minutes), remains the drug of choice. The majority of anesthesias (18 of 26, or 69%) were performed to remove snares from a limb or to treat snare-induced wounds. Four animals, or 15%, did not recover due to severe medical conditions. This would indicate that interventions with trained staff are a relatively safe undertaking, even in a high-risk or unhealthy subset of the population in uncontrolled field conditions.

Snare removals

Snares are the most frequent cause of human-induced injuries to mountain gorillas. Snares are set by poachers to trap antelope, but gorillas are either caught accidentally or because they are very curious. The rope and wire snares will often constrict around a limb, cutting off the blood supply to a hand or foot. If this tourniquet is not removed it can result in sloughing of skin and even loss of the whole hand or foot. This could impede the gorilla's ability to survive and/or result in death from bacterial infection and sepsis. Compiling injuries by poachers recorded by Karisoke Research Center, Mountain Gorilla Project, and MGVP between 1971 and 1998 revealed 67 reports of poacher-associated injuries: 50 (75%) were from snares from which 4 ended fatally and 46 survived, 6 (9%) were by firearms, 5 (7%) by spear or arrow, and 6 (9%) from unknown causes (N. Gerald-

Steklis & E.A. Williamson, personal communication). The fact that many gorillas have survived these snare injuries could be attributed to the prompt removal of the snares by the project veterinarians. Close attention to the management of the snare-induced wounds using surgical debridement, flushing, topical and parenteral antibiotics, and close monitoring of the healing process has almost certainly reduced the morbidity and mortality from secondary bacterial infections. Depending on the extent of tissue damage, the healing and therefore monitoring time has ranged from 2 weeks to 6 months.

Other treatments

Interventions involve the treatment of illness as well as trauma. Close observations of clinical signs can lead to successful treatment under field conditions of severely ill individuals, as is demonstrated by the following case (Hastings, 1991). A silverback male developed progressive clinical signs of purulent exudate draining from one eye and both nostrils, cough, shrunken abdomen, an elevated respiratory rate, inappetence, reduced movement as well as asymmetric enlargement of the laryngeal air sacs. He was treated with antibiotics administered by remote darting. Two days later his health condition was stable enough to allow anesthesia. Investigation in the field of aspirated material from the nasal cavity and air sac revealed a bacterial infection. The air sac was surgically opened, flushed with fluids and antibiotics, and parenteral fluids were also administered. The gorilla recovered completely.

A secondary benefit of emergency treatment under anesthesia is the opportunity to collect physiological data (although somewhat altered by anesthesia), morphometric data (including weights) and biological samples for further clinical, pathological and storage for biobanking (see below). Data is recorded in MedARKS.

Newly acquired field equipment, such as a pulse oximeter (Nellcor, Hayard) to measure pulse rate and percent oxygenation of the hemoglobin, and an ISTAT clinical analyzer (ISTAT Corporation, Princeton) to get immediate physiological blood results, along with a detailed physical examination, allows for on-site accurate assessment of the health of the patient. With this information, the clinician can determine what samples, and in what amounts, he/she needs to collect for further assessment of the individual clinical case, and for surveillance tests (genetic, nutritional, infectious disease, etc.) that would apply to the health of the population.

Sample collection is performed according to a standard operating procedure. Basic sample collection includes as much urine and feces as possible, ectoparasites if present, hair samples including roots, blood (up to

5 ml/kg safe volume in a healthy individual), and culture of the rectal and wound swabs. These stored samples hold a wealth of information about the health status of the population as a whole and allow investigators to access the dynamics of disease over time and with respect to events that occur to this population.

Preventive health care

Intensive detailed veterinary preventive medicine programs have been developed for domestic and wild animals in captivity. Zoos have long realized that maintaining a healthy population of animals is accomplished more easily and successfully by proactive prevention than by reactive emergency medicine. Preventive health care includes, but is not limited to, health monitoring, disease surveillance, pathology, and immunization. The goal is early detection and management of the disease to minimize negative impacts on the gorilla population.

Health monitoring and disease surveillance

Health monitoring means regular visits to the groups to assess health status by observation, as well as collecting and screening biological samples collected noninvasively. It is advantageous to plan management measures prior to the occurrence of a disease and implement preventive measure and regulations in order to diminish the impact. Veterinarians make regular visits to the habituated groups to observe clinical signs of disease and collect data in order to determine if the population is healthy or not. Every observable individual is screened for problems such as discharge from body orifices, abnormal respiration rates, injuries, poor hair condition, swollen navel (in newborns), scratching and abnormal manipulation behavior (transport of infants, plant manipulation, grooming). Groups are observed for composition (newborns and die-offs), social structure, and the spatial behavior of group members. These observations allow the program to gather data on the prevalence of different clinical entities such as wounds, respiratory disease, and diarrhea.

Disease surveillance is the proactive testing for high-risk diseases to which a population is susceptible. An example of this is tuberculosis in captive primates. Unfortunately, many of the factors necessary for the success of a preventive medicine program, which are controllable in a captive situation, cannot be manipulated or modified on a practical level in the wild, but the principles and goals can be applied (Foster, 1993; Sleeman, 1998). Such a program necessitates knowledge of baseline data of which diseases and events influence morbidity and mortality within the population. Even this relatively simple task of inventorying disease entities

Table 13.1. *Parasites of the different gorilla subspecies*

	Gorilla g. beringei	*Gorilla g. beringei*	*Gorilla g. beringei*	*Gorilla g. graueri*	*Gorilla g. gorilla*
Country	Rwanda	Rwanda/DRC	Uganda	DRC	Gabon
Study site	Virungas	Virungas	Bwindi	Kahuzi	Lopé
Year	1997	1989/92	1990/96	1993	1995
Total prevalence	*97%*	*96/100%*	*100/98%*	*67%*	*36%*
Helminths					
Anoplocephala gorillae	85%	51/×%	85/89%	27%	18%
Strongyloides fullebo/sp.	10%	96/100%	24/16%	8%	
Strongyles					
(*Oesophagostomum* sp.)	100%	84/×%	100/89%	43%	
Probstmayria gorillae/sp.	9%		100/×%	11%	
Trichostrongylus sp.	7%	×	×		
Gongylonema (possible)					8%
Capillaria hepatica					
Paralibiostrongylus kalinae			×		
Hyostrongylus kigenziensis			×		
Impalaia sp./possible		×			
Murshidia devians/sp. (*Trichuris trichiura*)	×				
Protozoa					
Giardia lamblia	3%	×			
Entamoeba coli	20%	×			
Enteromonas hominis	1%				
Entamoeba histolytica	27%	×			
Entamoeba hartmanni	16%	×			2%
Iodamoeba buetscheli	44%	×			
Endolimax nana	1%				
Chilomastix					

× = identified but not quantified. Blank spaces represent no data available.

is difficult, in this case, because of the lack of access to biological material on which to perform such disease surveys. Literature searches and questionnaire surveys have delineated some important zoonotic diseases between primates and man.

The first long-term parasitology study was carried out at Karisoke Research Center by Ian Redmond between November 1976 and April 1978 (Redmond, 1983). Monitoring feces for the prevalence of parasites such as fecal helminths together with clinical signs can give helpful insights into changes of immune competence of the host. Since that time, the parasitological screening program has expanded to include external liver and blood parasites as well as protozoa. Most of the data have come from opportunistic sampling and autopsies.

Table 13.1 compares data from different studies in the Virunga Volcanoes region to data on eastern and western lowland gorillas (Landsout-Soukate *et al.*, 1995; Eilenberger, 1998). While all investigated mountain gorillas in the Parc National des Volcans in Rwanda have tolerated the ecto- and endoparasites detected so far, without showing signs of illness, "new parasites" introduced by humans or domestic animal could have potentially devastating effects on the population dynamic of the host (Grenfell & Gulland, 1995). Three parasites *Trichuris trichiura, Chilomastix,* and *Endolimax nana*, which have not been found during prior investigations, were discovered through a survey of fecal samples (Mudakikwa *et al.*, 1998). Although it is difficult to document the source of these parasites, it is possible that they were introduced by humans, since they are commonly found in humans at the park borders (Ashford *et al.*, 1990; Eilenberger, 1998). Another example of possible transfer of parasites from man to gorillas was described by Kalema *et al.* (1998) when the cause of a neonatal gorilla death was delineated as scabies at the same time that there was an outbreak of scabies in the surrounding human population. Nizeyi *et al.* (1999) reported the prevalence of cryptosporidia to be 11% in the gorilla population of Bwindi Impenetrable National Park.

A study carried out in the Parc National des Volcans from August 1996 to January 1997 showed that it is feasible to collect urine from habituated free-ranging mountain gorillas for diagnostic purposes. The study established urinary reference intervals (Mudakikwa & Sleeman, 1997). Analysis of urine can be used to monitor endocrine functions (Czekala *et al.,* 1994) as well as carbohydrate metabolism, kidney and liver function, and acid–base balance and urinary tract infections. In addition, erythrocytes or leukocytes, neoplastic cells, epithelial cells, crystals, parasitic ova, and bacteria give further information about the health of the investigated gorilla.

Table 13.2. *ELISA titers of mountain gorilla sera*

Animal	Sex	Age (yrs)	Group	HSV-1[a]	HSV-2	SA8[b]
Tiger	F	20	NA[c]	<100	<100	<100
Peanuts	M	30	NA	<100	<100	<100
Josie	F	7.5	NA	<100	<100	<100
WGM	M	11	3	1600	6400	ND[d]
Kato	F	3	2	6400	25000	1600
Guhuma	F	NA	NA	12 800	12 800	3200
Mukanza	F	4.5	1	3200	3200	800

[a] HSV = Herpes Simplex Virus
[b] SA8 = Herpes virus from baboons and other African monkeys
[c] NA = Not Available
[d] ND = Not Determined
Source: After Eberle (1992).

Routine monitoring of the habituated mountain gorillas has also in-cluded morphological traits that might indicate genetic homogeneity. Syn-dactyly, or webbed digits, was first described in mountain gorillas by Schulz (1934) and Fossey (1983). A survey in 1996 of syndactyly in four habituated groups in the Parc National des Volcans (Routh & Sleeman, 1997) revealed a high prevalence of 52.7% (39/74) of this trait. Only one animal showed syndactyly of the fingers, whereas other individuals show syndactyly of the toes, of which the majority (36/39) involved the digits III and IV. Approximately half of the individuals (21/39) have some degree of bilateral syndactyly. Although this trait has been used in taxonomic de-scriptions of the mountain gorilla, it is possible that this characteristic is inherited. However no conclusion can be made regarding the etiology and significance of syndactyly until molecular genetic studies and behavioral observations have established the family trees and genetic relatedness of the population.

Serological epidemiology is the investigation of disease by the measure-ment of variables present in serum. Antibodies provide evidence of current or previous exposure to infectious agents. This is useful in identifying major health problems in the population, the distribution of disease, and diseases to which the animals are immunologically naïve. A study by Eberle (1992) showed serologic evidence of an alpha herpes virus in 58% of seven individuals tested, mostly in the younger individuals (Table 13.2). Blood samples were obtained from animals either found dead or when anesthetized for life-threatening disease. This virus appears to be most closely related to, but distinct from, human herpes simplex virus type 2. Suspicious facial macules and papules were observed on adult female and infant/juvenile gorillas and may have represented the clinical manifestation of this virus. No mortality has been linked to this viral infection and its zoonotic potential is yet unknown. Similarly, the observed antibody response in the gorillas may not confer immunity to the human herpes simplex viruses that may in turn represent a serious health threat to the mountain gorilla population.

In 1995 MGVP began utilizing ARKS and MedARKS (International Species Inventory System), which are computerized animal record keeping programs. Each individual habituated gorilla in Rwanda now has an identifying accession number with historical and medical information. It is hoped that this will expand to include Democratic Republic of Congo and Uganda. These programs allow for efficient collection, entry, and maintenance of data pertaining to live collections. Although developed in America for zoos, they are easily applied to wild populations. Utilization of ARKS and MedARKS has standardized record keeping, allowing for trends in health concerns to emerge, and for the system to be used as a management tool. MedARKS data, in conjunction with other databases such as Geographic Information Systems, will allow for the epidemiological analysis of disease by looking at the temporal and spatial patterns of the different clinical signs (Byers & Hastings, 1988; Eilenberger, 1998). Archived long-term data can also be used for retrospective studies.

Pathology

One of the most powerful tools in any preventive program is pathology, which includes complete macroscopic post-mortem results from every dead gorilla body recovered. This is complemented by histopathologic tissue examination and sample collection for storage for further investigation (see "Future research", below). Pathology provides information on many levels: (1) diagnosis during a disease outbreak so that effective treatment or management changes can be applied to the remaining susceptible animals, minimizing the negative impact on the population, (2) incidental findings which may not cause mortality, but cause significant morbidity within a population, and (3) defining high-risk subgroups within the population, allowing targeting of resources.

Even though post-mortems are often done under adverse conditions on less than fresh tissues, important data are compiled. Active CITES and CDC permits assure rapid shipping of fixed tissues to experienced primate pathologists after receipt of export papers from the host country. The review of the histopathology of all available tissues further delineates lesions and causes of morbidity and mortality (L.J. Lowenstine, unpublished data). There have been 89 documented examinations of dead

Table 13.3. *Parasitic lesions in juvenile and adult mountain gorillas found from histopathology (necropsy)*

Lesim	Frequency
Small intestinal cestodiasis (*Anoplocephala*)	6/8
Peritoneal adhesions (abdominal nematodiasis)	5/8[a]
Proliferative gastritis (trichostrongyles)	4/8
Periportal hepatic fibrosis (*Capillaria hepatica*)	3/8[b]
Colonic mural nodules (*Oesophagostomum* sp.)	3/8
Ectoparasites (lice)	1/8[c]

[a] All adults.
[b] 3 others with just fibrosis.
[c] Juvenile.

gorillas since Fossey's first visit. Fifteen are listed in her book (Fossey, 1983). Many were performed by investigators with varying medical knowledge and before protocols were developed. There are 42 detailed postmortem reports, and 25 cases for which representative tissue exits. The 89 deaths represent only a fraction of gorilla deaths since the 1960s, thus much information has been lost due to dead animals not being located or inaccessibility to the park.

Post-mortems reveal the prevalence of pathogens, which can lead to morbidity, such as the parasites listed in Table 13.3.

Skeletal and dental pathology can reveal helpful additional information to explain retrospectively clinical observations by field vets such as draining of exudate, swellings, difficulties in chewing, movement, and behavior. Furthermore, skeletal pathology gives insights into developmental abnormalities, aging processes, systemic bone disease, and bone neoplasm, which cannot be discovered otherwise. An investigation by Lovell in 1990 revealed that 22% of mountain gorilla skeletons show either trauma, inflammation, arthritis, dental abnormalities, or periodontal diseases. Developmental anomalies (6%) and bone neoplasm (6%) were also detected. Trauma mainly occurred in the postcranium and longbones (8/11), and some longbones showed signs of natural healing. A higher percentage of males showed broken canine teeth and skull fractures, which Lovell attributed to physical aggression between individuals. Table 13.4 lists causes of mortality and Table 13.5 lists the distribution of mortality by age and sex.

Human/gorilla exposure

Introduction of human disease is a real threat to the gorillas for two reasons: (1) due to a close genetic relationship with humans, gorillas are

352 *Antoine B. Mudakikwa* et al.

Table 13.4. *Causes of mortality by age group*

Infant <1 year of age		Juvenile 1–8 years of age		Adult >8 years of age	
Cause	Number	Cause	Number	Cause	Number
Trauma	1	Trauma	3	Trauma	3
Pneumonia	2	Pneumonia	3	Pneumonia	4
Dystocia	1	Bacterial septicemia	1	Bacterial hepatitis	1
Maternal neglect	1			Colitis	1
Infanticide	4			Heart failure	1
				Undetermined	6

Table 13.5. *Age and sex distribution of mortality of 32 gorillas for which data exist, 1978 to 1990*

Age	Males	Females	Total
Infant <1 year of age	4	5	9
Juvenile 1–8 years of age	3	4	7
Adult >8 years of age	5	11	16
Total	12	20	32

especially susceptible to human pathogens; (2) a newly introduced pathogen can have a devastating effect on the gorilla population because of the lack of immune protection against these agents (Grenfell & Gulland, 1995).

There are four distinct groups of humans which pose unique disease risks for gorillas: (1) large numbers of local people farming around the park; (2) gorilla conservation personnel, including trackers, guides, porters, park managers, biologists, and veterinarians; (3) ecotourists; (4) intruders who during the recent war were living on all sides of the Virunga Conservation Area without sanitation, proper nutrition, medical care, and who left behind many open latrines and uncovered garbage.

Contact between local people (excluding park staff) and the gorillas is limited to occasions when people enter the park illegally, or when gorillas leave the park and enter fields. Crop raiding by gorillas happens more often in Uganda (Bwindi Impenetrable National Park) than around the Virungas, where it seems to be incidental. With the burgeoning human population in the Virunga region, it is probable that there will be an increase in contact between the local human population and mountain gorillas. The increasing human population will require more land for agricultural use, and this pressure may lead to more encroachment into the forest for various activities, also increasing the likelihood of disease transmission. Thus, monitoring of the diseases present in the local population is essential to identify diseases that could be a threat to the gorillas. This monitoring could be done in collaboration with the local human health authorities. Recent investigations in Uganda (M. Rooney & J. Sleeman, 1997, unpublished data) and Congo (Eilenberger, 1998) in hospitals near to park borders and with local populations identified the occurrence of a variety of diseases. These investigations identified respiratory tract infections, intestinal parasites, skin disorders, and measles as the most likely types of anthropozoonotic diseases. This information is very helpful in devising appropriate disease prevention measures. Local government Health Monitoring Information Systems could contribute greatly to monitoring disease trends in the local human population, and serve as a warning system for detecting potential disease threats to the gorillas.

During times of stability, the most frequent close contact with the gorillas is by park staff and tourists, who possibly then pose the greatest threat of introduction of novel disease to the gorillas. Education of these people is an important part of the work of veterinarians. Veterinarians take part in training sessions for the park staff where they explain disease transmission, how to report signs of illness, and how to assist during emergency treatment. Ideally, humans would be quarantined and a health check performed before visiting the gorillas, but quarantine cannot be accomplished accurately with wildlife. In Bwindi Impenetrable National Park, it is even more difficult to limit contact because of roads and trails passing through the forest.

Ecotourists travel from many different areas of the world, potentially carrying endemic diseases. They may be stressed by exhausting travel agendas and foreign routines that might alter the immune system. Tourists often have a one-time close encounter with gorillas within 72 hours of leaving home. All of these factors can enhance disease transmission. The following regulations have been implemented in gorilla tourism programs to reduce the exposure of the gorillas to pathogens:

- Sick persons are denied access to the gorillas
- Children under 15 years are not allowed to visit
- Visits are limited to one visit to each group per day
- Visits are limited to eight persons and for one hour only
- A distance of more than 7 meters from the gorillas is maintained
- Touching gorillas is strictly prohibited: if a gorilla approaches, then the visitor must slowly back away
- Visitors who have to sneeze or cough are asked to turn away from the gorillas

- Visitors who break the regulations are asked to leave the group
- Human waste is buried at least 30 cm deep
- Littering, smoking, and eating are prohibited during the visit

Visitor regulations were first developed 20 years ago by the Mountain Gorilla Project, and were revised by conservation organizations, field veterinarians, and park managers during an evaluation in February 1999. The regulations are based on basic epidemiological principles. Although during the last 15 years no disease transmission between mountain gorillas and humans has been proven, the distance regulation has been adjusted according to modern research on the transmission of airborne diseases (Homsy, 1999). Nevertheless gorillas that approach humans out of curiosity sometimes decrease the effectiveness of this program.

Immunization
The ability of vaccines to reduce the prevalence of disease within human and animal populations is well documented and this tool is widely utilized. In 1988, a higher than usual mortality rate was observed in the Virunga gorillas, and an apparently novel respiratory disease spread to three of seven habituated gorilla groups. Although not yet definitively diagnosed, measles was implicated, and a human attenuated live vaccine was delivered by remote injection to 65 gorillas, a large portion of the habituated gorilla population (Hastings, 1991). The disease outbreak subsided. Although efficacy was not established at the time, this procedure proved that vaccines could be effectively and safely delivered to enough individuals within a population for efficacy to be established. At the present time, vaccines are not utilized due to adherence to the non-intervention policy; however, it might be considered in the case of a life-threatening epidemic. Data on efficacy and safety of vaccines have been established in captive lowland gorillas. MGVP has the equipment and expertise to use this powerful tool if policies change or if another outbreak of a disease occurs for which a proven vaccine exists.

Future research
For most wild species, few data exist about the etiology and pathogenesis of their diseases, or about basic physiological parameters such as normal gut fauna and flora. Many wildlife diseases have multifactorial causes due to the variety of parameters (e.g. diet, climate, vectors) in the natural habitat, which are not under human control. Research is necessary to assess the significance of a disease on gorilla individuals or subpopulations

and to identify factors that influence the vulnerability of certain population to pathogens.

The major goal of research by the MGVP is to seek knowledge that allows better management of the gorilla population and its habitat from a medical perspective. It usually involves new approaches to prevention, innovative and improved diagnostics and treatments of disease entities. Modern wildlife disease management may succeed by manipulating the agent, the host population, the environment, human activities, or a combination of these. Priorities are usually governed by the severity of existing or perceived health threats. One of the focuses of research is the risk of human disease and the consequences for gorilla health.

Park staff have frequent close contact with the gorillas and possibly pose the largest threat of the introduction of novel disease to the gorillas. A recent investigation of the human population at the park border and the guards in Kahuzi-Biega National Park in Democratic Republic of Congo (DRC) has shown that 54% of the people studied carried helminths and 24% protozoa, all of which have an anthropozoonotic potential for primates (Eilenberger, 1998). As sanitation and health care is comparable to the conditions in the Parc National des Virunga (DRC) and Parc National des Volcans (Rwanda) the implementation of a health program which includes pre-employment as well as yearly physical examinations is a high priority. These should include blood and TB tests, a vaccination program, regular fecal screening, and anti-parasitological treatment. Sanitation of park border areas is also an important aspect of a preventive medicine program. The effect of reduced sanitation can be documented through, for example, a higher prevalence of the helminth *Capillaria hepatica* in gorillas near human settlements (Graczyk *et al.*, 1999). It is thought that *Capillaria* prevalence might correlate with a higher prevalence of rodents, the intermediate host of capillariasis in these areas. Knowledge of these aspects may help with policy-making and management of the parks.

The mountain gorilla population is small. A serious disease outbreak combined with their long inter-birth intervals and high infant mortality could threaten the survival of the population. Reduction in population size has very likely contributed to a loss of genetic diversity. In order to store information, the Morris Animal Foundation has established a new biological resource center. Biological specimens from mountain gorillas are being centralized, inventoried, and correctly preserved for future research. The center will store hair, feces, urine, blood, and frozen and fixed tissue in a diverse range of preservatives and conditions to provide samples that will be available for the widest range of research techniques. These stored samples hold a wealth of information about the health status of the population as a whole and will allow retrospective studies to assess the dynamic of diseases over time and with respect to events that occur in this population. The biological resource center is open to researchers from all over the world who would like to investigate tissues. Researchers will submit proposals to the Wildlife Scientific Advisory Board of Morris Animal Foundation. Data from these studies will provide additional data for the long-term management of the mountain gorilla.

Conclusions

The Population Viability and Habitat Assessment for mountain gorillas, held in 1997, stressed that disease is one of the biggest potential threats to the mountain gorilla (Werikhe *et al.*, 1998). Veterinary input is therefore one of the major activities contributing to the gorillas' survival. Clinical medicine will help save individuals suffering from poacher injuries or life-threatening disease. But given the difficulties associated with intervention and long-term treatment in wild animals, a preventive approach to health care is advantageous. An early detection of disease will help to minimize negative impacts on the remaining gorilla population. Complete health surveys have been accomplished for the safety and benefit of a variety of endangered species (Karesh & Cook, 1995). Only when such data on population dynamics are correlated with baseline data on health will the information necessary to guide management plans be obtained. The non-intervention policy limits the collection of data on existing pathogens, although modern methods in sample collection, preservation, and identification of pathogens from living and dead animals have increased the capacity of health care programs. The question of whether disease management in wild gorillas is an interference in natural selection can be countered by the fact that mountain gorillas now live in an environment modified by humans to a certain degree. Many wildlife diseases are the result of, or at least influenced by, human activities, and all over the world health management plans are being implemented to counteract extinction of endangered species (Wobeser, 1994). The threat of the introduction of human disease seems considerable because of the burgeoning human population in the Virunga Volcanoes region, where there is poor health care (e.g. immunization) and limited education about disease transmission. The risk has also been enhanced by the recent war, with settlements and human refuse left in the park, as well as the growth of tourism with tourists coming from all continents carrying foreign pathogens. Identifying practical preventive measures to reduce potential anthropozoonotic diseases is therefore a major goal of current veterinary care. Research as well as the results of clinical and pathological investigations have revealed the existence of previously undescribed pathogens such as cryptosporidia, amoeba, and capillaria in the mountain gorillas. Veterinarians have shown how to treat successfully a variety of wounds and infections, as well as safe methods to anesthetize and vaccinate gorillas in the wild. The work has helped to identify vulnerability and causes of mortality. It was observed that various parasites are tolerated and belong to the normal flora of the gorillas but that some others also can cause signs of ill health. Veterinary research has revealed spatial and temporal patterns in disease frequency. Preventive health care, treatment, and regulations have been improved over the last 15 years and appear to have effectively protected the mountain gorillas from catastrophic decline. But the lack of information on critical health-related factors shows that data collection and research on health-related issues will remain a priority for the future.

Acknowledgements
We are grateful to Dr. Linda Lowenstine for the summary of the post-mortem examinations she provided. We thank the Office Rwandais du Tourisme et des Parcs Nationaux for the privilege of working in the Parc National des Volcans, Rwanda. We thank project staff and former veterinarians for their dedication to the conservation of the mountain gorillas. MGVP is funded by the Morris Animal Foundation and Mrs. Ruth Morris Keesling. We also thank Dr. Eric Miller for his review and Peter Blank for his help in preparing the manuscript.

References

Ashford, R W, Reid, G D F & Butynski, T M (1990) The intestinal fauna of man and mountain gorillas in a shared habitat. *Annals of Tropical Medicine and Parasitology*, **84**, 337–40.
Ashford, R W, Lawton, H, Butynski, T.M & Reid, G D F (1996) Patterns of intestinal parasitism in the mountain gorilla (*Gorilla gorilla*) in the Bwindi Impenetrable Forest, Uganda. *Journal of Zoology*, **239**, 507–14.
Butynski, T M & Kalina, J (1998) Gorilla tourism: a critical look. In *Conservation of Biological Resources*, ed. E J Milner-Gulland & R Mace, pp. 280–300. Oxford: Blackwell Scientific Publications.
Byers, A C & Hastings, B (1988) Mountain gorilla mortality and climatic factors in the Parc National des Volcans. *Mountain Research and Development*, **11**, 145–51.

Byers, A C & Hastings, B (1991) Mountain gorilla mortality and climate factors in the Parc National de Volcans, Ruhengeri Prefecture, Rwanda. *Mountain Research Development*, **11**, 145–52.

Czekala, N, Lance, V A & Sutherland-Smith, M (1994) Diurnal urinary corticoid excretion in the human and gorilla. *American Journal of Primatology*, **34**, 29–34.

Durette-Desset, M C, Chabaud, A G, Ashford, R W, Butynski, T M & Reid, G D F (1992) Two new species of the Trichostrongylidae, parasitic in *Gorilla gorilla beringei* in Uganda. *Systematic Parasitology*, **23**, 159–66.

Eberle, R (1994) Evidence of an alpha-herpes virus indigenous to mountain gorillas. *Journal of Medical Primatology*, **21**, 246–51.

Eilenberger, U (1998) Der Einfluss von individuellen, gruppenspezifischen und oekologischen Faktoren auf den Endoparasitenstatus von wildebended oestlichen Flachlandgorillas (*Gorilla gorilla graueri*). PhD thesis, Free University of Berlin.

Fossey, D (1983) *Gorillas in the Mist*. Boston MA: Houghton Mifflin.

Foster, J (1993) Health plans for the mountain gorillas of Rwanda. In *Zoo and Wild Animal Medicine, Current Therapy*, Vol. 3, ed. M E Fowler, pp. 331–4. Philadelphia PA: W.B. Saunders.

Garner, K J & Ryder, O A (1996) Mitochondrial DNA diversity in gorillas. *Molecular Phylogenetics and Evolution*, **6**, 39–48.

Graczyk, T K, Lowenstine, L J & Cranfield, M R (1999) *Capillaria hepatica* (Nematoda) infections in human-habituated mountain gorillas (*Gorilla gorilla beringei*) of the Parc National des Volcans, Rwanda. *American Society of Parasitology*, **85**, 1168–70.

Grenfell, B T & Gulland, F M D (1995) Ecological impact of parasitism on wildlife host populations. *Parasitology*, **111**, 4–14.

Hastings, B E (1991) The veterinary management of a laryngeal air sac infection in a free-ranging mountain gorilla. *Journal of Medical Primatology*, **20**, 361–4.

Hastings, B E, Condiotti, M, Sholley, C, Kenney, D & Foster, J (1988) Clinical signs of disease in wild mountain gorillas. In *Proceedings of the Joint Meeting of the American Association of Zoo Veterinarians and The American Association of Wildlife Veterinarians*, p. 107 (abstract).

Hastings, B E, Kenney, D, Lowenstine, L J & Foster, J (1991) Mountain gorillas and measles: ontogeny of a wildlife vaccination program. In *Proceedings of the American Association of Zoo Veterinarians Annual Meeting*, Calgary, pp. 98–205.

Hastings, B E, Gibbons, L M & Williams, J (1992). Parasites of free ranging mountain gorillas: survey and epidemiological factors. In *Proceedings of the Joint Meeting of the American Association of Zoo Veterinarians and the American Association of Wildlife Veterinarians*, Oakland, pp. 301–2.

Hill, A (1999) Defense in diversity. *Nature*, **398**, 668–9.

Holmes, J C (1996) Parasites as threats to biodiversity in shrinking ecosystems. *Biodiversity and Conservation*, **5**, 975–83.

Homsy, J (1999) *Ape Tourism and Human Diseases: How Close Should We Get?* Report to the International Gorilla Conservation Program. Rwanda.

Kalema, G, Kock, R A & Macfie, E J (1998). An outbreak of sarcoptic mange in free-ranging mountain gorillas (*Gorilla gorilla beringei*) in Bwindi Impenetrable Forest, Southwest Uganda. In *Proceedings of the Joint Meeting of the American Association of Zoo Veterinarians and the American Association of Wildlife Veterinarians*, p. 438.

Karesh, B W & Cook, A R (1995) Applications of veterinary medicine to *in situ* conservation efforts. *Oryx*, **29**, 244–52.

Landsout-Soukate, J, Tutin, C E G & Fernandez, M (1995) Intestinal parasites of sympatric gorillas and chimpanzees in Lopé Reserve, Gabon. *Annals of Tropical Medicine and Parasitology*, **89**, 73–9.

Loevinsohn, M E (1994) Climatic warming and increased malaria incidence in Rwanda. *The Lancet*, **343**, 714–18.

Lovell, N C (1990) Skeletal and dental pathology of free-ranging mountain gorillas. *American Journal of Physical Anthropology*, **81**, 399–412.

Lowenstine, L J (1990) Long distance pathology, or will the mountain gorilla fit in the diplomatic pouch? In *Proceedings of the American Association of Zoo Veterinarians Annual Meeting*, pp. 178–85.

Mudakikwa, A B & Sleeman, J M (1997) Analysis of urine from free-living mountain gorillas (*Gorilla gorilla beringei*) for normal physiological values. In *Proceedings of the American Association of Zoo Veterinarians Annual Meeting*, p. 103 (abstract).

Mudakikwa, A B, Sleeman, J M, Foster, J W, Meader, L L & Patton, S (1998). An indicator of human impact: gastrointestinal parasites of mountain gorillas (*Gorilla gorilla beringei*) from the Virunga Volcanoes Region, Central Africa. In *Proceedings of the Joint Meeting of the American Association of Zoo Veterinarians and the American Association of Wildlife Veterinarians*, pp. 436–7.

Nizeyi, J B, Mwebe, R, Nanteza, A, Cranfield, M R, Kalema, G N N & Graczyk, T K (1999) *Cryptosporidium* sp. and *Giardia* sp. infections in mountain gorillas (*Gorilla gorilla beringei*) of the Bwindi Impenetrable National Park, Uganda. *American Society of Parasitology*, **85**, 1084–8.

Redmond, I (1983) Karisoke parasitology research: summary of parasitology research, November 1976 to April 1978. In *Gorillas in the Mist*, D Fossey, pp. 271–8. Boston MA: Houghton Mifflin.

Rooney, M & Sleeman, J (1988) Identifying potential disease threats to the mountain gorillas (*Gorilla gorilla beringei*) and chimpanzees (*Pan troglodytes*) of Uganda by establishing the diseases endemic to the human population living in close proximity to the great ape habitats. Report to the Center for Conservation Medicine, Tufts Veterinary School.

Routh, A & Sleeman, J (1997) A preliminary survey of syndactyly in the mountain gorillas (*Gorilla gorilla beringei*). In *Proceedings of the Spring Meeting of the British Veterinary Zoological Society*, Howletts and Port Lympne Wild Animal Parks, June 14–15, 1997, pp. 22–5.

Schultz, A H (1934) Some distinguishing characters of the mountain gorilla. *Journal of Mammalogy*, **15**, 51–61.

Sleeman, J M (1998) Preventive medicine program for the mountain gorillas

(*Gorilla gorilla beringei*) of Rwanda: a model for other endangered primate populations. In *European Association of Zoo and Wildlife Veterinarians, Second Scientific Meeting*, Chester, UK, May 21–4, 1998, pp. 127–132.

Sleeman, J M & Mudakikwa, A B (1998) Analysis of urine from free-living mountain gorillas (*Gorilla gorilla beringei*) for normal physiological values. *Journal of Zoo and Wildlife Medicine*, **29**, 432–4.

Sleeman, J M, Cameron, K, Mudakikwa, A B, Nizeyi, J B, Anderson, S, Richardson, H M, Macfie, E J, Hastings, B E & Foster, J W (1998) Field anaesthesia of free ranging mountain gorillas (*Gorilla gorilla beringei*) from the Virunga Volcano Region, Central Africa. In *Proceedings of the Joint Meeting of the American Association of Zoo Veterinarians and the American Association of Wildlife Veterinarians*, pp. 1–4.

Werikhe, S, Macfie, L, Rosen, N & Miller, P (1998) *Can the Mountain Gorilla Survive? Population and Habitat Viability Assessment for* Gorilla gorilla beringei. Apple Valley MN: IUCN SSC Conservation Breeding Specialist Group.

Wobeser, G A (1994) *Investigation and Management of Diseases in Wild Animals.* New York: Plenum Press.

14 Conservation-oriented research in the Virunga region

ANDREW J. PLUMPTRE & ELIZABETH A. WILLIAMSON

Park/farmland boundary in Parc des Volcans, Rwanda. (Photo by Andrew J. Plumptre.)

Threats to gorillas in the Virungas during the 20th century

Gorilla (*Gorilla gorilla*) populations are threatened throughout their range, primarily by human activities. As human populations increase in Africa the available room for gorillas is decreasing and gorillas are coming into contact with humans much more than they used to. Each of the three subspecies (see chapter 1 for discussion about possible changes in taxonomy), *G. g .gorilla* (population of about 110 000), *G. g. graueri* (population of about 17 000), and *G .g. beringei* (population of about 600) are threatened by slightly different factors that result from increasing human populations, and all gorilla populations are considered under IUCN criteria as either vulnerable or endangered (Harcourt, 1996). The major threats to gorillas can be categorized as: (1) habitat loss, modification, or fragmentation; (2) hunting or poaching; (3) disease transmission from humans; and (4) war or political unrest. The mountain gorillas in the Virungas have been exposed to each of these threats over the period during which western scientists have known about them. Despite the international publicity about these animals and despite their potential to earn foreign currency for the countries where they are found, mountain gorillas still face several of these threats. This chapter reviews the threats gorillas have faced in the Virungas over time and summarizes some of the research that has aimed to address these threats and provide information to park managers to better conserve the gorillas. We focus primarily on Rwanda and the Parc National des Volcans because the Karisoke Research Center, where most researchers have been based, is in this country. However we do make comparisons with the rest of the Virungas in Uganda and the Democratic Republic of Congo (DRC) and with the Bwindi Impenetrable National Park in Uganda.

Habitat loss, modification, or fragmentation

This category of threat includes loss or fragmentation of forest due to agriculture and burning, and degradation of forest through human modification such as logging, fuel wood and non-timber forest product collection, and grazing domestic animals. Today the whole Virunga region, including Ugandan, Rwandan, and Congolese sectors, covers an area of about 430 km², although much of the high-altitude vegetation is not particularly suitable for gorillas (Weber & Vedder, 1983). For this population of gorillas to survive they can not afford to lose any more terrain.

When the mountain gorilla was first discovered for western science in 1902 by Oscar von Beringe, its range and available habitat were considerably larger than that which remains today. The land around the Virungas is very fertile, as the soils derive from volcanic deposits, and consequently it can support a high human population density. Human populations have increased drastically since 1902 and now the land around the Virunga forests supports 300–600 people per square kilometer, most of whom are subsistence farmers. In 1959, George Schaller carried out the first extensive research on mountain gorillas in the wild. At this time he considered habitat loss to be the greatest danger to the survival of the mountain gorillas (Schaller, 1963). He estimated that there were 400–500 gorillas in the Virungas. Within the next 10 years the total area of the Parc National des Volcans (the Rwandan part of the Virungas – see map in chapter 1) was gradually reduced to 46% of its original size (Weber, 1987*a*), so that only 125 km^2, less than one-third of the Virunga forest, now remains in Rwanda. The bulk of this habitat loss in Rwanda was due to a pyrethrum project that excised 10 000 hectares of land in 1968. All forest between the altitudes of 1600 m and 2600 m has been removed from Ruhengeri prefecture, Ruhengeri being the largest town in this region. Weber & Vedder (1983) estimated a 40%–50% decline in the number of gorillas remaining in 1976–78 following this habitat loss.

Removal of natural forest for agriculture destroys gorilla habitat, but there are additional ways that habitat has been affected in the Virungas. Humans have used the forest extensively in the past as a source of building materials, firewood, water, and grazing land for cattle. Regeneration of the vegetation at high altitudes (2600–4500 m) in these mountains is slow and consequently even low levels of human activity can have a long-lasting impact. During the 1950s, 1960s, and 1970s, herds of several hundred cattle were grazed in the Virungas and their activities are known to have modified the habitat (Schaller, 1963). Herds of cattle in the Virungas were described by Schaller (1963) as an insidious danger, as their numbers were uncontrolled. In Zaire (now the Democratic Republic of Congo) park authorities shot cattle to deter the pastoralists from entering the park (Schaller, 1963), and in the 1970s Fossey used the same tactics in Rwanda (Fossey, 1983). Fossey worked at a time when the administrative structures she was responsible to in Rwanda were minimal. This allowed her a certain autonomy, and although these measures were drastic, they played a role in the eventual removal of all cattle from the Parc National des Volcans by 1976 (Harcourt, 1986; Ruhengeri residents, personal communications).

In 1979 when plans were announced to clear another 52 km^2 for cattle grazing, parkland conversion was considered the greatest threat to the gorillas' survival. A means of making the gorillas "pay for themselves" and protecting the park was needed, and thus a tourism program was planned and developed (Vedder, 1989).

Today the risk of forest conversion to agriculture is low, although not completely eliminated. There have been suggestions by local political leaders in northwestern Rwanda that parts of the Parc National des Volcans be converted to agriculture or that cattle should be grazed in the park; however, the government has not taken these suggestions seriously. In early 2000, a proposal was made to resettle some refugees from Gishwati forest into the Parc National des Volcans but this decision was reversed by a Government Commission. However, these two incidents do show that without constant vigilance, more land could be lost in future.

Hunting or poaching

Hunting can be direct or indirect, depending on whether gorillas are targeted or whether they are unintentionally trapped by snares. The major reasons for killing gorillas are (1) meat, (2) capture of animals for zoos, and (3) body parts for trophies or religious rites.

The bushmeat trade in central Africa is currently one of the main threats to western lowland gorillas (*G. g. gorilla*). An estimated 1.2 million metric tonnes of bushmeat are consumed each year in the range of this animal (Wilkie & Carpenter, 1999), although gorillas form less than 1% of this trade. However, where gorillas are hunted regularly, their populations are usually drastically reduced or eliminated (Wilkie & Carpenter, 1999). It has been estimated that in 1999 about 48% of the Grauer's gorillas in the montain sector of Kahuzi-Biega National Park were killed for meat as a result of civil strife in DRC (O. Ilambu, personal communication; see Yamagiwa, this volume). Hunting of mountain gorillas for meat has also occurred but rarely. It was thought that many gorillas were killed for food in Zaire (now DRC) during fighting for independence in the 1970s (Weber & Vedder, 1983). Rwandans and Ugandans traditionally do not consume primate meat but in DRC primates are often eaten.

Infant gorillas were also captured for foreign zoos. For example, in 1968–69, 18 gorillas were killed in attempts to capture of gorilla infants (Fossey, 1983). There have been no known deliberate gorilla killings in Rwanda since 1982; however, gorillas are still very much at risk from poaching even where gorilla tourism has brought in foreign currency. For instance in Uganda in 1995 four gorillas were killed in the Bwindi Impenetrable National Park, allegedly to obtain an infant which was smuggled into DRC for sale.

Poaching of gorillas for trophies (skulls, hands, and feet) and skins to sell to collectors has been another motive to kill them. Mountain gorillas were hunted directly for trophies until quite recently. Von Beringe shot the first one to be discovered and Carl Akeley shot several for the American Museum of Natural History in the early 1920s. Akeley then went on to help establish the Virunga National Park, the first park in Africa. Religious beliefs and rites were also behind some killings of gorillas in Africa in the first half of the 20th century and even today mountain gorillas are the only subspecies that is not hunted for their body parts. According to Fossey (1983), poachers hunting for trophies were responsible for two-thirds of gorilla deaths during the late 1960s and 1970s in the groups she monitored.

Direct killings of mountain gorillas have been very rare in the last 20 years but indirect killing of mountain gorillas still occurs. Wire and rope snares are set to trap antelopes in the Virungas, and animals that get caught struggle to get free, creating deep cuts to their limbs. If these become infected the animal can easily die.

To combat poachers, Fossey initially tried to thwart their activities by cutting trap-lines and herding gorillas away from areas where snares had been set. In 1978–79 there was an eruption of gorilla killings in the habituated gorilla groups in Rwanda. Prior to these killings anti-poacher patrols were organized *ad hoc* from Karisoke Research Center. The creation of the Digit Fund following the slaughter of Fossey's favourite gorilla, Digit, gave Fossey a source of funds to employ anti-poaching teams and establish regular patrols. The formation of the Mountain Gorilla Project in Rwanda in 1979 increased and improved patrols and law enforcement and thus discouraged interest in gorilla infants and body parts (Harcourt, 1986; Vedder & Weber, 1990). The subsequent creation of the regional International Gorilla Conservation Program from the Mountain Gorilla Project led to greater collaboration in ranger patrolling and sharing of information between countries.

Disease transmission from humans

Gorillas are susceptible to human diseases and many in zoos are vaccinated against the common human ailments. With small populations of gorillas, any infectious disease could devastate the population. Diseases transmitted to immunologically naïve populations have resulted in massive mortality in other species – up to entire populations (Thorne & Williams, 1988; Macdonald, 1996) and primates are especially vulnerable due to their slow reproductive rates (Young, 1994). Disease outbreaks and subsequent deaths of habituated mountain gorillas include respiratory outbreaks and scabies (Hastings *et al.*, 1991; Kalema *et al.*, 1998). In the closely related chimpanzee (*Pan troglodytes*) polio and respiratory diseases have occurred (Goodall, 1986).

The gorillas in the Virungas have probably been exposed to human parasites from the local human population for decades. For many years people used the forest to graze cattle, collect firewood and building poles, and hunt and will have defecated and urinated whilst in the forest. More recently, gorilla ecotourism has increased the potential threat of disease transmission. Whilst most of the international tourists visiting Rwanda are fairly fit, having been inoculated against certain diseases, they may be carrying new viruses for the region, such as influenza. It is these illnesses to which the gorillas have never been exposed that are potentially the most dangerous. If poorly controlled, tourism can also lead to increased stress in the animals, which can increase susceptibility to disease (Hudson, 1992; McNeilage, 1996). While this was recognized as a risk at the start of the tourism program, the loss of habitat was considered a far greater threat to the gorillas at the time and the tourism program was implemented with rules in place to regulate tourist visiting times and the number of tourists per group.

During the war and civil unrest from 1991 to 1998, large numbers of people moved through the Virungas, basing themselves for long periods in the forest, and these people were not subject to any of the controls imposed on tourists. They were poorly nourished, living in harsh conditions, and many died in the park. The magnitude of risk from thousands of people streaming through or living in the park is obviously far greater than that posed by the relatively small numbers of tourists who spend a short time in the park. Disease transmission from both the local human populations and from tourists is treated in greater depth in chapter 13 (Mudakikwa *et al.*).

War and political unrest

Civil wars are not a new threat to the conservation of protected areas; however, the participants are much better armed than in the past. The ever-shrinking forests are ideal hiding places/retreats for armed opposition groups. Many national parks in Africa are associated with the presence of rebels, and conflicts extend over a much larger arena than they used to. Currently there are rebel groups in Kahuzi-Biega National Park, Garamba National Park, Upemba National Park, Salonga National Park, and Virunga National Park in DRC. In Uganda, rebels are found in the Ruwenzori National Park and Semuliki National Park. Protected areas that straddle international borders are particularly at risk as people can move back and forth between countries more freely.

Civil war erupted at independence in both Rwanda and DRC during the 1960s and probably had an impact on the gorillas. It is known that gorillas around the volcano Mikeno in DRC dropped in number between 1960 and 1971–73 (Weber & Vedder, 1983). In 1990, the Rwandan Patriotic Front (RPF), a movement created by a group of refugees who fled the civil war in Rwanda in the 1960s, invaded Rwanda from Uganda and based some of its troops in the forest between the volcanoes Sabyinyo and Muhabura. The Rwandan army at the time used to launch shells into the forest to try to dislodge the RPF. In 1994, the RPF took power following the genocide, at which point many of the local people fled into the forest and into Zaire (now DRC). Refugee camps housing hundreds of thousands of people were established near the Virunga Volcanoes and firewood collection led to the destruction of the forest near these camps before international attention was focused on this issue. The bulk of the refugee population did not return to Rwanda until late 1996. Between late 1994 and late 1996 the security situation in the volcanoes was relatively calm. During 1997–99 there were regular incursions by the militias (called the "Interahamwe") that were involved in the genocide, and retaliatory strikes by the Rwandese army. In 1997 and 1998, the Interahamwe used the Parc National des Volcans as a base to launch attacks into the Ruhengeri and Gisenyi prefectures. Local people stopped farming along the park boundary and moved to "internally displaced peoples' camps". Consequently there were many people living in the park with little access to food except what they could find or grow in the forest. Much of the population moved back and forth between the forest and their homes to escape the killings and the crossfire. Many established huts and even small farms in parts of the forest. Since early 1999 the security on the Rwandan side has improved greatly but the DRC sector of the Virungas is still insecure and military escorts are needed to enter this part of the forest.

Research assessing and monitoring threats

What research, either based from Karisoke Research Center or taking place in and around the Virungas, has addressed some of the threats highlighted above and how did it contribute to better conservation? It can be argued that all research information can potentially be used to improve conservation in some way but we will confine ourselves here to summarizing projects we consider to have addressed issues directly pertinent to conservation.

Several of the research topics described deal with socioeconomic studies of the local human population. Karisoke Research Center historically has not been directly involved in all of the socioeconomic research but has collaborated closely with the projects that have undertaken these studies. Although we present the results of some socioeconomic studies, this is an area that needs much more research effort and could be expanded in future.

Projet RRAM – research on water catchment and the ecological role of the Parc National des Volcans

A project called "Ruhengeri and its Resources" (Projet RRAM) operated in Ruhengeri in the mid 1980s, supported by USAID. This project was established by Bill Weber following some of his work at Karisoke Research Center, and it provided a wealth of data on the ecological value of the forest (Weber, 1987*a*). The Parc National des Volcans is protected by a dense cover of vegetation that controls flooding, landslides, and erosion. A high level of rainfall is generated in the volcanoes, and this rain is captured in a way that does not provoke erosion – there is a gradual release of water into the surrounding areas, and thus a perennial supply of clean water. In a survey carried out by the Projet RRAM, 50% of farmers reported that agricultural productivity in Ruhengeri was declining, and 19% of them attributed this to erosion in the years following a loss of 54% of the surface area of the park. The Parc National des Volcans forms only 0.5% of Rwanda's surface area but contributes an estimated 10% of the water catchment (Weber, 1979).

Weber documented that the regional land use could not increase in any significant way in the Ruhengeri prefecture. Even converting the entire park to agricultural land would only provide extremely marginal land for the equivalent of one year of Rwanda's human population growth. The slopes of the volcanoes would be too steep for sustainable agricultural production and could only provide short-term benefits rather than the longer-term benefits of gorilla tourism. This ecosystem approach to research in the Virungas, rather than a focus on gorillas, provided valuable information to bolster the arguments for keeping the forest cover on the volcanoes.

Gorilla population censuses

Population monitoring of any rare species is essential if conservation practitioners are to be able to assess the effectiveness of their management. Consequently regular censuses were carried out from the start of the gorilla research. In 1963 Schaller estimated that 400–500 gorillas lived in the Virungas. Schaller developed a census technique using nest counts and measurements of dung to estimate population size (Schaller, 1963; Weber & Vedder, 1983; McNeilage *et al.*, 2001). Each night an adult or juvenile gorilla constructs a nest and by following fresh trail until they find the most recent nests, researchers can obtain a total count of animals in a protected area. Usually three counts are made for each group in case any nests are missed. By measuring the size of the dung the gorillas deposit in the nest a researcher can obtain a measure of the structure of the population because size of the dung is related to size of the animal (Schaller, 1963). Subsequent censuses in 1971–73 and 1976–78 raised the alarm by clearly showing drastic declines in the gorilla population to about 252 animals (Harcourt & Fossey, 1981; Weber & Vedder, 1983). Following the launch of the Mountain Gorilla Project and ecotourism to the gorillas, subsequent censuses in 1981, 1986, and 1989 showed numbers increasing again to 310 individuals (Sholley, 1990). To make an accurate count in the Virungas, all three habitat countries must be surveyed simultaneously, since gorilla groups move freely between Rwanda, DRC and Uganda. A follow-up to the most recent 1989 census was made impossible by the civil war and the laying of anti-personnel land mines on the international border between Rwanda and DRC. In 1994 a Rwandan demining team began operating in the Parc National des Volcans with financial support from the International Gorilla Conservation Program, and cleared all the mines, but insecurity in the region remained an obstacle to carrying out a census.

Ten years after the last census, we can not give a precise number for the current gorilla population in the Virungas. The habituated groups monitored on a daily basis between the volcanoes Karisimbi and Sabyinyo, and in Uganda, are in good health and reproducing, although the impact of recent insecurity on groups ranging elsewhere in the Virungas is unknown. An update on the status of the gorilla population is much needed and is a top priority of the park managers and conservation organizations.

Socioeconomic research

Socioeconomic studies are important because the needs of the members of the local community who live around the Virunga Volcanoes and their attitudes towards the forest and conservation can determine the success or failure of any conservation project. People's attitudes towards protected areas and conservation will be strongly affected if they suffer costs as a result of the presence of the park, such as crop-raiding by wild animals or increased risk to themselves from large animals.

Attitudes towards the forest by the local political authorities will also affect whether further land is excised from the forest. Given that the entire Virunga conservation area (Rwanda, DRC, and Uganda) is currently protected by National Park status, it would seem that there is little danger of habitat loss; however, even now, the integrity of these national parks is not assured. Prior to and during the civil wars in Uganda in the 1980s, people encroached on the then Mgahinga Game Reserve and farmed inside its boundaries (Butynski *et al.*, 1990). In 1994, the siting of the world's largest refugee camp on the edge of the Virunga National Park enabled people to enter the forest and harvest fuelwood illegally, resulting in the deforestation of about $113 km^2$ of the park in DRC (Henquin & Blondel, 1996). In Rwanda, local authorities suggested use of part of the park around the volcano Sabyinyo for human resettlement in 1996. This was prevented through prompt action by the government in Kigali. None the less, illegal encroachment remains a serious threat at the edges of the park.

Influencing attitudes so that the likelihood of habitat loss is lessened is therefore an important task for a park warden and his staff. Local attitudes can be influenced positively via perceived benefits, and negatively through losses attributed to having a national park on their doorstep. Perceived benefits or losses are usually, but not exclusively, based on economic measures.

Local attitudes

People living around the Virungas resent being prevented from obtaining certain products that they obtained in the past, such as firewood, building materials (bamboo, vines, poles), medicinal plants, bushmeat, and water. In the 1980s, Weber (1981, 1987*b*) researched people's attitudes towards the park. Using a questionnaire survey he showed that people strongly objected to the presence of the park because of these perceived losses.

In addition negative attitudes towards the parks are exacerbated when people suffer crop damage caused by animals venturing out of the forest at night to feed in their fields. In 1996, a survey of farmers' attitudes to crop-raiding animals also examined measures taken to combat them (Plumptre & Bizumuremyi, 1996). Local people were asked to identify which animals were the greatest threat to their crops by interviewing households within 2.5 km of the park boundary. Buffalo (*Syncerus caffer*) were considered to be the greatest problem up to 2.0 km from the park (Figure 14.1). Whether this species is responsible for the greatest economic loss is doubtful, but people may view buffalo as the main threat because they are dangerous to human life and can reduce their freedom of movement at night. In the past elephants (*Loxodonta africana*) were more of a threat than they are today because their numbers were once much greater, whereas they have now been drastically reduced in the Virungas.

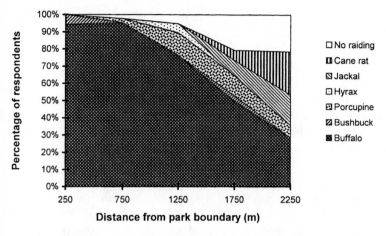

Figure 14.1. The percentage of respondents who labeled a particular species as the worst crop-raider. The figure shows how responses varied with distance (m) of the household from the park boundary.

Responding to negative attitudes

In the 1980s, the Mountain Gorilla Project initiated an education program in Rwanda to increase awareness of the international importance of the gorillas, the economic benefits of wildlife from tourism, and the ecological benefits of the forest. Weber (1987*a*) carried out a follow-up survey to the one mentioned above and showed that people's attitudes changed significantly following this education program, despite the fact that little revenue from tourism went directly to people who lived around the park. The main benefit to the local population was from employment opportunities. Currently about 80 government-employed guides and guards and 20 Karisoke Research Center trackers are employed in Rwanda with a few additional administrative staff for the national parks agency. Employment by the tourist industry also increased with the hiring of porters and the employment of hotel staff, drivers of vehicles, and other associated spin-offs. A few people profit by selling crafts to tourists.

In the 1990s, the conservation community placed great emphasis on the economic benefits that accrue to local communities living around national parks. In Uganda several Integrated Conservation Development Projects have been established including one around a park with gorillas, the Bwindi Impenetrable National Park. Sharing revenue from tourism and the collection of non-timber forest products is being practiced around this park in southwest Uganda but the dollar value that each individual receives is very small when averaged across the population. Here the revenues are used to fund community projects around the park rather than provide an individual with direct employment. With human densities of 300–600 people per km^2 around the Parc National des Volcans and Bwindi, it is going to be very difficult to generate benefits of financial significance, if the profits are shared equally. However, it is better that some revenue sharing takes place than none at all.

Another tool that has been used to improve local people's attitudes in Uganda has been to allow limited use of certain areas of the forest through the creation of multiple-use zones, where people are allowed to place beehives for honey and harvest medicinal plants. Multiple-use zones are also being proposed for Mgahinga National Park in the Virungas. A recent census of gorillas in Bwindi found that gorillas were not present in the multiple-use zones (McNeilage *et al.*, 2001), and although there is no direct evidence that human use is preventing gorillas using this area, it is possible that such use of the forest may have the effect of reducing the habitat available to gorillas. This practice should be monitored carefully. Of greater concern is a recent report (CARE, 1998) which showed that despite the creation of multiple-use zones in Bwindi, 68% of people still felt that the costs of living close to the park outweighed the benefits. This shows the importance of research in monitoring people's attitudes towards conservation when implementing community conservation projects.

Poaching

Research into several aspects of poaching includes assessing the frequency and location of snares in the Parc National des Volcans to determine whether more effective patrolling techniques can be developed. An examination of the number of snares found in different regions of the Parc National des Volcans by Office Rwandais du Tourisme et des Parcs Nationaux (ORTPN) guards between 1982 and 1988 (T. Lawrence, unpublished data from Mountain Gorilla Project) shows the markedly higher numbers of snares taken annually from sector III (Figure 14.2). This may be partly a result of higher patrol effort (both ORTPN and Karisoke Research Center staff patrolled this sector) but joint patrols did not take place until 1986 before which patrol effort was similar for each sector. This part of the park may have had higher ungulate populations, because it had been protected for longer with the presence of the Karisoke Research Center, and so consequently may have attracted more poachers. Analysis of the impact of patrol effort shows that as the number of patrols increased more snares were found, but that the number of snares per patrol dropped significantly after about 20 patrols per month (Figure 14.3). Data on snares collected at Karisoke Research Center show that during the year there are peaks in snares found around Christmas and Easter (N. Gerald-

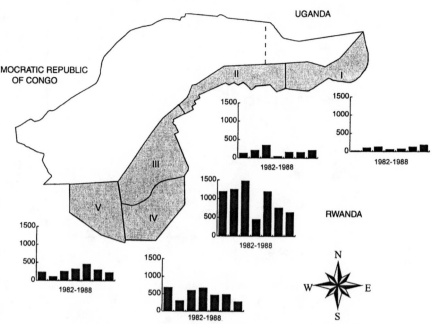

Figure 14.2. The number of snares collected annually in different sectors of the
Parc National des Volcans, Rwanda between 1982 and 1988. Sector III includes
snares from the Karisoke Research Center and ORTPN (Rwandan parks)
guards from 1986. Prior to 1986 only data for snares collected by ORTPN
guards could be found.

Steklis, personal communication), times when households often hope to
eat meat (A. Plumptre, unpublished data). The number of snares found in
the Virungas in the 1989 gorilla census was far higher than the number
found in Bwindi in 1997 despite a much higher guard density in the
Virungas. Why this was the case is difficult to understand but given that
ungulate sightings and dung were far less abundant in Bwindi it may not be
as worthwhile to set snares (McNeilage et al., 1998).

A recent survey measured the socioeconomic status of poachers in the
community adjacent to the Parc National des Volcans (Plumptre &
Bizumuremyi, 1996). This research showed that poachers were mainly
targeting bushbuck (Tragelaphus scriptus) and black-fronted duiker
(Cephalophus nigrifrons) although a few also set snares for buffalo and
hyrax (Dendrohyrax arboreus). Gorillas are not targeted, probably because
traditionally Rwandans do not eat primate meat, and because these people
had been exposed to an environmental education program. However,
snares set for other animals can easily trap a gorilla's hand or foot and
have led to the death or injury of many gorillas. Responses to questions

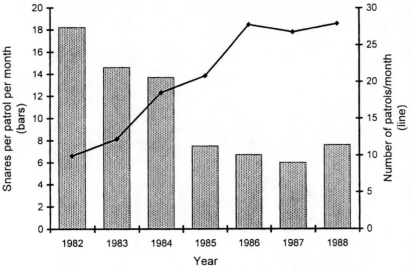

Figure 14.3. The mean number of snares recovered per patrol each month (bars) in the Parc National des Volcans and patrol effort (line) between 1982 and 1988.

about hunting wild animals in their fields showed that 28.7% of people living within 2.5 km of the boundary of the Parc National des Volcans admitted to hunting animals when they came into their fields and 27.6% admitted to setting snares or pitfall traps (Table 14.1). Additionally, 33.1% of people interviewed admitted to buying bushmeat although nobody said they had killed or eaten gorillas (Plumptre & Bizumuremyi, 1996). A preference for the taste of bushmeat will always be important in the decision about whether to buy it or not, but often it is the cost that really determines a sale. Data from Ruhengeri show that bushmeat is consistently cheaper than domestic meat. In fact domestic meat prices rose between 1990 and 1996 from about 350 Rwandan francs (FRw) to 450 FRw/kg, whereas bushmeat prices dropped over the same period from 60 to 40 FRw/kg (Plumptre & Bizumuremyi, 1996; Plumptre *et al.*, 1997). The rate of buying bushmeat increased following the genocide and the rate of selling of bushmeat per hunting effort also increased (Plumptre *et al.*, 1997), partly as a result of these price changes. Farmers who admitted to poaching in the park had significantly smaller fields, and fewer domestic livestock than the average household (Table 14.2). In addition, those people who admitted to poaching lived nearer the edge of the park. Given this fact and their small field area, people who poach will be most vulnerable to crop-raiding losses and this may be one of the factors that drive

Table 14.1. *The percentage of respondents who admitted to "illegal" activities concerned with hunting of bushmeat around the Parc National des Volcans*

	West of park	East of park	Combined
Percent admitting to snaring (fields or park)	25.6	29.7	27.6
Percent admitting to hunting in fields	20.0	37.4	28.7
Percent admitting to hunting in park	7.8	14.3	11.0
Percent admitting to buying bushmeat	18.9	47.3	33.1

The responses of people in the west (between Karisimbi and Sabyinyo) are compared with those in the east (between Sabyinyo and Muhabura). This split was made because people in the east tended on average to have more wealth and access to job opportunities. It was felt it was important to look at those people who admit to setting snares as well as those hunting in fields/park because of the impact snares can have on gorillas and because people might admit to using snares if they did not have to state where they snared.

Table 14.2. *The results of Mann–Whitney U tests and χ^2 tests between measures of wealth and whether a person admits to snaring (in their fields or in the park) or hunting in their fields or the park*

	Setting snares	Hunting in fields	Hunting in park
Distance from park	ns	ns	− **
Number of children	ns	+ *	ns
Number of sheep	ns	ns	− *
Number of goats	ns	ns	− *
Number of cattle	ns	ns	ns
Livestock biomass	ns	ns	− *
Area of fields	− *	ns	− **
Residence time	ns	+ *	ns
Employed	− *	ns	ns
Owns radio	ns	ns	ns
Owns bicycle	ns	ns	ns

+ = positively associated with the measure, − = negatively associated. ns = not significant, * = $p < 0.05$, ** = $p < 0.01$.

them to poach. It is probable that the frustrations people sense with crop-raiding animals lead to negative attitudes towards the park, increasing the probability that they will buy bushmeat. Consequently conservation managers need to address the crop-raiding problem for long-term success in conservation. Community conservation programs established around the Parc National des Volcans should target the poorer members of society and find alternative methods of income generation. This will not solve the whole problem because some people admitted to hunting for cultural reasons and because they liked the taste of the bushmeat, but it could help reduce the pressure.

Impacts of tourism

In the early 1980s, tourism to visit habituated gorillas was organized as an economic alternative to cutting the forest for other uses, thus providing a persuasive argument against the threat to excise yet more land for cattle pasture (Vedder, 1989). Gorilla tourism became a great success in terms of revenue for Rwanda, increased protection of the park, and close surveillance of additional gorilla groups (Harcourt, 1986; Vedder, 1989; Weber, 1993).

Benefits of tourism

Tourism enabled ORTPN, the National Parks authority in Rwanda, not only to become self-financing by the end of the 1980s, but also to subsidize salaries, patrols, and operating costs in the other protected areas of Rwanda, Akagera National Park and the Nyungwe Forest Reserve. Following the international release of the film "Gorillas in the Mist", the price of gorilla permits was increased from $120 to $170 (Sholley, 1988) and more recently to $250.

International publicity surrounding the gorillas and the advent of organized tourism brought many visitors to Rwanda and made tourism the third highest foreign-currency earner for this country after tea and coffee, at nearly $1 million per annum (Weber, 1993). Monitoring of habituated groups for tourism increased surveillance in the park by leading to the presence of several teams of parks staff in the forest (McNeilage, 1996). Daily observations facilitated rapid intervention when necessary, for example, to remove snares. With increased protection from poachers, more infants per adult female were recorded in groups that were habituated and visited by tourists (Harcourt & Fossey, 1981), and in 1986 the population was seen to be increasing for the first time in three decades (Vedder & Weber, 1990). International awareness and concern for the plight of gorillas enhanced through tourism also generated funds for conservation activities and research. The gorilla became a national symbol and was depicted everywhere in Rwanda: on postcards, stamps, cloth, murals, and carvings in tourist shops. Today the Rwandese passport, visas for foreigners, and bank notes all feature gorillas.

Some of the proceeds of the tourism benefit the local community around the parks by providing employment or boosting the economy (food, hotel bills, etc). In Uganda, revenue sharing of tourism receipts operates around Uganda's Mgahinga Gorilla and Bwindi Impenetrable National Parks, and has probably led to greater support for these parks by the local community. It has been agreed in principle, although yet to be put into practice in Rwanda, that profits generated through tourism should also be shared with people living around park, or used to fund schools and health centers for the local population.

Dangers linked to tourism
Pressures from tourism are high as it is a lucrative business, and some people question the continued justification for gorilla viewing as a conservation measure. Butynski & Kalina (1998) examined various aspects of gorilla tourism, including the role of money and politics. They conclude that gorilla tourism is likely to be sustainable only:

- where conservation is given priority over economic and political concerns
- where decisions affecting tourism are based on sound and objective science
- where scientifically formulated regulations governing visits are rigorously controlled.

Even before tourism began, impacts of visits on the gorillas' behavior were feared, for example changing the gorillas' ranging patterns, impeding the transfer of females to other groups, and hindering reproduction (Vedder, 1989). As tourism was being developed the groups were monitored by researchers as the size of groups of tourists was increased to evaluate whether the gorillas were changing their behavior drastically. Stress to the animals can be provoked during the habituation process or through regular contact with unfamiliar humans, which could potentially result in immunosuppression or reductions in reproductive success. It is acknowledged that these impacts have never been adequately evaluated. However, extrapolation from research on population demographics carried out at Karisoke Research Center, indicates that tourism has not been deleterious to the gorillas' overall health, behavior, and ecology. Any negative impacts seem to have been outweighed by improved monitoring and protection.

Habituation, the loss of fear of humans, could render gorillas extremely vulnerable in dangerous situations. Sadly this was the case, when habituated gorillas in the Virunga national park were shot at point-blank range in 1995. Also, in the poaching of eastern lowland gorillas in the Kahuzi-Biega National Park, habituated gorillas were some of the first killed during 1999–2000. An entire group was wiped out with a few rounds of machine gun fire (Yamagiwa, 1999).

Until recently we have relied on speculation, extrapolation, and common sense to evaluate the risks of disease transmission from humans to gorillas. Tourism introduced a new element to that threat. A study commissioned by the International Gorilla Conservation Program (Homsy, 1999) reviewed tourism regulations in light of epidemiological data and the risk of disease transmission between people and gorillas. Studies of captive gorillas show they have a definite susceptibility to human diseases, but not the same resistance as humans (Homsy, 1999). As a result of this shared susceptibility, certain human pathogens can affect gorillas, not only respiratory (measles, herpes, pneumonia) but also equally important enteric diseases (polio, salmonella). Homsy concluded that: "together with the high population pressure surrounding the parks, disease exposure ironically makes tourism one of the single greatest threats to mountain gorilla survival" and that "the best hope for a least damaging tourism program resides in the widespread sensitization, awareness and understanding of the catastrophic consequences of unconscious gorilla tourism".

Despite the dangers inherent in tourism, it provides a mechanism for ensuring that the parks and the gorillas are valued for many reasons, and has probably saved the gorillas in the Virunga Volcanoes from further habitat loss or degradation.

Armed conflict

Research from Karisoke Research Center has investigated two aspects of the civil war in Rwanda. These have focused on the effects of the fighting on animal populations and the effects of the war on staff of conservation organizations.

Effects of civil war on ungulate populations

To date it has not been possible to measure accurately the impact of the civil war on the gorilla population, but assessments have been made for the more abundant ungulates in the Parc National des Volcans (Plumptre *et al.*, 1997). This study, undertaken in 1996 during a period of calm in northwestern Rwanda, showed that the bushbuck and buffalo populations had not changed greatly in the region of the Karisoke Research Center between 1988–89 and 1996, whilst the black-fronted duiker population had increased. Buffalo counts showed no significant change in any use of habitat type, but bushbuck and black-fronted duikers increased significantly in the alpine habitat at high altitude (Plumptre *et al.*, 1997). However a questionnaire survey of people's impressions of population changes across the entire park, based on crop-raiding frequency, indicated that

Table 14.3. *Crop-raiding frequencies (number of days per year) within 500 meters of the park boundary*

Animal	Time	West of park	East of park	Combined
Buffalo	pre 1994	209 **	175 **	193 ns
	post 1994	72	353	208
Bushbuck	pre 1994	268 **	283 ns	275 **
	post 1994	15	176	163
Black-fronted	pre 1994	106 *	261 *	181 *
duiker	post 1994	44	132	87
Porcupine	pre 1994	137 ns	268 ns	200 ns
	post 1994	104	258	179

Wilcoxon signed ranks tests were used to detect differences before and after 1994. As above the respondents were separated depending on whether they lived in the west or east.
* = $p < 0.05$, ** = $p < 0.01$, ns = not significant.

Figure 14.4. Buffalo, one of the large ungulates found in the Virungas, also raid crops. (Photo by A. J. Plumptre.)

certain animal populations had changed and that these had changed in different ways in the west and east of the Virungas (Table 14.3). People's perceptions should be regarded with care and not necessarily equated with reality, but they do give a measure of how people feel crop-raiding is affecting their livelihoods. It is interesting that people in the east reported greater crop-raiding by buffalo following the war whilst those in the west reported lower crop-raiding in 1996 (Figure 14.4). At this time a wall was being constructed in Uganda to prevent buffalo crop-raiding which may have led to increased raiding on the Rwandan side.

Impacts of insecurity on staff

Civil war in Rwanda has made the headlines as a result of the genocide. What lessons can be learned from the attempts to conserve the gorillas and other protected areas in Rwanda that could be applied elsewhere? Why was it that despite the risks to their lives people continued to work to protect these animals? One of us (A. Plumptre) carried out a survey of the staff of the Karisoke Research Center and the Nyungwe Forest Conservation Project (PCFN) in south west Rwanda to determine (1) what had been the effect of the war on these people, (2) what motivated them to continue working, and (3) what international conservation organizations could do better during times of war to help their staff.

Insecurity has been high for longer around the Virungas than in other protected areas in Rwanda such as the Nyungwe Forest. How had the insecurity affected the staff working for these two projects? As a result of the fighting in this region between 1990 and 1999 there have been at least ten deaths of ORTPN staff in Rwanda in the Virungas, 12 deaths of Karisoke Research Center staff, 44 of parks staff of the Institut Congolais pour la Conservation de la Nature (ICCN) in the Virunga National Park, four of ICCN staff in the Kahuzi-Biega National Park in the DRC (N. Mushenzi, personal communication), and one of an ORTPN employee in Nyungwe Forest. Of the Karisoke Research Center staff, 94% had been robbed (compared with 41% of PCFN staff) at some point during the war and 88% had lost a member of their family (compared with 43% of the PCFN staff). Few of the PCFN staff believed their work had increased the risk to their life (14%) but 69% of the Karisoke Research Center staff believed this was the case. This was because people around the Parc National des Volcans were targeted by the Interahamwe if they worked with the government or with organizations that collaborated with the government.

When asked what motivated them to continue working despite the risks to their lives, salary was obviously a prominent response. The regularity of payment was often noted as being as important as the amounts that were paid. Other aspects of the work that provided motivation for many interviewed included the love of nature/work and the importance of their work nationally and internationally. This latter reason is interesting because it means that the education programs that explained the international uniqueness of the mountain gorilla had led to people being prepared to try to protect them at great risk to themselves. The Karisoke Research Center staff valued the fact that the project looked after their families during the war and during 1997–99 because life was so much more dangerous for them during this time.

When asked what projects could do for their staff during war situations, by far the greatest response was to try to reduce the risks to their staff. The specific suggestions included helping staff find somewhere safe to live, or another location where they could work more safely. If this was not possible a few of the PCFN staff suggested stopping the work but on the whole most people wanted to try to continue working despite the dangers. Other aspects that were considered to be important were to collaborate with whichever authorities were in power, rather than obviously taking a political stance. Improving communications (such as providing two-way radios for within-park communications and trying to meet with staff wherever possible), and educating the local population during any peaceful interludes about the importance of the forest for their long-term livelihoods, were also considered to be important. Since this study was carried out two-way radio systems have been provided to staff at both the Parc National des Volcans and the Nyungwe Forest.

In Rwanda, appealing to people's national pride in their gorillas has not been anywhere near as prevalent as economic arguments, and yet national pride in gorilla conservation may be more important than any of the hoped-for economic incentives. In this survey, parks staff in the PCFN and Karisoke Research Center stated that one of the factors that kept them working during the civil war was pride in protecting part of their national heritage and the global importance of their forests for conservation. This is often the motivation for conserving animals in the industrialized countries also. Creating positive attitudes is something that the national parks agencies in Rwanda (ORTPN), DRC (ICCN), and Uganda (Uganda Wildlife Authority) should all aspire to, and education programs are very important in addressing these issues.

Effects of insecurity on the gorillas

Despite civil unrest from 1991 to 1998, it was only during 1997–98 that trackers and park staff were prevented from monitoring the gorillas for a prolonged period. While the habituated gorillas in Rwanda have fared extremely well (one gorilla was shot dead in 1992), it is known that since 1995 at least 18 gorillas have been killed in the Virunga National Park (DRC) either as victims of poachers with firearms or getting caught in crossfire. Remarkably, almost all habituated individuals were accounted for in Rwanda in 1999 after 14 months without any monitoring. Despite an

382 Andrew J. Plumptre & Elizabeth A. Williamson

acute increase in illegal activities by the local population when Karisoke Research Center and ORTPN staff suspended patrols, the gorillas were visibly in good physical shape – there had been no loss of limbs or life to snares. Many births had occurred in the research groups and there seemed to be a good potential for growth of the population (E.A. Williamson, personal observation). There may have been other less obvious impacts, resulting from the stress of confronting armed combat or fleeing people. The worry now is that diseases that may have been contracted from the human refugees inhabiting the Parc National des Volcans in 1997–98 will have serious long-term impacts on the gorilla population in Rwanda (see Mudakikwa *et al.*, this volume).

Application of behavioral and ecological research to conservation
The wealth of knowledge gained through studies at Karisoke Research Center forms the basis for this book, yet it is not always evident that "pure" research can provide essential information for practical conservation. Ecological functions of the forest must be demonstrated, management needs have to be established, tourism and other impacts on the ecosystem should be monitored, to enable us to develop and improve tools with which to protect the park and the gorillas. However the importance of behavioral research is becoming increasingly acknowledged for conservation management (Clemmons & Buchholz, 1997). The following section provides some examples of how various aspects of "pure" behavioral and ecological studies can be "applied" to conservation.

Long-term monitoring of gorillas at Karisoke Research Center
The long-term research from Karisoke Research Center is one of few continuous studies of an animal population that has spanned several decades. Data on inter-birth intervals and other reproductive parameters are all-important when it comes to assessing rates of change (e.g. Harcourt *et al.*, 1981; Watts, 1991*a*; Robbins, 1995). Census results and population statistics not only show changes in the actual numbers of gorillas, but reproductive health and potential are indicated by the age–sex composition of the population (e.g. Harcourt & Fossey, 1981; Weber & Vedder, 1983). Thus demographic and life history data are especially important for park managers to assess the effectiveness of their activities. Steadily increasing gorilla numbers in the 1980s indicated that conservation actions were having a positive impact.

Similarly, understanding aspects of feeding ecology, nutrition, and ranging behavior (Waterman *et al.*, 1983; Watts, 1987 & 1991*b;* Plumptre, 1995, 1996) is important in determining whether the Virunga population could increase in size, and for improving management practices in areas which are not used by gorillas (McNeilage, 1995). For example, ecological data will be crucial to evaluate "underuse" of forest south of Karisimbi (Watts, 1998*a*, *b*). If we are able to assess avoidance of certain areas of the Virungas, park managers may be able to alleviate detrimental conditions and improve the gorillas' chances of survival.

Park managers are able to use the results of research at Karisoke Research Center to interpret dramatic instances of natural behavior, such as infanticide (Fossey, 1984; Watts, 1989), while knowledge of the natural processes of male emigration and female immigration explains transfers and "disappearances" of individuals (e.g. Sicotte, 1993).

One pertinent concern for the mountain gorillas' future is whether the size of their gene pool has been reduced to a level where inbreeding may become an serious problem. Analysis showed that habitat loss is a greater danger to the gorillas than inbreeding, and such studies can guide park managers in consolidating their resources (Harcourt, 1996).

Designing tourism programs and providing baseline information for comparisons with "tourist" groups

Karisoke Research Center played a critical role in developing the techniques of gorilla habituation. Researchers with experience gained from Karisoke Research Center initiated the tourism program in Rwanda. Part of the success of this program was based on our knowledge of gorilla diet, daily travel distance, and ranging (e.g. Watts, 1984), making it possible to predict group movements and locate the gorillas with relative ease. Predictability of daily activity rhythms was also important for the tourism program, and visits were timed to coincide with gorillas' rest periods when possible, facilitating excellent observation conditions for the visitors.

Continued long-term monitoring of gorillas at Karisoke Research Center provides a baseline from which to judge the impacts of tourism. These data allow park managers to assess whether new or altered behaviors observed in groups visited by tourists might result from stress caused by tourism. Tourism guidelines have been developed through our understanding of gorilla behavior and our awareness of what disturbs them. It is important for park managers to minimize stress to these animals because no one can afford to put at risk such a small population if tourism revenues are to be maintained. Monitoring should be coupled with targeted research to evaluate the impacts of tourism and provide information for managers to ensure that tourism is implemented sustainably. Watts (1998*c*) demonstrated a strong correlation between rainfall and mortality, and recommended that extra stringent measures to limit disease transmission should be taken by anyone, tourists and researchers, visiting gorillas during the rainy season.

The positive integration of research and management will ensure that management decisions are based on sound scientific data. Since 1997, the International Gorilla Conservation Program has organized regional meetings to bring together managers of protected areas from the three Virunga parks, the Parc National des Volcans, the Parc National des Virunga, and Mgahinga National Park, and their partners from non-governmental organizations. During these meetings, tourism regulations have been reviewed to provide managers with input on what is and is not sustainable. Researchers and managers were able to develop together recommendations based on the findings of Homsy (1999).

Providing information for conservation education, fund-raising, and zoos

The "Karisoke Research Center" gorillas have attracted world-wide attention, fueled through documentary films and magazine articles about their lives. Education programs, both nationally and internationally, rely on information from research projects and Karisoke Research Center has generated the bulk of the information that currently exists about mountain gorilla behavior and ecology.

Public commitment to try to save the mountain gorillas provides funds for the activities of the conservation non-governmental organizations, which benefit the parks in Rwanda and other parks in the region. This high profile attracts visitors to the gorillas, bringing revenue to the country and helping Rwanda develop its pride in its national heritage. A positive image enhanced by a well-managed tourism program stimulates tourism further and generates publicity.

Additionally, lessons learned from the mountain gorillas have helped enormously in the care and management of the large captive population of western lowland gorillas. The gorillas themselves have benefited as their physiological and psychological needs are better understood and catered for (e.g. Watts, 1990).

Future research and linkages at Karisoke Research Center

One aspect of long-term monitoring that has been lacking is detailed monitoring of gorilla groups visited by tourists. Documenting the dynamics of a larger number of groups would expand our knowledge of the natural variation between groups in the Virunga population. In addition most research has been concentrated in Rwanda and much more work needs to be done in the Virunga National Park in DRC. This park, which forms two-thirds of the Virunga forest, is altitudinally and ecologically quite distinct from the Rwandan side, and so data would be more representative of the whole population (McNeilage, this volume).

Whilst the continuation of ecological and behavioral research on the mountain gorillas is important, it is vital that other aspects critical to management of the park are addressed. A park warden faces many issues, including education of the local community, problems with crop-raiding animals, illegal hunting, and community use of certain forest products. Research is an essential requirement for the development of management plans and monitoring actions implemented by park management. The importance of coordinating research and management and closely linking research to management has been emphasized (e.g. Harman 1994; Alexander 1995). Reinforcing links between Karisoke Research Center and park managers will underscore the value of the research center and help address management problems.

Acknowledgements
The Wildlife Conservation Society and Dian Fossey Gorilla Fund International funded the collection of the new data summarized in this chapter, and we thank the following for help in the collection of some of these data: Jean Bosco Bizumuremyi, Tom Lawrence, Jean Damascène Ndaruhebeye, François Javier Ngarukiyintwari, and Leonidas Ngabayisonga. Several conservation organizations have been involved in the work reported here: Fauna and Flora International, World Wide Fund for Nature, and African Wildlife Foundation supported the Mountain Gorilla Project and now the International Gorilla Conservation Program; Dian Fossey Gorilla Fund International supports the Karisoke Research Center and the Wildlife Conservation Society has supported all the censuses of the gorillas and some of the research projects. We would like to thank the Government of Rwanda and the Office Rwandais du Tourisme et des Parcs Nationaux for permission and support to work in the Parc National des Volcans.

At the end of the 1990s, while the gorillas thrived, many experienced gorilla trackers lost their lives. We would like to dedicate this chapter to Vatiri André, as a representative of those who have died in tragic circumstances in recent years. Vatiri died in 1997, the most difficult period in the lives of the Karisoke Research Center staff, and he symbolized the long-term commitment of the trackers to the fight to save the gorillas.

References

Alexander, M (1995) Management planning in relation to protected areas. *Parks*, 5, 2–11.

Butynski, T M & Kalina, J (1998) Gorilla tourism: a critical look. In *Conservation of Biological Resources*, ed. E J Milner-Gulland & R Mace, pp. 280–300. Oxford: Blackwell Scientific Publications.

Butynski, T M, Werikhe, S E & Kalina, J (1990) Status, distribution and conservation of the mountain gorilla in the gorilla game reserve, Uganda. *Primate Conservation*, 11, 31–41.

CARE (1998) *Survey of Knowledge and Attitudes of People from Communities Bordering BINP and MGNP*. CARE/Development Through Conservation, no. 98/391.

Clemmons, J R & Buchholz, R (1997) *Behavioural Approaches to Conservation in the Wild*. Cambridge: Cambridge University Press.

Fossey, D (1983) *Gorillas in the Mist*. Boston MA: Houghton Mifflin.

Fossey, D (1984) Infanticide in mountain gorillas *(Gorilla gorilla beringei)* with comparative notes on chimpanzees. In *Infanticide: Comparative and Evolutionary Perspectives*, ed. G Hausfater & S B Hrdy, pp. 217–35. New York: Aldine Press.

Goodall, J (1986) *The Chimpanzees of Gombe: Patterns of Behaviour*. Cambridge MA: Harvard University Press.

Harcourt, A H (1986) Gorilla conservation: anatomy of a campaign. In *Primates: The Road to Self-Sustaining Populations*, ed K Benirschke, pp. 31–46. New York: Springer-Verlag.

Harcourt, A H (1996) Is the gorilla a threatened species? How should we judge? *Biological Conservation*, 75, 165–76.

Harcourt, A H & Fossey, D (1981) The Virunga gorillas: decline of an island population. *African Journal of Ecology*, 19, 83–97.

Harcourt, A H, Fossey, D & Sabater Pi, J (1981) Demography of *Gorilla gorilla*. *Journal of Zoology*, 195, 215–33.

Harman, D (1994) Coordinating research and management to enhance protected areas. In *Proceedings of the 4th World Congress on National Parks and Protected Areas*, Caracas, Venezuela, IUCN.

Hastings, B E, Kenny, D, Lowenstine, L J & Foster, J W (1991) Mountain gorillas and measles: ontogeny of a wildlife vaccination program. In *Proceedings of the Annual Meeting of the American Association of Zoo Veterinarians*, pp. 198–205.

Henquin, B & Blondel, N (1996) Etude par télédétection sur l'évolution récente de la couverture boisée du Parc National des Virunga. Unpublished report, Laboratoire d'hydrologie et de télédétection, Gembloux, Belgium.

Homsy, J (1999) *Ape Tourism and Human Diseases: How Close Should We Get?* Report to the International Gorilla Conservation Program. Rwanda.

Hudson, H R (1992) The relationship between stress and disease in orphan gorillas and its significance for gorilla tourism. *Gorilla Conservation News*, 6, 8–10.

Kalema, G, Kock, R A & Macfie, E (1998) An outbreak of sarcoptic mange in free-ranging mountain gorillas (*Gorilla gorilla beringei*) in Bwindi Impenetrable National Park, Southwestern Uganda. In *Joint Proceedings of the American Association of Zoo Veterinarians and American Association of Wildlife Veterinarians Annual Meeting*, Omaha, Nebraska, p. 438 (abstract).

Macdonald, D W (1996) Dangerous liaisons and disease. *Nature*, **379**, 400–1.

McNeilage, A J (1995) Mountain gorillas in the Virunga Volcanoes: ecology and carrying capacity. PhD thesis, University of Bristol.

McNeilage, A (1996) Ecotourism and mountain gorillas in the Virungas. In *The Exploitation of Mammal Populations*, ed. V J Taylor & N Dunstone, pp. 334–44. London: Chapman & Hall.

McNeilage, A, Plumptre, A J, Brock-Doyle, A & Vedder, A (1998) *Bwindi Impenetrable National Park, Uganda Gorilla and Large Mammal Census, 1997*. Wildlife Conservation Society, Working Paper No. 14.

McNeilage, A, Plumptre, A J, Brock-Doyle, A & Vedder, A (2001). Bwindi Impenetrable National Park Uganda: mountain gorilla census. *Oryx*, in press.

Moss, C (1988) *Elephant Memories*. London: Elm Tree Books.

Plumptre, A (1995) The chemical composition of montane plants and its influence on the diet of the large mammalian herbivores in the Parc National des Volcans, Rwanda. *Journal of Zoology*, **235**, 323–37.

Plumptre, A J (1996) Modelling the impact of large herbivores on the food supply of mountain gorillas and implications for management. *Biological Conservation*, **75**, 147–55.

Plumptre, A J & Bizumuremyi, J B (1996) Ungulates and hunting in the Parc National des Volcans, Rwanda: the effects of the Rwandan civil war on ungulate populations and the socioeconomics of poaching. Unpublished report to the Wildlife Conservation Society.

Plumptre, A J, Bizumuremyi, J B, Uwimana, F & Ndaruhebeye, J D (1997) The effects of the Rwandan civil war on poaching of ungulates in the Parc National des Volcans. *Oryx*, **31**, 265–73.

Robbins, M M (1995) A demographic analysis of male life history and social structure of mountain gorillas. *Behaviour*, **132**, 21–47.

Sarmiento, E E, Butynski, T M & Kalina, J (1996) Gorillas of Bwindi Impenetrable Forest and the Virunga Volcanoes: taxonomic implications of morphological and ecological differences. *American Journal of Primatology*, **40**, 1–21.

Schaller, G B (1963) *The Mountain Gorilla: Ecology and Behavior*. Chicago: University of Chicago Press.

Sholley, C R (1988) Annual report of the Mountain Gorillas Project. Unpublished report.

Sholley, C R (1990) Census of the mountain gorillas in the Virungas of central Africa. Unpublished report.

Sicotte, P (1993) Inter-group encounters and female transfer in mountain gorillas: influence of group composition on male behavior. *American Journal of Primatology*, **30**, 21–36.

Thorne, E T & Williams, E S (1988) Disease and endangered species: the black-footed ferret as a recent example. *Conservation Biology*, **2**, 66–74.

Vedder, A (1989) In the hall of the mountain king. *Animal Kingdom*, **92**, 30–43.

Vedder, A & Weber, A W (1990) The Mountain Gorilla Project. In *Living with Wildlife: Wildlife Resource Management with Local Participation*, ed. A Kiss, World Bank Technical Publication 130, pp. 83–90, Washington DC: World Bank.

Waterman, P, Choo, G, Vedder, A L & Watts, D P (1983) Digestibility, digestion-inhibitors and nutrients of herbaceous foliage and green stems from an African montane flora and comparison with other tropical flora. *Oecologia*, **60**, 244–9.

Watts, D P (1984) Composition and variability of mountain gorilla diets in the Central Virungas. *American Journal of Primatology*, **7**, 323–56.

Watts, D P (1987) Effects of mountain gorilla foraging activities on the productivity of their food plant species. *African Journal of Ecology*, **25**, 155–63.

Watts, D P (1989) Infanticide in mountain gorillas: new cases and a reconsideration of the evidence. *Ethology*, **81**, 1–18.

Watts, D P (1990) Mountain gorilla life histories, reproductive competition, and sociosexual behavior and some implications for captive husbandry. *Zoo Biology*, **9**, 185–200.

Watts, D P (1991*a*) Mountain gorilla reproduction and sexual behavior. *American Journal of Primatology*, **24**, 211–18.

Watts, D P (1991*b*) Strategies of habitat use by mountain gorillas. *Folia Primatologica*, **56**, 1–16.

Watts, D P (1998*a*) Long-term habitat use by mountain gorillas (*Gorilla gorilla beringei*). I. Consistency, variation, and home range size and stability. *International Journal of Primatology*, **19**, 651–80.

Watts, D P (1998*b*) Long-term habitat use by mountain gorillas (*Gorilla gorilla beringei*). II. Reuse of foraging areas in relation to resource abundance, quality, and depletion. *International Journal of Primatology*, **19**, 681–702.

Watts, D P (1998*c*) Seasonality in the ecology and life histories of mountain gorillas (*Gorilla gorilla beringei*). *International Journal of Primatology*, **19**, 929–48.

Weber, A W (1979) Conservation of the Virunga gorilla. *Wildlife News (African Wildlife Foundation)*, **14**, 7–9.

Weber, A W (1981) Conservation of the Virunga gorillas: a socio-economic perspective on habitat and wildlife preservation in Rwanda. MSc thesis, University of Wisconsin.

Weber, A W (1987*a*) *Ruhengeri and its Resources: An Environmental Profile of the Ruhengeri Prefecture*. Kigali: Environmental Training and Management in Africa/USAID.

Weber, A W (1987*b*) Socio-ecological factors in the conservation of afromontane forest reserves. In *Primate Conservation in the Tropical Rain Forest*, ed. C Marsh & R Mittermeier, pp. 205–29. New York: Alan R. Liss.

Weber, A W (1993) Primate conservation and ecotourism in Africa. In *Perspectives on Biodiversity: Case Studies of Genetic Resource Conservation and Development*, ed. C S Potter, J I Cohen & D Janczewski, pp. 129–50. Washington DC: American Association for the Advancement of Science Press.

Weber, A W & Vedder, A L (1983) Population dynamics of the Virunga gorillas

1959–1978. *Biological Conservation*, **26**, 341–66.

Wilkie, D S & Carpenter, J (1999) The impact of bushmeat hunting on forest fauna and local economies in the Congo Basin: a review of the literature. *Biodiversity and Conservation*, **8**, 927–55.

Yamagiwa, J (1999) Slaughter of the gorillas in the Kahuzi-Biega park. *Gorilla Journal*, **19**, 4–6.

Young, T P (1994) Natural die-offs of large mammals: implications for conservation. *Conservation Biology*, **8**, 410–18.

15 Status of the Virunga mountain gorilla population

H. DIETER STEKLIS AND NETZIN GERALD-STEKLIS

Ginseng and infant son Bwenge in 1991. (Photo by Pascale Sicotte.)

Introduction

The mountain gorilla subspecies is considered to include two geographically isolated populations in east–central Africa (Garner & Ryder, 1996). One population inhabits the Bwindi Impenetrable Forest National Park of southwestern Uganda and comprises about 300 individuals (McNeilage *et al.*, 1998), while the other, similarly sized, population is restricted to the Virunga Volcanoes area (Figure 15.1). This chapter examines the conservation status of the Virunga mountain gorilla population and its vulnerability to extinction factors.

The Virunga gorillas are limited to three national parks bordering three nations: Parc National des Virunga, Parc National des Volcans, and Mgahinga Gorilla National Park. Together, they comprise an approximate 430 km² protected habitat that is commonly referred to as the "Virunga Conservation Area". The long history of research in the region (see Introduction to this volume) has provided several decades' worth of behavioral, ecological, and demographic data on this population, including the results of several censuses. The dramatic decline of the Virunga population in the 1970s (Harcourt & Fossey, 1981) raised the distinct possibility that this population was at risk of becoming extinct in the same century in which it had been discovered (Fossey, 1983). This concern has provided a strong rationale for subsequent population status assessments.

The question of what is a viable population, or minimum viable population, has been answered in different ways and is an area of continuing debate (e.g. Woodruff, 1989). There is broad agreement, however, on the importance of population attributes (e.g. size and structure) and environmental factors (e.g. disease, habitat loss) that affect viability, and such variables are often included in population viability (or vulnerability) analyses (PVA). Accordingly, we will examine data on population size and structure and review population viability models that have to varying degrees incorporated population characteristics in modeling the impact of environmental factors. In a final section, we discuss the implications of these data for the mountain gorilla's conservation status and conservation management. We will also briefly examine the conservation implications of whether or not to include the Bwindi population in the mountain gorilla taxon (see box "Gorilla taxonomy" in the Introduction to this volume).

Population size and structure

The size and structure of a population are critical in assessing whether it might be at risk of extinction. It has been recognized for some time that a small, isolated population with a low reproductive rate, as is the case for the Virunga gorillas, is at greater risk of extinction, largely because it is

392

Mountain Gorilla
Conservation Areas

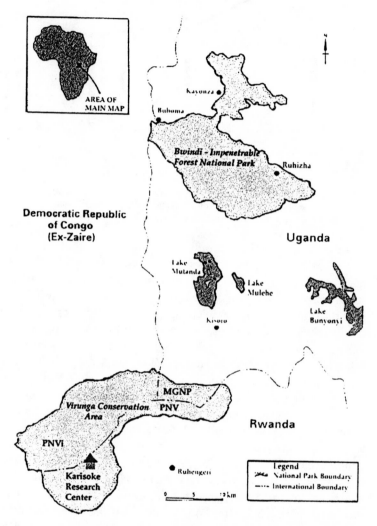

Figure 15.1. Locations of the two mountain gorilla populations in the Virunga Conservation Area and Bwindi Impenetrable National Park (adapted from Sarmiento *et al.*, 1996). See text for discussion on mountain gorilla taxonomic status.

394   H. Dieter Steklis and Netzin Gerald-Steklis

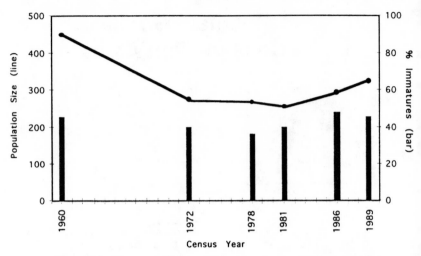

Figure 15.2. Virunga gorilla population size and percent of immatures from
censuses conducted between 1960 and 1989. Immature % calculated from animals
counted within groups (solitary males excluded). (After Harcourt, 1995.)

more vulnerable to the naturally occurring effects of random fluctuations
in birth and death rates, or demographic stochasticity (Soulé, 1987; Burgman *et al.*, 1993). It has less often been appreciated that population
structure, particularly social organization, plays an equally important role
in extinction risk. For example, populations of primates that live in small
family groups (e.g. orangutans) seem more prone to demographic stochasticity and thus more vulnerable to extinction than those that live in more
extended social groups (e.g. baboons) (Dobson & Lyles, 1989). Accordingly, in this section, we will review available data on the Virunga mountain
gorilla population's size and structure (group composition, group size, and
number of groups), including trends in these variables over time.

   Beginning in 1959 (Schaller, 1963), the total Virunga population size
and its structure has been estimated on six occasions, with the last census
conducted in 1989 (Sholley, 1991) (Figure 15.2; Table 15.1). Schaller based
his total population size estimate on actual counts of individuals in groups
in areas of intensive study, evidence of groups in areas of foot surveys (e.g.
nests and dung), and, based on this information, additional numbers of
individuals that could be supported in areas not surveyed. Census techniques were improved beginning with the 1971–73 census (Harcourt &
Groom, 1972; Groom, 1973) in an attempt to provide a total count of the
population. Improvements included a sweep technique, whereby the entire
conservation area was systematically surveyed from west to east. Whenever possible, actual counts of individuals were made, otherwise counts

Table 15.1. *Virunga mountain gorilla population size and structure based on census results from 1959 to 1989*

Census	Total estimated	Total counted	Number of groups	Count in groups	Lone males	Mean group size
1959–60	450	~279	19	~279	0	~15
1971–73	274	261	31	246	15	7.9
1976–78	268	252	28	246	6	8.8
1981	254	242	28	237	5	8.5
1986	293	279	29	268	11	9.2
1989	324	309	32	303	6	9.1

Census data from Schaller, 1963 (1959–60); Harcourt *et al.*, 1981 (1971–73); Weber & Vedder, 1983 (1976–78); Harcourt *et al.*, 1983; Aveling & Harcourt, 1984; Aveling & Aveling, 1989 (1981); Vedder, 1989 (1986); Sholley, 1990 (1989).

were based on nests and size of dung, which correlate with age (as established previously by Schaller, 1963). The total estimated population size is an average of the minimum population size (i.e. total counted individuals) and maximum population size (i.e. total counted plus 10% estimated error in the count). The results of the six censuses show that the population declined from about 450 in 1959–60 to about 254 in 1981 (Harcourt & Fossey, 1981), and after 1981 consistently increased to about 324 by 1989 (Vedder, 1989; Sholley, 1991).

Previous studies have examined these changes in population size by Virunga sub-region, indicating that the overall population size changes illustrated in Figure 15.2 have not been uniform across the entire conservation area. Human disturbance (principally poaching) and habitat loss differentially affected certain Virunga sub-regions. The large drop in population size from 1959–60 to the 1971–73 census was primarily due to massive killing of gorillas in the area around Mount Mikeno in the Parc National des Virunga (for details of location, see map in the Introduction to this volume), which was likely the result of direct hunting brought on by widespread civil war in the mid 1960s in the Democratic Republic of Congo (DRC, then Zaire) (Weber & Vedder, 1983). The remainder of the population decline was caused by the conversion of approximately 100 km^2 of Rwanda's Parc National des Volcans for agriculture in 1968–69, as well as a significant reduction in the Ugandan portion of the conservation area (Weber & Vedder, 1983).

The further but lesser decline of the Virunga population between 1973 and 1981 has been attributed to several factors, including cattle grazing inside park boundaries, gorilla deaths from snares set for antelope, and continuing human disturbance (Harcourt & Fossey, 1981; Aveling & Harcourt, 1984). Again, most of this decline occurred in the Mikeno sector subpopulation (Aveling & Harcourt, 1984; Harcourt, 1986; Vedder, 1989), as a result of human disturbance coupled with little protection.

The increase in overall population size after 1981 is most likely a direct effect of conservation efforts. In 1979, the Mountain Gorilla Project was launched in Rwanda as a multifaceted conservation program that included education, ecotourism, and, importantly, increased numbers of park guards for clearing cattle from park lands and for anti-poaching patrols (Harcourt, 1986), especially in sectors of the Parc National des Volcans that were not monitored by the Karisoke Research Center. A gorilla tourism program, modeled after the one begun earlier in Rwanda, was launched in the neighbouring DRC after 1984 (Aveling & Aveling, 1989). We can assume that these effective gorilla protection programs primarily contributed to the overall population increase by 1989.

More direct evidence of the positive effects of protection efforts is provided by studies comparing recruitment (i.e. percent or number of immatures) to the population in the DRC (unprotected) versus the Rwandan (protected) portion of the Virungas. Comparison of percent immatures between the early 1970s and 1981 showed a decrease in the unprotected zone from 37% to 33%, while the percentage increased in the protected zone from 36% to 44% (Aveling & Harcourt, 1984). More striking was the change in number of juveniles and infants per adult female per group: in the early 1970s, the two zones did not differ in this regard, but by 1981, the groups on the Rwandan side of the Virungas had twice as many juveniles and infants per female per group as did the groups on the DRC side (Harcourt, 1986). Thus, by 1981, even though the Virunga population as a whole was still declining, effective protection efforts in Rwanda actually resulted in substantial recruitment to the groups on the Rwandan side. Subsequent improved protection in DRC showed that by 1986 the proportion of immatures was equal on both sides (Aveling & Aveling, 1989), with parity between the two populations in percent immatures of over 40% also evident in 1989 (our own calculation, using 1989 census data). Indeed, as shown in Figure 15.2, in parallel with the population size increase, percent immatures was consistently above 40% after 1981 for the entire Virunga population, suggesting that by then the population overall was on a healthy growth trajectory (Vedder, 1989).

We might expect the population size increase from 1981 to 1989 to be expressed through proliferation of groups (e.g. by lone silverbacks forming new groups, fission of groups) and/or by increasing the size of existing groups. Analysis of the total count in groups from the 1973–1989 censuses

(Table 15.1) indicates a trend for total count in groups to be associated with both mean group size ($r = 0.72$, $0.10 < p < 0.25$, $n = 5$, Spearman rank-order, one-tailed) and number of groups ($r = 0.77$, $0.10 < p < 0.25$, $n = 5$, one-tailed). Since number of groups and mean group size are not correlated ($r = 0.17$, n.s., $n = 5$), they are independently related to total population size. The best we can conclude from the five censuses is that increases in the total count tended to be associated with both increased group size and increased number of groups.

While we have no census data on changes in the Virunga population after 1989, data from groups in the Rwanda sector that is regularly monitored by the Karisoke Research Center and park staff (i.e. groups habituated for research or tourism) indicate that monitored groups continued to increase in size after 1989. The Karisoke long-term demographic records show that group size for groups monitored between 1989 and 2000 increased from a 1989 median of 17 ($n = 4$) to 18 ($n = 5$) in 1994 and to 24 ($n = 5$) by the year 2000. Indeed, at the time of this writing, one of the groups followed for research is the largest ever recorded in the Virungas at 41 individuals. Any inference about the whole population from the monitored one in the post-1989 years must be treated with caution, however. This is because following the outbreak of civil war in 1990 in the Virunga region, much of the monitored population on the Rwandan side enjoyed a continued, higher level of protection than was the case for the rest of the Virunga Conservation Area. Indeed, since 1989, the majority of civil-war-related killing of gorillas has occurred in the DRC (see Plumptre & Williamson, this volume).

The increase in group size after 1989, at least in the monitored population, raises the question of the social and ecological conditions that might favor larger groups. Both census data and monitored group data indicate that larger groups tend to be multimale (i.e. with more than one silverback), suggesting that advantages of multimale group membership may offset the reproductive costs otherwise incurred in large groups. For example, for the 1989 census (from Sholley, 1991), multimale groups had significantly larger mean group sizes (multimale: 15.6 ± 3.4 SE, $n = 8$; single male: 8.0 ± 1.0 SE, $n = 21$; $t = -2.14$, d.f. $= 8$, $p = 0.03$) and more adult females than single male groups (multimale: 5.2 ± 1.3 SE, $n = 8$; single male: 2.6 ± 0.4 SE, $n = 21$; $t = -1.93$, d.f. $= 9$, $p = 0.04$). [The latter calculation of adult females used the method employed in previous censuses (e.g. Weber & Vedder, 1983), wherein unsexed adults are equally divided between adult females and blackback males, with any remaining unsexed adults assigned to adult females.] Similarly, examining four monitored groups that alternated between single and multimale states, Robbins (1995) found female group size was positively related to number of silverbacks in a group. Finally, the actual proportion of multimale to single male groups has gone up from 8% in 1986 to 27% in 1989 [our calculations from 1986 and 1989 census data, Vedder (1989) and Sholley (1991), respectively]. It should be recognized, however, that the 1976–78 census data show a substantially larger percent (42%) of multimale groups (Watts, 2000).

Several social factors may contribute to the formation and maintenance of large, multimale groups, although there may also be reproductive costs to these large groups. Watts (1990) found that increasing group size tends to be associated with decreased reproductive rates (i.e. births per female per year and surviving births per female per year), but this trend was reversed at the largest group size (about 17) and highest number of females per group (11). One benefit for females of living in multimale groups is improved infanticide protection (Watts, 1989), which is consistent with females preferring to transfer to multimale groups (Watts, 2000). As well, females begin to reproduce earlier in multimale vs. single-male groups (Gerald, 1995), which may lead to higher lifetime reproductive output. For males, the advantages are principally increased mating opportunities and infant survival (Robbins, 1995). Watts (2000) modeled the fitness payoffs for males remaining as part of multimale groups (as "followers") or leaving such groups to form their own (as "bachelors"), and found substantially higher fitness payoffs for followers vs. bachelors. Finally, large, multimale groups may be able to better retain female members since silverbacks can cooperate in the "herding" of females, which is effective in preventing female transfer (Sicotte, 1993).

Habitat quality and level of protection for monitored groups may also contribute to the particularly large group sizes observed after 1989. The largest groups' ranges fall inside the rich habitat of the Karisimbi–Visoke saddle (Harcourt & Fossey, 1981; Harcourt et al., 1981; see map detail in Introduction, this volume). McNeilage (1995) found a correlation between food density in a group's home range and group size for his sampled groups. As well, the regular monitoring of these groups leads to better protection, and presumably to larger group sizes because well-protected groups do not suffer the depletion effects of poaching-related deaths nor the effects of habitat degradation due to human disturbance. Lower habitat quality and higher levels of human disturbance have been argued to account for the consistently lower mean group sizes observed in the eastern sector of the Virungas (Fossey & Harcourt, 1977; Harcourt & Fossey, 1981; Weber & Vedder, 1983). Similarly, poaching-related declines in the Mikeno sector (in DRC) were associated with decreased group sizes, while habitat loss and degradation were linked to decreased group sizes outside the Mikeno sector (Weber & Vedder, 1983). Also supportive of a direct connection between group size and level of protection is the finding from the 1986 census (Vedder, 1989) that monitored groups (i.e. groups in areas patrolled or visited frequently) were significantly larger than unprotected groups. Thus, while poor habitat or human disturbance may effectively limit group size, the large groups found in the Rwandan monitored sector are likely the result of the combined effects of a food-rich habitat, relatively high levels of protection, and advantages gained from multimale group membership.

Population viability analyses
The data reviewed in the last section on population size and structure indicate that 1981 marked a positive turning-point in the Virunga mountain gorilla population's size and growth potential. Despite this positive turn, however, many questions and concerns remain about the population's long-term viability. All populations fluctuate in numbers as a result of random variation (or stochasticity) in their demography and environment. Population viability analyses (PVA) model the responses of a population to stochastic variation in demographic and environmental processes (Boyce, 1992) and predict the probability of a population going extinct (or, conversely, the probability of persistence) over a specified time period. PVAs rely on life history traits and may be improved by including details of population structure and behavioral characteristics. The results of a PVA may be used to draw up population management recommendations. In this section we will review and evaluate the several PVAs that have been performed on the Virunga mountain gorilla population, with a particular focus on those conducted since Harcourt's (1995) review.

Harcourt (1995) provides a quick assessment of the Virunga population's viability by examining the genetic consequences of small population size. A small population may lose genetic variation as a result of stochastic demographic processes, which can increase its extinction risk either by reducing its ability to respond adaptively to environmental circumstances (Franklin, 1980), or by increasing its susceptibility to disease (O'Brien & Evermann, 1988). An estimate of genetic viability can be derived from calculating the size of the effective breeding population (Franklin, 1980). For the Virunga population at its smallest size in 1981, Harcourt (1995) calculated an effective population size (N_e) of 70–100 individuals. While this small size, as Harcourt (1995) suggests, may provide reason for alarm, it is offset by the gorillas' relatively long generation time of 20 years, which yields a low risk of extinction from loss of genetic heterozygosity within 400 years. Following Harcourt's (1995) method, using 1989 census data

(Table 15.1), we arrive at an N_e of about 80–120 for the 1989 population. Therefore, from a genetic standpoint, the 1989 population continued to have an at least equally low risk of extinction for several hundred years.

Harcourt (1995) also reviewed the results of three PVAs that modeled the effects of demographic stochasticity on population viability: Weber & Vedder (1983), Akcakaya & Ginzburg (1991), and Durant & Mace (1994). Of these, only the Durant & Mace (1994) analysis incorporated social behavioral factors, such as realistic conditions relating to dispersal. The results of these analyses, Harcourt (1995) points out, lead to the same conclusion: while the Virunga mountain gorilla effective population size is small, and the population is isolated, its healthy annual growth rate (about 3%) since the early 1980s appears sufficient to buffer it against both demographic and natural environmental stochasticity for at least several centuries.

The aforementioned studies, however, did not include environmental stochasticity (e.g. disease) as part of their viability analyses. Subsequent analyses by Durant (1998, 2000) and by Miller *et al.* (1998) do incorporate environmental factors. Durant (1998) performed a sensitivity analysis in which she investigated which demographic parameters have the greatest impact on population growth rate and percent effective population size. Based on a deterministic growth rate equation (from Charlesworth, 1980) that included five demographic parameters (adult survival rate, juvenile survival rate, annual reproductive rate, longevity, age at first reproduction), she found that the deterministic growth rate and percent effective population size were most sensitive to changes in adult or juveniles survival rates. Based on these results, and the assumption that juvenile survival is more amenable to management, she suggested that intervention practices should target juveniles to improve their survival rates.

Using a computer population simulation model (POPGEN: Durant *et al.*, 1992), Durant (1998) also examined the effectiveness of a juvenile survival management strategy within a modeled meta-population (i.e. a population containing 10 breeding or social groups, of initially 10 individuals with a stable age/sex structure). Her model also included several behavioral factors that determine the sex-specific conditions and consequences of migration between groups (e.g. a female could only migrate to join another breeding group or a solitary adult male; an adult male would not migrate if he was the only adult male in his group), and two levels of migration probability (0.01 and 0.1). An intervention strategy at either low or high risk consisted of raising juvenile survival rates to 1 until the group no longer showed indicators of extinction for 10 years. She found that the effectiveness of this intervention strategy depended on how many of the 10 groups were managed, the migration rate, and at what level of risk the intervention strategy was implemented. At the higher migration rate, the most economical management strategy was to intervene as soon as low-risk indicators appeared and to manage four of the groups. At the lower migration rate, no management strategy was reasonably economical.

In an extension of her previous analyses, Durant (2000) used similar demographic and behavioral factors but added environmental stochasticity. In her models the groups were subjected to frequent reproductive disasters (occurring with a mean frequency of 1 every 5 years, reducing birth rate to 0) and infrequent survival disasters (similar to a severe epizootic occurring at a mean frequency of once every 50 years, reducing survival across all ages by 50%), that occurred independently across all groups. While frequent reproductive disasters had a negligible effect on population persistence, infrequent survival disasters, at the low migration rate, resulted in a persistence of less than 50% by 750 years, compared to about 90% persistence at the high migration rate. In this latter scenario, groups depleted by the disaster presumably are more easily repopulated through migration. In addition, a subdivided population achieved better persistence than an undivided panmictic one because, under the conditions modeled, there is a long-term spreading of risk benefit in a subdivided population.

The overall conclusion from these further models echoes one reached earlier, namely, that the Virunga population is a fairly robust one that is not particularly vulnerable to modeled stochastic events for several centuries. Thus, while some factors (i.e. epizootic-like disasters, intervention strategies) had significant effects on persistence, these did not become apparent until after 200 years. At the same time, the effects modeled may become apparent sooner if, for example, baseline adult survival rates are decreased, reflecting more realistic estimates, or if the overall population ceiling is reduced to a more realistic level. Durant's models all used demographic data from an early analysis of Harcourt *et al.* (1981). More reliable demographic parameters are now available because of a longer-term database (about 10 years vs. about 30 years: Gerald, 1995). Given the model's sensitivity of population growth rates to adult survival, using a more up-to-date baseline adult survival rate of 0.87 (compared to 0.96) may result in a substantial decrease in population growth rate and hence slower recovery from disasters. Similarly, Durant's initial meta-population size was a reasonable 300, but the population ceiling (i.e. a size at which reproduction is suppressed in order to prevent numbers from exceeding this level) was set at 1200. This population ceiling seems unreasonable in that it far exceeds any estimate of population size likely to be supported by the Virunga habitat. For example, McNeilage (1995) arrived at a carrying capacity of 650, defined as the number of gorillas the habitat could support, with current patterns of habitat use maintained and without detrimental effects on either the population or the habitat. Thus, with a more realistic population ceiling, the conditions of decreased persistence modeled by Durant may become evident in fewer years.

An altogether different PVA was developed in a 1997 workshop held in Uganda to evaluate the survival prospects of the mountain gorilla (Werikhe *et al.*, 1998). The PVA used a simulation software package called VORTEX v.7, which is a Monte Carlo simulation of the effects of demographic, environmental, and genetic stochasticity as well as the effects of deterministic forces on wildlife populations (Lacey, 1993; Lacey *et al.*, 1995). The PVA included a sensitivity analysis followed by population simulations, both based on up-to-date estimates of demographic variables obtained from the long-term Karisoke records for the monitored population.

The sensitivity analysis focused on those demographic parameters for which a range of values was available from the long-term demographic data (age at first reproduction and last reproduction for females, age at first reproduction for males, proportion of annual female breeding success, and extent of male reproductive pool). Of the parameters included, the sensitivity analysis showed that population growth rate was most affected by annual proportion of females breeding and age of onset of female reproduction. Using the best estimates of each demographic parameter, the model produced a mean stochastic population growth rate of 3.80% ($r = 0.038$, 0.023 SD), which is broadly consistent with the estimated observed growth rate of 3.05% between 1981 and 1989, indicating a strong growth potential. The low variance in the mean growth rate indicated low demographic stochasticity.

However, because the model did not include the effects of any natural environmental stochasticity, such a growth rate is likely inflated, reflecting more the maximum inherent population growth potential rather than a realistic long-term average growth rate. In an effort to bring the growth rate down to a more realistic, long-range, baseline value, a natural environmental variable was introduced into the model. Disease was modeled as a realistic candidate of natural environmental stochasticity, since naturally occurring respiratory illnesses and viral disease, for example, have been documented for the Virunga gorilla population (Homsy, 1999; Wallis & Lee, 1999). Two disease scenarios were used in this baseline model, which attempts to represent a population with naturally occurring demographic and environmental stochasticity: (1) influenza-like disease with no reproductive decrease (10% annual probability of occurrence, 5% reduction in survivorship, no effect on reproduction), and (2) viral disease with some reproductive decrease (10% annual probability of occurrence, 25% reduction in survivorship, 20% reduction in proportion of females breeding). The resulting mean stochastic growth rate indicated a population that was only slightly growing at 0.3% ($r = 0.003$, 0.095 SD), with a zero probability of extinction and high heterozygosity after 100 years. The large standard deviation around the growth rate suggests that the simulated population fluctuated substantially in size over time caused by disease events, but that the population's strong growth potential enabled it to recover, resulting in a slightly increasing population. Thus all subsequent effects of additional environmental factors (i.e. more severe disease, war, and a forced population subdivision) were examined relative to this baseline model, with the probability of extinction based on 100 years and 500 simulations (Miller *et al.*, 1998).

First the effects of a severe disease were modeled. The disease simulation, based on discussion with gorilla-specialist veterinarians at the workshop, was an infrequent but severe viral disease with high reproductive decrease (4% annual probability of occurrence, 25% reduction in survivorship, 100% reduction in proportion of females breeding). When this severe disease was added to the baseline model, the resulting growth rate was negative. None the less, the probability of extinction was low (0.018).

Since civil war has plagued the countries surrounding the Virunga conservation area periodically since the late 1950s, the second analysis modeled the effects of war. Different war scenarios, reflecting historical knowledge of the frequency and impact of war in the region, were modeled. These included different combinations of the following effects: 10% reduction in proportion of females successfully breeding (due to stress), 5% mortality increases imposed on infants and adults of both sexes (due to direct and accidental killings), post-war chronic effects on reproduction and mortality of full or 50% recovery of baseline rates, 25% or 50% permanent reduction in carrying capacity (as a proxy for loss of habitat and/or increase of habitat rendered unusable due to human impact), and impact of the different diseases previously modeled (assuming the stress of war conditions makes individuals susceptible to disease). In all scenarios, the simulation began under conditions of war, which would last for 10 years, with recurrence every 30 years. The highest impact was seen when chronic post-war effects were included in the simulation. These effects produced the highest probabilities of extinction and worst (negative) growth rates. In comparison to the chronic war effects, the effects of reducing carrying capacity were substantially milder.

A final analysis examined the effects of severe disease and war, as specified earlier, in a population divided into two unequal subpopulations with no migration between them. While the probability of extinction was highest for the smaller of the two subpopulations, it was consistently less for the meta-population than for the panmictic one under the same conditions (between 40–80% decrease in extinction probability). This result is consistent with the spread of risk benefit shown in Durant's (1998) meta-population analyses.

Like previous PVAs, the results of this one show a robust population growth rate. This growth rate is slowed, and indeed reversed, under several modeled deleterious environmental conditions. In assessing the relative impact of the different environmental disasters on probability of extinction, the chronic scenarios of war have the most impact (all about a 3% extinction probability), followed by severe disease (about 2%), with reductions in carrying capacity under war conditions having the least impact ($< 1\%$), with effects more pronounced under panmictic than subdivided population conditions. However, in these models the probability of extinction never reaches 10%. This is likely because of the slow growth rates projected over a relatively short time span (100 years). As pointed out by Miller *et al.* (1998), when the simulations are extended to 150 years, extinction probabilities increase greatly, largely because population size at 100 years usually drops to well below 100 animals.

Compared to the 1997 PVA analyses, Durant's (1998, 2000) simulations, even for her worst-case scenarios, consistently show considerably longer time intervals (at least 200–250 years) before persistence decreases to 90% (or 10% probability of extinction). While the models used roughly similar demographic parameters, they differed in many respects that complicate direct comparison. Nevertheless, a few obvious differences between the models may be responsible for the differences in outcome. For example, as mentioned earlier, Durant's model included a population ceiling of 1200, compared to the more realistic 650 figure (McNeilage, 1995) used in the 1997 PVA, which may contribute to her unrealistically optimistic persistence probabilities. On the other hand, Durant's more realistic modeling of a gorilla meta-population, compared to either the single panmictic one or the divided, no-migration one used in the 1997 PVA, suggests the relatively shorter time course of extinction generated by the latter model is unrealistically pessimistic.

Results from population modeling confirm what is evident from census data, namely, that the Virunga population has a strong positive growth rate. Further, its population structure (i.e. subdivision into social groups) leads to low probabilities of extinction from loss of genetic variation. The high and robust inherent stochastic demographic growth rate of over 3%, with little variance, suggests that the rate is at best a purely demographic one and unlikely to include environmental stochasticity. The value of modeling is that it allows the simulation of environmental stochasticity and its combination with the observed demographic stochasticity to yield a more realistic long-term forecast. Even with the addition of modeled environmental stochastic factors, however, the growth rate, although significantly reduced, remains positive. Though the effects of environmental stochasticity may be better modeled with more accurate, empirically derived information, the outcome of adding any environmental stochasticity will likely be the same: a reduced, but probably not negative growth rate because of the inherent strong growth trajectory the population exhibits. As a result, the population has a low probability of extinction at least for several hundred years based on natural demographic and environmental stochasticity.

Given demographic and environmental stochasticity, it is therefore unlikely that the high 3% growth rate observed only in the recent past will characterize the population in the long term. In that case, the carrying capacity, estimated as about 650 (McNeilage, 1995) may not be reached in the short time-span McNeilage suggests (by year 2011). For example in the 1997 PVA simulation, which incorporates demographic and environmental stochasticity, the growth rate is reduced to less than 1%, at which rate carrying capacity is not attained within the 100 years modeled, and in fact remains well below it.

Discussion

The IUCN (1994) provides broad criteria for assigning an animal taxon to a level of extinction risk. The criteria include population size and structure, and rate of growth, in recognition of their empirically demonstrated importance for estimating extinction risk (Mace & Lande, 1991). Because the criteria refer to taxa of animals (i.e. ideally, meaningful genetic–evolutionary groups), it must first be decided what constitutes the taxonomically defined population of mountain gorillas. For example, recent past practice has included both the Bwindi and Virunga gorilla populations in the same subspecies taxon *Gorilla gorilla beringei*. This results in a combined estimated total population size (for the taxon) of about 616 [see McNeilage *et al.* (1998) for results of 1997 Bwindi population census] with about 290 mature individuals. This adult population size is just larger than the cutoff (250) for inclusion in the "critically endangered" category and thus places them into the next lower risk category of "endangered".

This classification is at odds, however, with IUCN's, which places the mountain gorilla within the "critically endangered" category. This indicates, for one, that the IUCN has since 1996 (*IUCN Red Data Book*) excluded the Bwindi gorillas from the mountain gorilla taxon, probably in light of recently reopened debate about its morphological and ecological uniqueness (Sarmiento *et al.*, 1996; T.M. Butynski, personal communication). More importantly, however, it suggests that even when the mountain gorilla taxon is reduced to the population in the Virunga conservation area, IUCN, inexplicably, is not adhering to its own threat classification criteria. To fall into the "critically endangered" category, the Virunga population would have to show an adult population size of less than 250 (which it does) with all individuals in a single population (which they are), *and* show evidence of a continuing decline (*which it does not*). In other words, the Virunga population, regardless of taxonomy and contrary to IUCN's classification, does not satisfy the criteria for being included in the "critically endangered" category, but rather should be classified as "endangered" (see also Harcourt, 1996).

While this lower risk classification (than IUCN's own) of the Virunga gorillas, we believe, more accurately reflects available data for the population (both census and PVA), it must be tempered by the fact that the population has not been counted in more than a decade. And while the monitored (and well-protected) part of the population has fared well in the intervening years, we cannot say the same with certainty for the population as a whole. This is especially the case given the renewed outbreak of civil war in the Virunga region beginning in 1990 and with continuing repercussions, the total effects of which on both gorillas and habitat are as yet unknown. Some might well argue, therefore, that the lower risk categorization may not accurately reflect the current state of affairs and that it should not justify a lowered degree of vigilance.

Although the PVA results all indicate a generally low extinction risk for several hundred years in response to the environmental factors modeled, this essentially healthy report might easily blind us to the population's vulnerability to large deterministic effects that could in a relatively short time-span decimate population numbers, placing the population at higher risk of extinction than the models indicate. For example, human-introduced diseases could exert such a large effect. Two outbreaks of respiratory illness in the population, one in 1988, the other in 1990, may have been human-introduced (Homsy, 1999). In each case, the disease caused substantial morbidity and some mortality. The worst of the two was the 1988 outbreak (suspected as measles), which affected 20 individuals in different groups, and two juveniles and three adult females died. The true impact of the outbreak is unknown as a massive vaccination effort in the habituated groups curtailed the spread of the disease (Mudakikwa *et al.*, this volume). It is entirely probable that a human-introduced disease, to which gorillas lack immunity, in the absence of intervention could have a one-time catastrophic effect in excess of any disease scenario modeled.

Large-scale habitat destruction, of which there already is a history in the Virungas, is another potential catastrophic environmental threat. For example, using human birth rates in the countries surrounding the Virunga conservation area to project the rate of forest clearance in relation to human population density, Harcourt (1995) estimated that all remaining forests outside national parks in these countries will be cleared in less than 100 years. This would result in such intense pressure on the remaining habitat that it could mean extinction of the gorilla population within the same time-frame. Although the present population is not likely anywhere near carrying capacity, habitat loss or degradation could produce regional (e.g. eastern vs. western part of the Virungas) large-scale, density-dependent negative effects, in light of the significant heterogeneity in quality of the Virunga habitat that affects the distribution and size of groups (Weber & Vedder, 1983; McNeilage, 1995).

These observations about the population's vulnerability suggest that early warning signs of a potential environmentally induced catastrophe should be recognized and attended to. We recommend that, in addition to the continued use of gorilla groups for research, gorilla ecotourism should be seen as playing a larger role in conservation than its more obvious economic one. Though Durant's (1998) models indicate that the most cost-effective population management strategy consists of intensive monitoring of a few groups, such a strategy will only work for the detection of risk indicators caused by normal demographic and environmental stochasticity. The larger deterministic environmental effects we are concerned with merit a different management approach. This "catastrophe avoidance" approach requires, we believe, monitoring of a large proportion of the population over a wide area and early intervention.

The monitoring that occurs alongside a successful tourism program makes such a strategy feasible and economically viable. Gorilla tourism in all three countries is considered critical to the populations' continued protection and valuation in general. In the Virungas, about 70% of the population (in about 13 groups) is now habituated either for tourism or research purposes (Butynski & Kalina, 1998). In evaluating the benefits and problems of gorilla tourism, the latter authors suggest that, with so large a percent of the population regularly visited by people, introduction of disease and as-yet unknown effects of frequent visitation on gorilla behavior and reproductive health pose the most serious problems. We suggest, however, that these potential problems, while real, can be addressed with proper management and precautions (see Homsy, 1999), and, if this is done, the benefits of the extra benevolent eyes and ears in the forest will outweigh the (yet unknown) costs of their presence.

There is clearly a need for systematic research on the effects of tourism on gorilla behavior and biology, but there are data that indirectly give us some idea. For example, the proportion of immatures was higher for protected groups (research and tourist groups) compared to unprotected groups for the past three censuses (McNeilage, 1996). Further, using females in the Karisoke long-term records (1967–94) whose reproductive careers were exclusively in either tourism or research groups, we compared reproduction rates for research groups (0.28 births/female/year; 195 female years, 55 births) and tourist groups (0.23 births/female/year; 139 female years, 32 births) and found no significant difference between the two. These results suggest that daily visitation by larger numbers of novel visitors (research groups are visited by fewer and familiar visitors/researchers) has no effect on reproductive health. The evidence so far, therefore, indicates that, in addition to the direct economic benefit, the collateral benefits of monitoring and protection may outweigh any of the potential costs, especially if proper precautions against disease transmission are followed (Homsy, 1999).

In general, we can conclude the following about the status of the Virunga mountain gorilla population. Although it shows a low extinction risk over the next century or more on genetic and demographic grounds, its small adult population size means it is vulnerable to potentially large-scale deterministic environmental effects, such as human-introduced disease or habitat loss, and thus deserves to be regarded as endangered. On these grounds, continued intensive and extensive monitoring and protection are recommended.

Summary
Based on six censuses between 1959 and 1989, after an initial decrease, the population of mountain gorillas has increased since 1981. It is likely that the post-1981 increase is a direct effect of protection efforts. Mean group size in the monitored population has increased since 1989, which may indicate a similar trend in the general population. Largest groups tend to be multimale, suggesting that, when local habitat quality permits it, the reproductive advantages of multimale group membership outweigh the competitive costs of large group size. Population viability models support the conclusion from census data of a robust growth rate and resilience to natural demographic and environmental stochasticity. Depending on the

model and the demographic parameters, sensitivity analyses indicate that changes in adult and juvenile mortality have the largest effects on growth rates, followed by more moderate effects of variability in age of first reproduction and annual proportion of females breeding. A meta-population with migration between groups has a higher persistence than a single panmictic one. In modeling different war scenarios that included disease, chronic post-war effects, and reductions in carrying capacity, the highest probabilities of extinction and worst (negative) growth rates were seen when chronic post-war effects were included in the simulation. By comparison, the effects of reducing carrying capacity were substantially milder. None of the simulations showed a high probability of extinction before 100 years. Larger deterministic factors acting on a shorter time-scale (e.g. human-introduced disease, massive habitat destruction) may thus play a greater role in this population's viability. To guard against such catastrophic events, intensive and extensive monitoring is required, which can be an important collateral benefit of ecotourism. While the IUCN has classified the Virunga mountain gorilla as "critically endangered", the evidence overall indicates it should be categorized as "endangered".

Acknowledgements
We gratefully acknowledge the numerous researchers and Rwandan field staff for their contributions to the Karisoke Research Center long-term demographic database. We owe a special thanks to the editors for their patience and to the editors and anonymous reviewers for their many helpful suggestions for improvement. We also thank Drs. Sandy Harcourt and Tom Butynski who provided valuable input on the IUCN classification criteria. Finally, we are grateful to the Dian Fossey Gorilla Fund International for supporting our research.

References

Akcakaya, H R & Ginzburg, L R (1991) Ecological risk analysis for single and multiple populations. In *Species Conservation: A Population-Biological Approach*, ed. A Seitz & V Loeschke, pp. 73–87. Basel: Birkhauser Verlag.

Aveling, C & Aveling, R (1989) Gorilla conservation in Zaire. *Oryx*, **23**, 64–70.

Aveling, C & Harcourt, H (1984) A census of the Virunga gorillas. *Oryx*, **18**, 8–13.

Boyce, M S (1992) Population viability analyses. *Annual Review of Ecology and Systematics*, **23**, 481–506.

Burgman, M A, Ferson, S & Akcakaya, H R (1993) *Risk Assessment in Conservation Biology*. London: Chapman & Hall.

Butynski, T M & Kalina, J (1998) Gorilla tourism: a critical look. In *Conservation of Biological Resources*, ed. E J Milner-Gulland & R Mace, pp. 280–300. Oxford: Blackwell Scientific Publications.

Charlesworth, B (1980) *Evolution in Age-Structured Populations*. Cambridge: Cambridge University Press.

Dobson, A P & Lyles, A M (1989) The population dynamics and conservation of primate populations. *Conservation Biology*, **3**, 362–80.

Durant, S M (1998) A minimum intervention approach to conservation: the influence of social structure. In *Behavioral Ecology and Conservation Biology*, ed. T Caro, pp. 105–29. New York: Oxford University Press.

Durant, S M (2000) Incorporating behavior in predictive models of conservation. In *Behavior and Conservation*, ed. L M Gosling & W J Sutherland, pp. 172–97. New York: Cambridge University Press.

Durant, S M & Mace, R (1994) Species differences and population structure in population viability analysis. In *Creative Conservation: Interactive Management of Wild and Captive Animals*, ed. P G S Olney, G M Mace & A T C Feistner, pp. 67–91. London: Chapman & Hall.

Durant, S M, Harwood, J & Beudels, R (1992) Monitoring and management strategies for endangered populations of marine mammals and ungulates. In *Wildlife 2001*, ed. D R McCullough & R H Barrett, pp. 252–61. New York: Elsevier.

Fossey, D (1983) *Gorillas in the Mist*. Boston MA: Houghton Mifflin.

Fossey, D & Harcourt, A H (1977) Feeding ecology of free-ranging mountain gorilla (*Gorilla gorilla beringei*). In *Primate Ecology: Studies of Feeding and Ranging Behaviour in Lemurs, Monkeys and Apes*, ed. T H Clutton-Brock, pp. 415–47. New York: Academic Press.

Franklin, I R (1980) Evolutionary change in small populations. In *Conservation Biology: An Evolutionary–Ecological Perspective*, ed. M E Soulé & B A Wilcox, pp. 135–49. Sunderland, MA: Sinauer Associates.

Garner, K J & Ryder, O A (1996) Mitochondrial DNA diversity in gorillas. *Molecular Phylogenetics and Evolution*, **6**, 39–48.

Gerald, C N (1995) Demography of the Virunga mountain gorilla (*Gorilla gorilla beringei*). MSc thesis, Princeton University.

Groom, A F (1973) Squeezing out the mountain gorilla. *Oryx*, **2**, 207–15.

Groves, C P (1970) Population systematics of the gorillas. *Journal of Zoology*, **161**, 287–300.

Groves, C P & Stott, K W (1979) Systematic relationships of gorillas from Kahuzi, Tshiaberimu and Kayonza. *Folia Primatologica*, **32**, 161–79.

Harcourt, A H (1986) Gorilla conservation: anatomy of a campaign. In *Primates: The Road to Self-Sustaining Populations*, ed. K Benirschke, pp. 31–46. New York: Springer-Verlag.

Harcourt, A H (1995) Population viability estimates: theory and practice for a wild gorilla population. *Conservation Biology*, **9**, 134–42.

Harcourt, A H (1996) Is the gorilla a threatened species? How should we judge? *Biological Conservation*, **75**, 165–76.

Harcourt, A H & Fossey, D (1981) The Virunga gorilla: decline of an island population. *African Journal of Ecology*, **19**, 83–97.

Harcourt, A H & Groom, A F (1972) Gorilla census. *Oryx*, **11**, 355–63.

Harcourt, A H, Fossey D & Sabater Pi, J (1981) Demography of gorilla. *Journal of Zoology*, **195**, 215–33.

Harcourt, A H, Kineman, J, Campbell, G, Yamagiwa, G, Redmond, I, Aveling, C & Condiotti, M (1983) Conservation of the Virunga gorilla population. *African Journal of Ecology*, **21**, 139–42.

Homsy, J (1999) *Ape Tourism and Human Diseases: How Close Should We Get?* Report to the International Gorilla Conservation Program.

IUCN (1994) *The IUCN Red List Categories*. Gland: IUCN.

IUCN (1996) *1996 IUCN Red List of Threatened Animals*. Gland: IUCN.

Lacey, R C (1993) VORTEX – a computer simulation model for population viability analysis. *Wildlife Research*, **20**, 45–65.

Lacey, R C, Miller, P, Hughes, K & Kreeger, T (1995) *VORTEX* Users Manual, version 7: A Stochastic Simulation of the Extinction Process. Apple Valley MN: IUCN SSC Conservation Breeding Specialist Group.

Mace, G M & Lande, R (1991) Assessing extinction threats: toward a reevaluation of IUCN threatened species categories. *Conservation Biology*, **5**, 148–57.

McNeilage, A J (1995) Mountain gorillas in the Virunga Volcanoes: ecology and carrying capacity. PhD thesis, University of Bristol.

McNeilage, A J (1996) Ecotourism and mountain gorillas in the Virunga Volcanoes. In *The Exploitation of Mammal Populations*, ed. V J Taylor & N Dunstone, pp. 334–44. London: Chapman & Hall.

McNeilage, A J, Plumptre, A, Brock-Doyle, A & Vedder, A (1998) *Bwindi Impenetrable National Park, Uganda Gorilla and Large Mammal Census*. Wildlife Conservation Society, Working Paper No. 14.

Miller, P, Babaasa, D, Gerald-Steklis, N, Robbins, M, Ryder, O & Steklis, D (1998). Population biology and simulation modeling working group report. In *Can The Mountain Gorilla Survive? Population and Habitat Viability Assessment for* Gorilla gorilla beringei, ed. S Werikhe, L Macfie, N Rosen & P Miller, pp. 71–105. Apple Valley MN: IUCN SSC Conservation Breeding Specialist Group.

O'Brien, S J & Evermann, J F (1988) Interactive influence of infectious disease and genetic diversity in natural populations. *Trends in Ecology and Evolution*, **3**, 254–9.

Robbins, M (1995) A demographic analysis of male life history and social structure of mountain gorillas. *Behaviour*, **132**, 21–45.

Sarmiento, E, Butynski, T M & Kalina, J (1996) Ecological, morphological, and behavioral aspects of gorillas of Bwindi-Impenetrable and Virungas National Parks, with implications for gorilla taxonomic affinities. *American Journal of Primatology*, **40**, 1–21.

Schaller, G B (1963) *The Mountain Gorilla: Ecology and Behavior*. Chicago: University of Chicago Press.

Sholley, C (1991) Conserving gorillas in the midst of guerillas. In *American Association of Zoological Parks and Aquariums, Annual Conference Proceedings*, pp. 30–7.

Sicotte, P (1993) Inter-group encounters and female transfer in mountain gorillas:

influence of group composition on male behavior. *American Journal of Primatology*, **30**, 21–36.

Soulé, M E (1987) *Viable Populations for Conservation*. Cambridge: Cambridge University Press.

Vedder, A L (1989) Feeding ecology and conservation of the mountain gorilla (*Gorilla gorilla beringei*). PhD thesis, University of Wisconsin–Madison.

Wallis, J & Lee, D R (1999) Primate conservation: the prevention of disease transmission. *International Journal of Primatology*, **20**, 803–26.

Watts, D P (1989) Infanticide in mountain gorillas: new cases and a reconsideration of the evidence. *Ethology*, **81**, 1–18.

Watts, D P (1990) Ecology of gorillas and its relation to female transfer in mountain gorillas. *International Journal of Primatology*, **11**, 21–45.

Watts, D P (2000) Causes and consequences of variation in male mountain gorilla life histories and group membership. In *Primate Males*, ed. P Kappeler, pp. 169–79. Cambridge: Cambridge University Press.

Weber, A W & Vedder, A (1983) Population dynamics of the Virunga gorillas 1959–1978. *Biological Conservation*, **26**, 341–66.

Werikhe, S, Macfie, L, Rosen, N & Miller, P (1998) *Can The Mountain Gorilla Survive? Population and Habitat Viability Assessment for* Gorilla gorilla beringei. Apple Valley MN: IUCN SSC Conservation Breeding Specialist Group.

Woodruff, D S (1989) The problem of conserving genes and species. In *Conservation for the Twenty-First Century*, ed. D Western & M Pearl, pp. 76–88. New York: Oxford University Press.

Afterword: mountain gorillas at the turn of the century

BILL WEBER & AMY VEDDER

Two-year-old infant, Samvura. (Photo by Martha M. Robbins.)

We first arrived at Karisoke in early 1978. Digit, a young silverback made famous in films, had been killed a few weeks earlier and his death hung over the research station like a shroud. More deaths would follow in the months to come. Mweza, Quince, Uncle Bert, Macho, Kweli, Frito, Lee. Some were shot and gruesomely beheaded. Some suffered slow, painful deaths from trap wounds. Frito and Mwelu died of infanticide, but only after poachers killed Digit and Uncle Bert and destabilized the family structure. Quince died naturally, but just as the beautiful 8-year-old was entering her prime. We had come to help save mountain gorillas, but we buried far too many.

A dispassionate analysis might note that gorilla deaths around Karisoke in the late 1970s were just catching up with the rest of the Virunga range. Our census of 1978–79 confirmed earlier declines in the total gorilla population and highlighted the elimination of subpopulations on the forest's eastern and western extremes. The most dramatic losses between 1973 and 1978 were on Mount Mikeno, across the Congolese border within sight of Karisoke. These deaths could not be attributed to habitat loss, since they were in areas where none had occurred. Given no evidence of widespread disease, poaching must have claimed a high percentage of the 200 gorillas missing since George Schaller's pre-Independence survey in 1960. Yet serious habitat losses had occurred, too, especially in Rwanda. More than 50% of the Parc National des Volcans had been cleared for human settlement and agriculture, driving the gorillas higher into the mountains where they were exposed to greater cold and the effects of disease, with less food and shelter to sustain them.

Perhaps the greatest irony is that with all the death and destruction in the park, Rwanda was at peace in the late 1970s. If not the "Switzerland of Africa" that some proclaimed, Rwanda at that time was a center of calm in a region racked by violence and political upheaval. Western donors rewarded its perceived stability and studied neutrality with a rising tide of foreign investment. Some of these investments yielded clear rewards for the country and its people. Some rewarded the political elite. Some investments were killing the park and its gorillas. Ten thousand hectares were cleared from the Parc National des Volcans for pyrethrum production in 1969–70, subsidized by the European Development Fund. Less than ten years later, the Fund and others agreed to finance a cattle project for which another 5000 hectares – one-third of the remaining parkland and all of its rich bamboo zone – were to be cleared from the Parc National des Volcans. This decision, approved by the government and announced in early 1979, joined Digit's death as the primary catalyst for conservation action and the creation of the Mountain Gorilla Project.

414

Ten years later, in 1989, 320 mountain gorillas roamed the Virungas: an almost 25% increase in the population. More than 20 years later, at the turn of a new century and in the wake of horrific civil strife and human bloodshed, on-going monitoring indicates a further increase to more than 350 mountain gorillas. In a world where wildlife and wildlands are disappearing at an accelerating rate, good news is hard to find. When wildlife survives in a context of landlessness and extreme civil disorder, it is worth our while to consider why.

The mountain gorilla deserves much of the credit for its own survival. We have had the great fortune to work with elephants, bongo, and primates of many kinds; penguins and sea lions; and wolves, lynx, and loons in spectacular wildlands around the world. Yet our most powerful wildlife experiences remain those with mountain gorillas. No other creature is as compelling; no other draws us across that narrow, yet still deep divide between our species. The mountain gorilla has been its own best ambassador to the world of humans.

George Schaller's landmark study in 1960 provided a highly accurate and enduring portrait of mountain gorilla ecology and behavior. Dian Fossey added detail and personalities to that image, then held it up for all the world to see. Through magazine articles and films her work attracted the interest of millions and stimulated an insatiable appetite for more. Her growing audience formed a natural constituency for the mountain gorillas, yet political influence and financing were slow to follow. As the western world learned more about the gorillas, Rwandans remained largely uninformed about their famous neighbor. Barely half of those surveyed in 1979 could cite a single non-consumptive value of protecting either wildlife or the park. Not surprisingly, a majority of local farmers thought that the park should be abolished and opened to agriculture.

Glaring gaps in our understanding of the Virunga forest and its wildlife were revealed with the threat of further clearing for the cattle project. This was especially true for the ecological needs of gorillas and other species, which were largely unknown outside the Karisoke study area. Even less was known about socioeconomic factors in gorilla conservation and the source of human pressures on the park. Faced with increasingly complex problems in the late 1970s, Karisoke's narrow research base could not meet the challenge.

Ironically, it was the latent power of the global constituency which came to the rescue. Sparked by Digit's death and fanned by the reporting of American and especially British media, a ground fire of support spread around the world. The Digit Fund, established by Dian Fossey, brought monies directly to Karisoke to support gorilla monitoring and maintain a local anti-poaching presence. The Mountain Gorilla Preservation Fund took a broader view of its mandate. Fueled with unprecedented donations from the British public and guided by new ideas from the field, this awareness and fund-raising initiative was transformed into the Mountain Gorilla Project (MGP). With its comprehensive plan of attack, the MGP had sufficient merit, money, and momentum to overwhelm the proposed cattle-raising scheme. Ultimately, its three-pronged attack through anti-poaching, education, and ecotourism proved very successful. With expanded employment and income for local people, and a flood of foreign revenue for the central government, the politics of protection tipped in favor of the gorillas, despite increased controls on hunting and a halt to encroachment. Beyond Rwanda, the MGP established a model that has been widely copied and discussed. It was exciting to play a role in the MGP's creation and implementation. It was far more exciting to see it work.

Private money and private organizations drove the early operations of the MGP, and both have continued to play a central role in mountain gorilla conservation over the past two decades. The Digit Fund has maintained a steady flow of support to Karisoke, while a consortium of conservation non-governmental organizations has carried the MGP model across the border and adapted it to conditions in the Congolese and Ugandan sectors of the Virungas. The Belgian government also deserves credit for its long-term assistance to basic management of the Parc National des Volcans. In the mid 1980s, though, bilateral assistance increased dramatically. Hesitant at first to deal with the gorillas directly, the US Agency for International Development worked with us to establish an "integrated resource management project" that would address a broad range of land, water, wood, and health issues in the Virunga watershed, while giving some limited support to a new gorilla census and other data collection in the park. Later in the early 1990s, USAID–Rwanda provided direct funding to Karisoke for infrastructure expansion and operations, while USAID–Uganda supported parallel initiatives in Bwindi and the Mgahinga National Park.

Throughout the 1980s, Rwandan individuals and institutions were increasingly involved with the MGP and Karisoke. Gorilla guides were the face of a nation for the thousands of tourists who followed them into the forest. The Office Rwandais du Tourisme et des Parcs Nationaux grew in stature as its revenues swelled. Few of these revenues were shared with local communities, but more than 50 new guards and guides were hired with tourist entry fees. Fourteen years after Karisoke's creation, the first Rwandan students were recruited to conduct research at the station. A new field veterinary program, too, brought Rwandans into the world of gorilla conservation, where they worked with expatriate colleagues to care for gorillas that formerly would have died of their wounds or diseases. Rwandan involvement in all aspects of research and conservation grew larger over the years.

In 1991, George Schaller returned to see mountain gorillas for the first time in 30 years. On his return to the USA, he brought a tape-recording he called "sounds from the forest". It was filled with automatic rifle and artillery fire from a battle in which Schaller and other Karisoke staff had been inadvertently trapped. It would not be long before Karisoke itself was burned, looted, and abandoned during an intensified civil war – though both sides in that conflict remarkably pledged not to harm the gorillas. Other humans were not off-limits, though, and in 1994 the entire nation was plunged into the darkness of an unspeakable frenzy of genocidal killing. In the wake of this seismic event, new political leadership took control of Rwanda and a process of healing has slowly begun. But peace, fragile, remains elusive and the government is preoccupied with guerillas, not gorillas, in their midst.

With the outbreak of hostilities, bilateral agency conservation projects withdrew and have yet to reestablish a presence. With no foreign assistance and limited government involvement, private organizations have carried the primary responsibility for gorilla conservation throughout the 1990s and into the new century. Unable to maintain a permanent presence at Karisoke or even near the park, a few expatriate advisers and Rwandan staff nevertheless make regular visits to the gorillas and conduct patrols at considerable personal risk. They are the last and most vital links in a chain of conservation action that goes back more than 30 years.

There are many important lessons to be drawn from the long and complicated experience with mountain gorilla research and conservation. The first is that mountain gorillas are a powerful flagship species. Their sociality, intelligence, appearance, and their tolerance of humans make them incredibly attractive. They pique our interest and demand our concern in ways that few other species can approach. Even lowland gorillas seem more distant and are certainly less approachable.

Beyond their own intrinsic appeal, mountain gorillas have benefited from a very high public profile. Schaller's *Year of the Gorilla*, Fossey's popular articles and films, the book and movie versions of *Gorillas in the Mist*, and even Fossey's unsolved murder have all brought attention to the gorillas. So, too, has a tourism program through which tens of thousands of visitors have come to know the gorillas in a powerful, first-hand manner. There is no question that international development agencies were drawn to the gorillas by their "star" appeal. Within Rwanda, an active education program has exposed more than 100 000 people to gorilla films and presentations, while millions have seen the gorilla's image on everything from soap to their national currency – the latter image copied from an early MGP calendar. Image isn't everything, but a high profile has helped a critically endangered species survive despite the combined threats of poaching, habitat loss, and warfare.

The long-term record of gorilla monitoring and research at Karisoke has generated most of what is known about gorillas in the wild and stimulated world-wide popular interest in the species. Data from Karisoke study groups also provided a baseline from which demographic trends could be extrapolated from successive census counts. With the eventual creation of a tourism program, the same data sets provided a control against which potential impacts could be measured. This research and monitoring focus should remain at the heart of Karisoke's mission.

Applied multidisciplinary research, especially attention outside the park to the socioeconomic context in which conservation must occur, played a central role in the formulation of the MGP. Throughout the 1980s, Karisoke expanded its program of research on other species, such as buffalo and elephant, as well as basic forest ecology. Increasingly, research results informed conservation action, until civil war and its aftermath intervened. When conditions permit, Karisoke will have to reestablish a broad-based program of inquiry. The Center will also have to establish good lines of communication and collaboration with partner organizations in the private and government spheres. Conservation in Rwanda has worked best when the staffs from Karisoke, conservation non-governmental organizations, and government agencies are in regular contact with each other and working toward common goals.

One important role of interdisciplinary research is to identify the various factors, positive and negative, affecting mountain gorilla conservation. Once these factors are identified and given some sort of relative weighting, then the role of conservation is to actively alter those weights – increasing the positives, reducing the negatives – in a way that improves the gorillas' prospects. In Rwanda in the 1970s, local attitudes and economics worked against the gorillas, as did national economics and politics. Tourism jobs and salaries helped tip the local balance in favor of conservation, despite the perceived costs of prohibitions on hunting and clearing land. Foreign revenues of several million dollars per year accomplished the same goal at the national government level as did millions more in bilateral development assistance to projects in and around the park. Furthermore, both government and donors basked in the full glow of international acclaim for their efforts, where once they had been condemned. In these and other ways, the conservation balance in Rwanda shifted dramatically.

Non-governmental organizations have proven their durability, commitment, and flexibility with regard to mountain gorilla research and conservation over the past several decades. The original Mountain Gorilla Project ran its course, spinning off sister projects in neighboring Congo and Uganda. Eventually, the International Gorilla Conservation Program was formed to provide an umbrella for all of the regional tourism, education, and management initiatives concerning mountain gorillas. The Digit Fund evolved into the Dian Fossey Gorilla Fund and has continued to support the operations of the Karisoke Research Center and its personnel. Similarly, the Morris Animal Fund has sustained the Mountain Gorilla Veterinary Project and affiliated staff throughout the most difficult of times. The Wildlife Conservation Society has also provided direct support to gorilla conservation efforts, while maintaining its flagship program in Rwanda's Nyungwe Forest. As the government collapsed and then focused on other priorities, as bilateral assistance programs closed down and withdrew their personnel, these non-governmental organizations continued to operate on the ground throughout the worst of the 1990s. As Rwanda emerges from its long national nightmare, these groups are still there. More to the point, the forests and wildlife which they serve continue to exist. Much of the Akagera National Park has been cleared and many of its animals slaughtered. The Gishwati Forest to the south of the Virungas – which 20 years ago was larger and more diverse than the Parc National des Volcans – no longer exists. Yet the Parc National des Volcans and the Nyungwe Forest, the two sites with an active conservation presence of non-governmental organizations, are largely intact and appear to have lost little wildlife.

As much credit as these organizations deserve, they would have accomplished nothing over the past ten years without dedicated national staff. And the dedicated staff of the 1990s would not have been in place without the recruitment, training, and support activities of the 1980s. In the wake of the worst fighting in 1994, dozens of Rwandans continued core conservation activities. Gorillas were monitored and patrols were carried out, where possible. Most of the Rwandans stayed in place because they were local residents; most kept working, often without pay, because of a combination of commitment and the expectation that their long-term project would someday restart its activities. Some colleagues from non-governmental organizations were recruited to fill new government posts, where they have played important roles. The role of government institutions must also be recognized. The Rwandan park service and forest service are making extremely important decisions about the future of conservation in their country. The degree to which they listen, or even consult, outside advice depends to a very large degree on their prior record of relations with non-governmental and other expatriate organizations.

Not all lessons are good ones. The record of the past 30 years includes some significant failures and shortcomings. Karisoke was far too isolated in the late 1970s. It was unaware of the proposed cattle project, lacked the information and standing to oppose the scheme, and was actively hostile toward the Mountain Gorilla Project's efforts to present an alternative option that linked conservation and development. The MGP, too, had its shortcomings. Everyone spoke well of its education component, but tourism and anti-poaching always received more attention. Education activities, and the Rwandan personnel who ran them, were given less attention and followed no consistent program. Surveys still showed that local attitudes toward the park and gorillas became more positive over the years, but it is entirely possible that this reflected an appreciation of tourism-related jobs and income rather than attitudinal changes due to education.

Tourism was the MGP's crowning glory and may have single-handedly saved the gorillas, but it had its problems as well. Within a few years of the program's inception, Rwandans were priced out of the opportunity to see their own gorillas. With foreign tourists paying first $100, then almost $200 for a single day's ticket, Rwandan authorities abolished the two-tiered price system that allowed their own citizens to pay only a few dollars. A few years later, the number of daily visitors to several gorilla groups was increased from six to eight, expanding the risks of disturbance and disease transmission to the gorillas, and lowering the quality of the visitors' experience. Yet despite the volume of tourism revenue – an increase from a few thousand to more than a million dollars per year between 1979 and 1989 – the park service (ORTPN) consistently resisted any form of revenue-sharing with local communities, thus eliminating a potentially powerful incentive for increased local support for conservation.

Still, the positive lessons far outweigh the negative ones and much has been learned from past mistakes and difficulties. This learning has occurred because key groups and individuals have stayed the course and maintained their commitment over many years, even decades. In fact, one of the most basic lessons to be drawn from the mountain gorilla experience is the importance of a long-term commitment to a cause and a site. In the process, one not only learns more about the gorillas and their forest home, but also about the concerns and aspirations of the people and their country.

There are enormous challenges ahead. The Virungas are buffeted by the tides and swells of political and military conflict in the surrounding region. Peace and stability have thus far eluded the best efforts of diplomats, and they are certainly beyond the control of conservationists. Still, now is the time to plan for an eventual return to a relative state of normalcy and to identify a set of priority actions.

A first step is to convince the international donor agencies that conservation is an integral part of the development package. Bilateral and multilateral agencies currently resist involvement with gorillas or parks with the rationale that Rwanda has more pressing priorities. Without diminishing the importance of these other priorities, the conservation community must work with allies in the development community to stress the hard core financial importance of tourism to Rwanda's future, with the corollary recognition that there will be no tourism without viable wildlife populations and parks. The fact that Rwanda's mountain forests play a major role in water regulation and agriculture must also be stressed.

Ecotourism is a double-edged sword. It vanquished the cattle scheme of the late 1970s and was the cutting edge of the MGP. It provided the financial force behind the rise of the park service (ORTPN) from a minor institution to the major voice for conservation within the Rwandan government. Gorilla-based tourism informed and inspired the global ecotourism movement. But tourism always brings problems, most of which can be traced to greed. No matter how wonderful the hen that lays the golden egg, someone always wants to squeeze out a few more eggs. It is essential that the unique experience of spending an hour with wild mountain gorillas be rationed through pricing, with ticket prices for foreign visitors set as high as the market will bear. The strategy of increasing the number of visitors per day per gorilla family will only lower the quality of the experience, generate negative publicity, and ultimately cause prices to drop.

A related concern is disease transmission. Even though there is a self-selection for tourists in good health, it is absolutely necessary to take reasonable precautions to limit the potential spread of illness and disease. This requires maintaining a minimum distance between visitors and gorillas, as established by medical and veterinary experts. It also reinforces the need for limits on the number of visitors per group to a level – which we would place at six – at which guides can effectively control the vast majority of interactions. Finally, the majority of gorilla families must remain free from any contact with tourists or researchers, as in the last Virunga census when 21 of the 31 groups were still completely wild.

At the same time, it is essential that people remember why the gorilla tourism program was first initiated. It was started to counteract a cattle raising scheme that would have cleared one-third of the Parc National des Volcans and virtually all of the bamboo zone, from which the gorillas derive one of their most important food resources. Five thousand head of cattle and hundreds of attendant herders would have flooded the park and interacted regularly with the remaining gorillas. Gorilla tourism provided an alternative that protected the park's integrity, permitted the gorilla population to reverse its steady decline, and altered the economics and politics of conservation in favor of the gorillas. Tourism remains an extremely powerful weapon that may be crucial to the survival of the mountain gorilla. But like all powerful weapons, it must be wielded with care.

The Karisoke research station also faces critical challenges. Assuming that it will one day reestablish its full-time presence within the park, Karisoke needs to demonstrate its utility for applied wildlife and park management. As much as its leadership might want to be a force for conservation, an overly narrow focus on academic research topics could result in an unintended return to isolation, or irrelevance. This can be avoided through the development of a comprehensive research plan that seeks to address a broad range of biological and ecological subjects, as well as selected socioeconomic topics. That much said, Karisoke's strength will always lie in its gorilla research and monitoring program. This program informs the scientific community, inspires popular interest and support, and provides baseline information that is essential to understand the health of the population. Karisoke must also make sure that its doors are open to Rwandan students and researchers, and perhaps even those from neighboring countries. Finding room for an expanded mandate and staff will be a challenge and perhaps require a second station outside of the park. But the end result will be to establish Karisoke as the primary information source for the Virungas.

There is a pressing need to reestablish dialogue with local populations and political structures around the Parc National des Volcans. This population has been physically uprooted and its world view radically altered by the dramatic events of the past decade. Its composition has changed and human population densities near the park actually have declined. Yet this remains the group that interacts most with the park, that derives the greatest benefits and bears the greatest costs of conservation. While new political structures may take a while to emerge, there are opportunities to begin initial contacts with this group. Participatory surveys with follow-up feedback sessions may be the best way to learn about this changed population and see how it looks at the world – both the world of the forest and gorillas, and the world in which they must meet the needs of their families. Understanding these perceptions is the first step toward effective action.

Gorillas attract so much attention that it is often difficult to see beyond their immediate realm. Yet the Virunga forest cannot exist as an island fortress. Despite large losses to genocide, warfare, and AIDS, the Rwandan population continues to grow and seek more land. The conversion of the Akagera Park and the Gishwati Forest must be seen as harbingers of pressures to come. The recent proposal to declare the Nyungwe Forest a national park is a welcome move in the opposite direction. Still, conservationists must coordinate and reinforce their efforts to protect not just the Parc National des Volcans and contiguous parks in Congo and Uganda, but Nyungwe and the remainder of Akagera, too. Even more distant forests such as Bwindi, Kahuzi-Biega, and Kibira-Teza are important parts of the Afromontane archipelago. A regional perspective recognizes that each forest island that falls leaves the others more exposed.

A century has nearly passed since the mountain gorilla was discovered. It is safe to say that this isolated population would not have survived to this point without the exceptional efforts of a small number of researchers and conservationists, backed by a global network of support. The survival of the mountain gorilla in the 21st century will require even greater effort and support.

Index

aggressive behavior
  between females 220, 221, 223–5, 228–9, 232
    male interventions 227, 228
    reconciliation 226
  between males 36–7
    during inter-group encounters 76
    testosterone relationships 330
  male–female aggression 39–40
    following group fission 40–2
    influence on female mate choice 79–80
    sexual coercion 40, 44–5
  male–immature aggression 189
  see also agonism
agonism
  females 219–22, 223–5, 228–9
    agonistic support 218, 219–20
    asymmetries 221–2
  male–immatures 202–4
    agonistic support 202–4
  mother–infant 161, 172
  see also aggressive behavior
Akagera National Park 419, 423
alarm calls 245
Albert National Park 5
all-male groups 37–8
alpha herpes virus 349–50
androgens 320
  dominance rank relationships 330–1
anesthesia 343–4
anti-predator calls 245
antibiotic treatments 345
ARKS program 350
armed conflict see war

baboons 255
  female mate choice 60
  female–female relationships 231, 232–3
    estrogen excretion and 327
  food-processing skills 305
bamboo habitat 275–6, 285, 286
bites, to offspring 172
bonobos
  dietary strategy 296

  social organization 6, 49
bushmeat trade 364, 374
Bwindi Impenetrable Forest National Park 12, 19, 371, 392

calls see vocalizations
carrying capacity 402, 405
cattle grazing 363, 421–2
chacma baboons 231
chimpanzees 3
  extra-group mating 75
  female group transfer 64, 69–70
  food-processing skills 302, 306, 307
  foraging strategy 295, 296
  reproductive biology 326, 328, 329
  social organization 6, 49, 207
  tool-use 296
  vocalizations 247, 249, 258
  weaning 176
chorionic gonadotropin 320
  pregnancy detection 323
civil war see war
clinical medicine 343–6
  emergency treatments 343–4
  future research 354–6
  preventive health care 346–54, 355–6
    disease transmission from humans 348, 350, 351–4, 365–6, 378
    health monitoring 346–50
    immunization 354
    pathology 350–1
    snare removals 344–5
close-calls see vocalizations
cognitive niche 294, 306–7
colobine monkeys 294
communications 136, 256–7
  inter-group communication 245–6
  see also vocalizations
competition
  among females 229
  during inter-group encounters 60
  see also feeding competition; mating competition
Congo 9–10, 92, 353
consanguinity, vocal relationships and 252